GLOBAL PRODUCTION NETWORKING AND TECHNOLOGICAL CHANGE IN EAST ASIA

SHAHID YUSUF
M. ANJUM ALTAF
KAORU NABESHIMA
EDITORS

THE WORLD BANK
Washington, D.C.

A copublication of
the World Bank and
Oxford University Press

1818 H Street, NW
Washington, DC 20433
Telephone 202-473-1000
Internet www.worldbank.org
E-mail feedback@worldbank.org

First printing June 2004
1 2 3 4 07 06 05 04

The findings, interpretations, and conclusions expressed herein are those of the author(s) and do not necessarily reflect the views of the Board of Executive Directors of the World Bank or the governments they represent.

The World Bank does not guarantee the accuracy of the data included in this work. The boundaries, colors, denominations, and other information shown on any map in this work do not imply any judgment on the part of the World Bank concerning the legal status of any territory or the endorsement or acceptance of such boundaries.

Library of Congress Cataloging-in-Publication Data

Global production networking and technological change in East Asia / edited by Shahid Yusuf, M. Anjum Altaf, Kaoru Nabeshima.
p. cm
Includes bibliographical references and index.
ISBN 0-8213-5618-6
1. Manufacturing industries—East Asia. 2. Business networks—East Asia. 3. Production management—East Asia. 4. Technological innovations—East Asia. 5. Electronic industries—East Asia. 6. International business enterprises—East Asia. 7. Globalization—Economic aspects—East Asia. I. Yusuf, Shahid, 1949– II. Altaf, M. Anjum, 1950– III. Nabeshima, Kaoru.

HD9736.E18G55 2004
338.8'895—dc22

2004044087

Cover design by Debra Naylor of Naylor Design, Inc.

CONTENTS

PREFACE

This is the third volume in a series of publications from a study co-sponsored by the Government of Japan and the World Bank to examine the sources of economic growth in East Asia. The study was initiated in 1999 with the objective of identifying the most promising path to development in the light of global and regional changes.

The first volume, *Can East Asia Compete?*, published in 2002, provided a compact overview of the relevant strategic issues and future policy directions. *Innovative East Asia*, the second volume, analyzed each of main issues and consequent policy choices drawing comprehensively upon recent empirical research and the findings of firm surveys conducted for the study. Its principal message is that sustained economic growth in East Asia will rest on retaining the strengths of the past—stability, openness, investment, and human capital development—on overcoming the sources of current weaknesses—in the financial sectors, corporate, judicial, and social sectors—and on implementing the changes required by the evolving economic environment, particularly with regards to technology development.

This book, *Global Production Networking and Technological Change in East Asia*, is the first of two volumes of papers commissioned for the East Asia study. It provides detailed information, analysis and case studies further illuminating some of the topics covered in the earlier volumes. The contributors rigorously examine the effects of the changing global geography of production for the growth prospects of East Asian economies. They conclude that in the face of a global environment, economies in East Asia need to adapt to the changing character of global production networks and to nurture and develop technological capabilities in order to sustain their growth prospects. The companion volume, *Global Change and East Asian Policy Initiatives*, will include a second set of papers that explore regional and institutional policy options that East Asian economies face in a similarly in depth manner. Both volumes complement *Innovative East Asia* and are addressed to researchers, students and policymakers.

The financial backing of the government of Japan through its Policy and Human Resources Development Fund provided vital support for this project, as did senior public officials who gave generously of their time. We are deeply grateful to Haruhiko Kuroda, Naoko Ishii, Masahiro Kawai, Kiyoshi Kodera, Rintaro Tamaki, Junichi Maruyama and Takatoshi Ito. The staff of the World Bank's Tokyo office facilitated the reviews and seminars, and we greatly appreciate the assistance provided by Yukio Yoshimura, Shuzo Nakamura, Mika Iwasaki, Tomoko Hirai, and Hitomi Sasaki. We also deeply appreciate the support we received from Deepak Bhattasali and Jianqing Chen at the World Bank's Beijing office. We owe special thanks to K. Migara De Silva for his enthusiastic and tireless support in organizing and participating in seminars in Beijing and Tokyo.

The papers in this volume were presented at seminars and workshops in Beijing; Cambridge, Mass; Tokyo, and Washington, DC. The comments received helped the authors in revising their drafts. We would like to thank all of those who participated in these seminars, along with the many reviewers of the entire manuscript and, in particular, Jose Luis Guasch.

At the World Bank, the Development Research Group has for several years offered a base for the study. In addition, we are grateful for the support provided by East Asia and Pacific Region. We are especially indebted to Jemal-ud-din Kassum and Homi Kharas for their guidance and strong encouragement.

The study team was ably supported by the research skills of Soumya Chattopadhyay, Farhan Hameed, and Yifan Hu. The manuscript was prepared by Paulina Flewitt, Marc Sanford Shotten, and Rebecca Sugui; and we thank Susan Graham, Patricia Katayama, and Ilma Kramer of the Office of the Publisher for their expert management of the editorial and print production of the volume.

CHAPTER 1

COMPETITIVENESS THROUGH TECHNOLOGICAL ADVANCES UNDER GLOBAL PRODUCTION NETWORKING

Shahid Yusuf

Starting in the early 1990s, the emerging economies of East Asia accelerated the pace of deregulation and integration with the global economy.[1] Although the crisis of 1997–98 resulted in a brief hiatus, making countries wary of the risks from open capital accounts and skeptical of the gains, the process has continued into the early twenty-first century, spearheaded by China's entry into the World Trade Organization (WTO).[2] Regional and bilateral free trade areas, the continuing evolution

1. Since the late 1990s, globalization has been analyzed, chronicled, and debated in obsessive detail, with frequent backward glances at the earlier round of globalization that commenced around the turn of the nineteenth century. While the gains from trade and the sharing of technology and ideas have been acknowledged with qualifications (Bhagwati 2004), a legion of skeptics worry about cultural imperialism, the risks from volatile capital flows, the loss of jobs as a result of rapid shifts in comparative advantage, and the declining role of the nation-state. For a compact review of the trends signifying globalization and their implications for developing nations, see World Bank (2000). Micklethwait and Woolridge (2000) provide an engaging account for the general reader, while a close look at globalization in historical perspective can be found in O'Rourke and Williamson (1999) and Held and others (1999). Some of the tensions generated by globalization are brought out in Stiglitz (2002), and its impact on markets and on inequality is discussed by Bourguignon and others (2002). The interactions between globalization and the business sector are explored in Cairncross (2002) and Dicken (2003), while Yeung and Olds (2000) show how Chinese business groups are participating in and contributing to globalization.

2. A provisional summing-up of the considerable research on the effects of capital account openness would be as follows: middle-income countries with strong market and financial institutions are likely to benefit over the medium term from an open capital account. Low-income countries are less likely to derive much benefit, and their institutional gaps might make them prone to crises. Countries can contain the effects of crises by way of capital controls, although their utility, even in the cases of China, Chile, and Malaysia, is by no means rigorously established. See Arteta, Eichengreen, and Wyplosz (2001); Edison and others (2002); Eichengreen and Leblang (2003); Henry (2003), Klein (2003); Wyplosz (2001).

of international production networks, and substantial flows of foreign direct investment (FDI) within and to the region have tightened the economic strands binding East Asian firms to one another and to the developed nations of the Organisation for Economic Co-operation and Development (OECD). Together, these factors are defining the parameters impinging on the growth and profitability of firms.

This process of global integration is having five major consequences for East Asia. First, it has substantially widened the opportunities for firms, as tariff and nontariff barriers are stripped away, information circulates more freely, and transport costs decline.[3] The downside is that competitive pressures have also increased, and firms that do not meet the market test are quickly weeded out. Second, many more firms are now linked to global production networks, the outgrowth of ongoing efforts by many large multinational corporations (MNCs) to establish foreign subsidiaries, to deverticalize their activities, to outsource certain functions, and to subcontract the production of numerous components.[4] The rise of such global production networks has intensified both trade within industries—to almost a third of total world merchandise trade—and trade among affiliates of multinational companies—approximately 45 percent of total trade or about $3.2 trillion in 1999.[5] It is also differentiating the roles of firms that are original equipment manufacturers (OEMs),[6] firms that have risen into the ranks of original design manufacturers (ODMs), and firms that have become original brand manufacturers (OBMs) and stand at the apex of their global production networks. As discussed in this chapter, the prospects of OEMs are starting to diverge from those of ODMs and OBMs.[7]

3. For example, regional tariffs in the ASEAN (Association of South East Asian Nations) Free Trade Area fell to the 0–5 percent range for a wide range of goods, including auto parts ("ASEAN: Free Trade Fears" 2002; "Autos and AFTA" 2002).

4. The global production network, according to Ernst, provides the lead firm "with quick and low-cost access to resources, capabilities, and knowledge that complement its core competencies" (chapter 3 of this volume). For an application of the global production network—that is, the setting-up of decentralized worldwide production networks by firms that are either lead buyers or lead producers—to the apparel industry, see Gereffi and Memedovic (2003).

5. See also the analysis by Feinberg and Keane (2003) of the growth of trade related to multinational companies.

6. OEMs are firms that serve as subcontractors and produce according to the specifications of buyers. See Hanson, Mataloni, and Slaughter (2003); "Multinationals and Globalization" (2001); World Bank (2000).

7. Most East Asian firms from the industrializing countries fall into the ODM categories. Very few have progressed to the OBM category.

Third, whereas price competitiveness remains an important determinant of success in the marketplace, firms of all stripes are finding that the capacity to innovate is the key to productivity, growth, and greater profitability. This applies to OEM suppliers, to firms seeking to achieve greater independence and to enlarge their market share and profitability, using the additional leverage provided by innovative design, and to firms trying to establish a brand in the international marketplace.[8]

Fourth, the progressive leveling of trade barriers and the emphasis on lean production and frugal networking practices, supported by just-in-time delivery of components, are bringing about the geographic consolidation of certain industries such as automobiles, engineering, electronics, and white goods in closer proximity to their principal markets. For example, the production of one-ton trucks and, to a lesser extent, sedans for the Southeast Asian market is shifting to Thailand, and leading auto companies are likely to choose China and the Republic of Korea as the hub for their regional activities. Multinational companies favor China and Thailand because they have the lowest labor and overhead costs in the region. This and the quest for optimal scale is also leading to a shakeout of firms in major subsectors, such as hard disk drives (120 companies in the early 1980s—less than 10 remain out of 120) and personal computers, and to a coalescence of market power in the hands of relatively large firms with a regional, if not a global, footprint.

Fifth, auto assemblers and manufacturers of consumer electronics are streamlining their product lines to realize the economies from higher-volume production runs. They are also seeking to reduce the number of components to lessen the complexity of products, lower costs, and minimize the need for numerous suppliers.

STAGES OF EAST ASIAN DEVELOPMENT

How East Asian manufacturing firms respond to the changing market environment, harness the potential of global production networks, and buttress price competitiveness by building technological capability will strongly influence their growth prospects. The purpose of this introductory chapter is to underscore the significance of the global production networks and to indicate how their evolution requires East Asian firms to redouble their efforts to build technological capability, improve logistics, and harness foreign direct investment to increase competitiveness.

8. Hallward-Driemeier, Iarossi, and Sokoloff (2003) find that firms participating in export markets have higher productivity, although research reported by Tybout (2000) suggests that the causality runs from higher productivity to participation in trade rather than vice versa.

Labor-Intensive Light Manufacturing

Viewed through the lens of trade and technological change, development in East Asia can be divided into four stages. Starting in the 1960s, Hong Kong (China), Republic of Korea, Singapore, and Taiwan (China) began investing in light, labor-intensive manufacturing industries whose products were quickly able to penetrate the lower and most price-sensitive end of markets in Western countries. The American involvement in Vietnam at that time and the interest of the United States in strengthening the economies of its allies in the region led to policies that eased access to U.S. markets. Access was facilitated further by the efficiency of the retail sector and the capacity of American retail chains to source products from East Asia. Having established a beachhead, the East Asian exporters consolidated their market position by improving quality and broadening the range of manufactured exports. Within a matter of years, they established a reputation not just for competitiveness and quality but also for on-time delivery, flexibility, and readiness to enter new product lines.

Upgrading by Newly Industrialized Economies, Enter the Southeast Asians

This first stage of export-led growth in developing East Asia was followed by a second, extending from the early 1970s through the late 1980s, during which four newly industrialized economies were joined by several Southeast Asian countries and, after the early 1980s, by China.[9] The newly industrialized economies shifted into more capital- and skill-intensive manufacturing industries as well as into producer services, enabling newcomers to enter the subsectors they were vacating. East Asia's gathering reputation as a politically stable region with a demonstrated track record of industrial performance and export competitiveness attracted a rising flow of foreign direct investment from Japan, the United States, and some of the European countries. Having determined that East Asia had the manufacturing capability, the supplies of educated and trainable workers, and the policy environment conducive to rapid industrialization, Western and Japanese corporations began investing in local firms and establishing wholly owned facilities to leverage the advantages of lower labor and overhead costs.[10] Major corporations

9. This has been called the flying-geese pattern of development, initiated by Japan (Ozawa 2001). Each group of countries that has moved up the production-technology ladder in the wake of Japan has made way for a group of successful economies lower on the scale of development. See Ozawa (2003). Meanwhile, access to the U.S. market has stimulated export-led growth (Cutler, Berri, and Ozawa 2003).

10. Foreign direct investment from Japan to the East Asian economies increased as cost differentials widened following the Louvre Accord in 1985 that led to appreciation of the yen.

that were capable of managing geographically dispersed production facilities transferred components and machinery to overseas subsidiaries and subcontractors that engaged in labor-intensive assembly operations and exported the final product back to the parent company or to a third party. The cost advantages of parceling out production of parts and components to numerous specialized producers throughout East Asia increased as local manufacturing and intermediaries gained experience, tariff barriers fell, and logistics improved. This, in turn, led to rising trade flows within industries and within production networks led by MNCs. Starting with items such as garments, footwear, and consumer electronics, these patterns of production and trade spread to machinery, electronics, chemicals, and transport industries.

Production Networking Gathers Momentum

By the end of the 1980s, East Asia was being drawn into a web of relationships created by intertwined flows of trade and FDI. Hatch (2003, p. 29) states, "This export of Japanese capital and technology helped weave together the economies of Asia." He also cites the work of Tamara, who has remarked, "Japanese multinational companies are building a regional division of labor that emphasizes technology-intensive prototype production in Japan and mass production of standardized products in Asia" (Hatch 2003, p. 31). This third stage lasted through much of the 1990s and witnessed an unusually rapid expansion of trade. Between 1990 and 2000, world trade grew at an annual average rate of 6.6 percent. In the East Asian region, it grew nearly 14 percent annually. These years also saw the flowering of global production networks that were budding in the previous decade. Market deregulation, which had been gathering momentum, became widespread in conjunction with WTO-sanctioned trade liberalization, and it dramatically ratcheted up the level of market competition. An added twist was imparted by the intensifying of technological change, most notably, in the many manufacturing and service industries related to the provision, distribution, and use of information. Moreover, by simplifying the management of dispersed production and enhancing the efficiency of supply chains, information technology further encouraged firms to deverticalize and to outsource not just a variety of production tasks but also a multitude of back-office functions.

During the space of a few years, the combined forces of trade, FDI, and organizational restructuring elaborated a system of global networking that made the tradable segments of the light manufacturing sectors in East Asia's emerging economies almost coextensive with those of Japan, the United States, and, to a lesser extent, Europe. Large MNCs from these countries moved their labor-cost-sensitive activities to East Asia and, in addition, invested in other production facilities to access protected domestic markets. In industries such as autos, electronics, office equipment, and optical instruments, these moves by the large firms were matched by many of their local parts suppliers, which also set up subsidiaries in East Asia.

At the same time, the big retail chains in the Western countries and producers of commodity items began sourcing entire categories of products from East Asia, extending from men's shirts to personal computers. Some of these companies, such as Nike, retained control of design and research, while others limited themselves to brand management and quality control, leaving their overseas suppliers to take the initiative in submitting new designs and fresh products.

Technological Deepening

By about the mid-1990s, the manufacturing subsectors of the industrializing East Asian economies were linked to the economies of leading OECD countries through trade that was intermediated by buyer- and producer-driven supply chains and through a host of relationships arising from FDI. However, even as integration between the East Asian region and the OECD countries appeared to be tightening, the dynamics of continuing trade liberalization, intensifying competition from China, both in product markets and for FDI, and the apparent advantages of firm size in a globalizing world were ushering in a fourth stage of development in which competitiveness is predicated on technological deepening. A discussion of this most recent stage and what it demands of firms takes up the balance of this chapter. Its various facets are examined in detail by the contributors to this volume. There is no better exemplar of the leverage provided by steady technological capability than China, which has emerged in the short space of 10 years as the world's fourth largest trading nation in 2003 and second largest producer of information technology hardware, with close to 14 percent of the world market ("Link by Link" 2003).

SUSTAINING COMPETITIVENESS: THE ISSUES

How the East Asian economies respond to the challenges posed by this fourth stage will decisively influence their growth in the early twenty-first century.

Four issues differentiate this new and difficult phase of development from preceding phases. By exploring these issues, economies in the region can elucidate their major strategic choices. The issues revolve around changes in the global market brought by shifting dynamic comparative advantages; various strategies adopted by MNCs and the emergence of contract manufacturers; the characteristics and capacities of firms to face these changes; and government policies supporting firms' efforts.

Direction of Dynamic Comparative Advantage

The rapid industrialization of East Asian countries and their success at exporting have transformed the market for a host of manufactures. As the efficiency of production has risen and firms have expanded capacity, prices have dropped steadily and products once aimed at a limited number of high-income consumers have become commodities now sold on mass markets. For products ranging from bicycles to computer memory, entry barriers have fallen, supply elasticity has risen substantially, and, as a consequence, profit margins have narrowed.[11] Because wage costs are rising in the middle-income economies of East Asia, the pressure from lower-income countries in the region and in South Asia is growing. This is forcing producers in middle-income economies either to upgrade their offerings and differentiate their existing range of products, so as to accommodate higher costs and at the same time enhance earnings, or to diversify into other products and services where the competition is less severe. One widely pursued strategy, discussed by Dieter Ernst (chapter 3 in this volume), is for economies to upgrade to the assembly of high-technology products such as mobile handsets or digital cameras. However, most of the East Asian economies are already adept at assembling electronics equipment, and a shift to high-technology items does not necessarily increase local value added by much. Moreover, earnings only increase when the growth of supply for a more sophisticated line of products lags behind demand. A large and continuous increase in value added that is more likely to result in higher earnings is becoming tied to steady technological advance in certain commodity products and in complex products or services (Hobday 1998)[12] These include

---

11. By some accounts, excess capacity in the manufacturing sector is widespread because of high levels of investment in the 1990s, scale economies with associated falling marginal costs, and the high price of exit for firms with capital-intensive assets (see Crotty 2002).

12. For example, the profitability in the flat-panel display business has come to depend on the ability to move from one generation of manufacturing facilities to the next generation with greater production capacity and the technology to produce panels of larger size.

the machinery needed to produce the parts and components and to assemble the manufactures pouring out of East Asian factories, the conceptualization, design, integration, and maintenance of complex products such as plant equipment, and the supply of or high value–added services, most of which are now tradable and can command global markets.

For more than a decade, upgrading has been the mantra of East Asian middle-income economies. As they have collectively struggled to reach the next rung of the ladder, economies in the region have often only succeeded in extending the embrace of commodification. Upgrading can be achieved through incremental product or process innovation that can be the source of modest rents until the competition catches up, which can occur in a matter of months. However, when upgrading is based on significant design and product innovation protected by patent rights or the products themselves are difficult to imitate, substantial rents can accrue over several years.

Building a reliable base of technological capability is becoming increasingly urgent for all those East Asian economies seeking to avoid the trap of low-level growth associated with commodity production. The risk has certainly risen over the past five years, as low-cost producers in China have dramatically expanded capacity and flooded the markets for manufactured commodities.

Commodification of many mainstay tradable manufacturers is one reason for building technological capability, but there are others as well, including the changing requirements for participating in global production networks and the emerging threat from multinational contract manufacturers.

Evolution of Global Production Networks

During the third stage of East Asian development, global production networks offered efficient and enterprising firms a shortcut to international markets. A firm that was prepared to meet the product specifications and delivery schedules demanded by MNCs and buyers from the industrial countries gained overseas market access and some market intelligence. The firm also stood to benefit from steady demand for its products as well as a variety of technical assistance that could be obtained from client firms to boost skills, quality, production efficiency, and logistics.[13] Many firms gravitated toward membership in global production networks for these reasons. If they wanted to penetrate foreign markets and did not have the resources to engage in direct marketing, global production networks offered

13. See, for instance, Humphrey and Schmitz (2002). On the advances in and the importance of logistics, especially for time-sensitive products, see chapter 7.

an attractive intermediate alternative: the opportunity to export at an affordable transaction cost. Moreover, links with global production networks generated two-way flows of information that could be a means of inducing FDI in East Asian firms seeking a dose of capital, a partnership, or a transfer of ownership. Through the 1990s global production networks mediated a large increase in trade within industry and trade among MNCs, their subsidiaries, and affiliates.

Since the late 1990s, the criteria for participating in global production networks have become more exacting, especially for first-tier suppliers, because of pressures on the MNCs and the deepening imprint of information technology on transactions and business processes. In a remarkably short period of time, the large MNCs have assimilated and harnessed information technology and have become increasingly comfortable in using the new technologies to coordinate dispersed activities, manage the supply chain, synchronize production and marketing, and establish mechanisms for gathering market and technological intelligence and for sharing this information with the constituent parts of the corporation as well as with affiliated firms. Familiarity with information technology and a better understanding of its potential are inducing firms to seek leaner, often much flatter, organizational structures, to reconfigure and streamline inter-firm relationships, and to embark on innovative trading, process, and information-sharing arrangements with a wide range of business partners. Indeed, information technology is being woven into virtually every element of corporate strategy.

Information technology has served to reinforce the efforts of firms to enhance their competitiveness in a globalizing market environment by streamlining their own operations, retaining only those promising the highest returns, many of which can be overseas.[14] As firms subcontract with specialized suppliers, they are seeking to minimize the number of first-tier suppliers and to establish close long-term relationships with them, which can involve acquiring an equity stake or entering into long-term and binding contractual undertakings (Whittaker 2003). Toyota enlarged its equity in its three large parts suppliers—Aishin Seiki, Denso, and Toyota Gosei—in the late 1990s (Hatch 2003). By reducing the number of components used and purchasing from a smaller number of first-tier suppliers, assemblers can lower the cost of production and coordination in addition to simplifying supply chain management. More important, in a marketplace where competition is paced by technological advances, even the largest companies are

14. The global diversification of U.S. companies rose substantially between 1984 and 1997 (Denis, Denis, and Yost 2002).

finding it necessary to share the burden of technology development with suppliers or buyers.[15] Moreover, if technological advance is to be a shared activity with significant mutual exchanges of proprietary information among firms, lead corporations must choose and vet their potential partners with some care. Every relationship that assigns a major role in the development of a technology or the supply of a key component (or module) is fraught with risks and with reciprocal obligations. Hence, even as MNCs have deverticalized and sought to purchase what they once produced in-house from specialized suppliers in competition with one another, they are seeking dependable suppliers that can directly support their global operations. MNCs in the automotive sector, for example, are inclined to recruit first-tier suppliers from among their traditional suppliers, such as Bosch, Johnson Controls, Lear, or TRW, former subsidiaries now hived off as fully fledged independent companies, such as Delphi and Visteon, or entities in which they have acquired a sizable, if not a majority, stake. Most of the first-tier suppliers in Thailand are foreign owned, and this appears to be the trend in Korea as well.[16]

The imperative of finding trustworthy suppliers able to fulfill exacting requirements is continually sharpened by the worry that technological dependence and sharing with partners and suppliers can erode a firm's technological edge. A lead firm that has transferred technology to a supplier exposes itself to the risk that this supplier will permit a competitor access to the technology as well (the clients of wafer fabrication companies have expressed this concern). Even more threatening is the possibility that a subcontractor could acquire both the technological and the manufacturing capability to displace the lead firm in its principal product market (a possibility that worried Acer's clients).

In fact, the mixed experience with technology-sharing arrangements as well as with the ability of subcontractors to meet the worldwide needs of lead firms has underscored the advantage of selective reverticalization. Sony and Toyota have both moved in this direction in an effort to preserve vital

15. Knowledge-sharing networks, technology development partnerships, and research consortia have also proliferated as firms find it increasingly necessary to leverage the research and resources of other firms in order to arrive at desired outcomes. See Buchel and Raub (2002); Dyer, Kale, and Singh (2001); chapter 9 by Nabeshima. Dyer, Kale, and Singh (2001, p. 37) note, "Strategic alliances have become an important tool for achieving sustainable competitive advantage . . . . Currently, the top 500 global businesses have an average of 60 strategic alliances each."

16. These first-tier suppliers have gradually been enlarging their own roles. Johnson Controls, for example, has widened its functions to include cockpit engineering and added new items to those traditionally located in the cockpit, such as cell phones incorporated in the roof liners. Beyond that, the company also serves as the intermediary that deals with the second-tier suppliers providing parts for its modules ("How to Move" 2001).

skills. In countries where the local supply base is weak, it also argues for the acquisition of local firms. The most promising avenues, not infrequently, entail the acquisition of a start-up with an innovative technology or even an established firm with good local production, marketing, or brand assets.

As firms introduce these changes, they are altering their organizational structures and networking relationships. Some firms are deverticalizing and outsourcing components, subsystems, or services. Others are adding new production activities and entering new fields frequently related to their core business. These dynamics are continuously heightening the tempo of competition in East Asian markets. But, at the same time, competition and the growth of markets are presenting fresh opportunities. Local firms are struggling hard to become first-tier suppliers to the MNCs and thereby earn the higher profits accruing from close longer-term relationships and joint development of technologies. East Asian firms are also competing with foreign component suppliers that have followed lead MNCs to the region and set up production facilities in close proximity to those of the lead firms (Hatch 2003). Because MNCs are inclined to channel technology selectively to their own subsidiaries and affiliates and to demand just-in-time delivery, domestic firms are finding that they must either redouble their efforts at building technological capability (which is a stepping stone to fruitful research alliances) and exploit advances in logistics or risk being squeezed out of the global production networks.[17]

Inevitably, the more agile East Asian manufacturers, such as Haier and Hyundai, are laying the groundwork for their own global production networks. Such firms are first establishing a regional presence to solidify a brand image before venturing farther afield.

Emergence of Contract Manufacturers

Competition has emerged from another quarter as well. Coinciding with the deverticalization of leading MNCs is the growth of multinational contract manufacturers. Five firms, all based in North America (Celestica, Flextronics, Jabil Circuit, Sanmina/SCI, and Solectron), account for the lion's share of total output.[18] They have taken over

17. A firm's stock of patents can be the currency needed to acquire technology from others and to enter into partnerships.

18. The contract manufacturer phenomenon is discussed in detail by Sturgeon and Lester (chapter 2) and by Ernst (chapter 3).

the tasks of component sourcing, development, engineering, final assembly, and, to a limited extent, design for many of the major electronics and telecommunications firms. Contract manufacturers are responsible for myriad products from Microsoft's X-Box to Sony-Ericsson's mobile handsets to printers for Hewlett-Packard based on prototypes and design specifications provided by their clients. In many instances, they handle the entire supply chain, culminating in final assembly and delivery to either the client's warehouses or wholesale outlets.

Contract manufacturers have grown by acquiring the production facilities disposed of by the lead MNCs, by taking over firms in the industrializing economies, and by investing in new plants. Flextronics, the largest, had 87 plants in 27 countries and a turnover of $146 billion in 2002. Their competitive strength derives from economies of scale, scope, and geographic reach. Contract manufacturers can quote the lowest prices because of their high turnover; because they are now able to offer a wide range of electronics items, they can switch production from one category of manufacturers to another; they can pool the inventories of several customers and thereby cut total inventories; and, by maintaining production in close proximity to the major markets in North America, Europe, and East Asia, they can spread production over multiple countries and promise timely, often just-in-time, delivery (Lakenan, Boyd, and Frey 2001).

More recently, contract manufacturers are venturing into research and design activities with higher added value, possibly in response to the slowing world economy after 2001. This move could tighten their links with some of their clients, which might wish to focus even more narrowly on downstream marketing and brand management. Such capabilities also position the contract manufacturers to launch new products, possibly under their own brand name, although such a move risks alienating their key clients.

Moreover, in the lead up to the telecommunications bubble in 2000, firms such as Cisco discovered that subcontracting to contract manufacturers had substantially reduced their flexibility to boost production when demand for certain products surged. Their objectives of quick market penetration and growth of market share are likely to be at odds with those of contract manufacturers, which must manage inventory and costs with the utmost care to sustain what are often meager profit margins (Chapter 3 and Lakenan, Boyd, and Frey 2001).

In 2002 the multinational contract manufacturers, which have been joined by key East Asian ODMs such as Hon Hai Precision and Quanta, accounted for 20 percent of the market for electronics contract manufacturing. As they better integrate their expanding global production systems and strengthen their design capacity, the contract manufacturers are likely to be a force to be reckoned with.[19] The small- and medium-size firms in some of the Southeast Asian economies that lack technological capability and are faced with rising wage costs will need to devise strategies to cope with pressure from three sides: from the push by MNCs to reduce the number of first-tier suppliers and to enlarge their respective roles; from the continuing influx of foreign firms that subcontract with the multinational companies and are relocating some of their facilities to reduce costs and service their client operations; and from the regional and multinational contract manufacturers that are enlarging their economies of scale and scope and their links with MNCs. To meet these challenges, firms in East Asia need to make concerted efforts to build technological capabilities and to grow in size. In this they might be assisted by the size and unwieldiness of the contract manufacturers, which are frequently operating 50 to 70 plants in a score or more countries. Firms in East Asia might also have a little time to prepare for the challenge, because, as Dieter Ernst notes in chapter 3, relatively few of the plants operated by contract manufacturers are currently located in East Asia.

Building Technological Capability

Although, in the medium run, purchase of capital equipment embodying new technology and licensing can be the means of raising the technological threshold, over the longer term, there is no viable alternative to investment in research and development (R&D) by the firm.

R&D that facilitates technology transfer and some innovation is becoming a necessary stepping stone to enter the first rank of suppliers to MNCs, which now expect such firms to partner with them on the upgrading of existing products and to contribute to the next generation of technologies. Beyond this, investment in R&D can enable a firm to graduate from the status of an OEM supplier and credibly train its sights on design-based manufacturing as a prelude to launching its own branded products. The most likely escape from the treadmill of cutthroat competition in markets for commodities is via cultivating design skills and innovation. By enabling firms to differentiate their products or, in rare instances, create demand for entirely new products with a disruptive technology, research and design can generate rents, providing the financial wherewithal for growth.

R&D expenditure is high and rising throughout much of industrializing East Asia (relative to other industrializing regions), and more of it is now concentrated in the corporate sector, reflecting a trend observed in the advanced countries as well. The largest gains registered during 1991–2000 were by

19. Many risks associated with the contract manufacturer strategy arise from global market volatility in a low-margin business. These are discussed in chapter 3.

China, which raised R&D spending to more than 1 percent of GDP from 0.6 percent and whose purchasing power parity–adjusted spending is now the third highest in the world after the United States and Japan. China has also been able to increase its scientific publications, while the number of patent applications in China rose 26 percent annually between 1994 and 1999. The incentives offered and the lure of China's market have drawn most of the top 500 MNCs to China and induced them to establish 400 research centers ("China, R&D Spending" 2003; OECD 2003). On a smaller scale, China's efforts are being matched by those of Malaysia and Singapore, which are determined to become innovative economies. Estimated social returns to R&D range from 20 to 40 percent for the OECD countries to 60 percent for middle-income countries and to almost 100 percent for low-income ones ("Does Science Drive" 2000; Lederman and Maloney 2003). This investment is already yielding rich dividends in Korea and Taiwan (China), which rank fifth and third among economies with respect to U.S. utility patents registered between 1995 and 2001. These two economies also rank high when measured by the significance of the technologies represented by the patents. However, two-thirds of all patents registered by Korean entities were by the top five corporations, led by Samsung. In Taiwan (China), the five leading corporations registered one-quarter of all patents. Much of the R&D in Japan, Korea, and Taiwan (China) is by the larger firms (Whittaker 2003). Firms that seek to compete and to grow on the basis of innovation need to invest heavily in acquiring research capability and, beyond that, to spot promising innovations, develop them, and find ways to realize and profit from their full market potential. Serial innovation by small companies is relatively infrequent. Most small firms are lucky to have a single innovation to their credit, and significant innovations—or truly disruptive innovations—are rare.[20] Hence, where technological capability is underwritten by widespread innovation, entry barriers to new firms are low, and market and financial institutions support the growth of the more dynamic firms.

Technological dynamism has contributed to the global ambitions of Hyundai, LG, and Samsung and to the acquisition of a favorable brand image. Taiwanese (Chinese) firms have moved to the forefront of ODMs in the electronics subsector and have acquired a regional brand image. A few companies in China, Singapore (for example, Creative Technologies and

20. Geroski, Van Reenen, and Samiei (1996) find that few of the firms they studied in the United States and United Kingdom registered more than one patent or had more than one innovation to their credit. Research on 10 OECD countries shows that between 20 and 40 percent of the growth of total factor productivity is because of gains from the entry and exit of firms, with new innovation and productivity growth generated by new firms offsetting or more than offsetting the effects of exit (Scarpetta and others 2002). On the nature and potential of disruptive innovations that give rise to entirely new markets, see Christensen, Johnson, and Rigby (2002).

Chartered Semiconductor), and Thailand are also attempting to establish themselves through their technological prowess, but these countries still lack sufficient numbers of large firms in a position to sustain sizable research programs and to capitalize on their findings.

Becoming Bigger, Maybe Better

The global market environment now appears to favor firms that can operate on international scale. Such firms can combine the advantages of scale economies and greater learning by doing, with product or process innovation by way of in-house R&D combined with inter-firm collaboration. Larger firms are also better positioned to mobilize the finances, whether from their internal or market sources, to introduce process innovations and develop, test, and market new products.

Writing on industrial upgrading in Taiwan (China), Amsden and Chu (2003, pp. 2–3) observe,

> Small-scale firms may be the first in a latecomer country to introduce a world-class foreign technology. But typically, small firms take a backseat in upgrading to national large-scale firms. Whereas small firms in advanced countries may be technological pioneers, small firms in latecomer countries do not fulfill this role because they do not yet have technology at the world frontier. Many are [also] backward in terms of management skills . . . . Both small and large firms in Taiwan (China) moved their manufacturing operations to China, but only large firms had enough resources to maintain product development facilities in Taiwan, enabling them to upgrade concurrently. Both large and small firms in the electronics sector have R&D, but the content differs. Small firms dominate in the non-electrical machinery industry, but this industry has stagnated in terms of its share of GNP. Even the small suppliers of parts and components to electronics assemblers have been subject to one-stop shopping rationalization as each assembler has demanded a greater number of parts from a smaller number of suppliers, thereby pushing the importance of bigness down the supply chain. By the second or third generation of industrial development, most start-ups or promising small-scale firms are tied in one way or the other to an existing large firm.[21]

In an integrating world economy, the most profitable and growing firms are generally those able to market their products in several of the major international markets. Such firms are generally among a handful that dominate the global markets for a product or service, whether it is the software for mobile handsets (for example, ARM of Cambridge, United Kingdom),

21. Japanese small- and medium-scale enterprises were a less prolific source of innovation from the 1950s through the 1970s. Of 34 major innovations, they were responsible for only two (Hatch 2003).

the machinery for making frozen pizza, or the equipment for producing flat-panel displays (for example, Applied Materials). Large firms that are active on a global scale can leverage the advantages of size and international operations by creating their own global production networks; by acquiring promising innovative firms worldwide to augment their R&D capability and product range; and by investing in overseas subsidiaries that reap the benefits of proximity to important markets. This is the path taken by virtually all the leading Western and Japanese firms. It is a model being emulated with some success by a small number of Chinese, Korean, and Taiwanese (Chinese) firms.[22] In a world where the dynamics of competition are being redefined by innovation, information technology, and business relationships favoring firms that have honed their research and design skills and have the capacity to operate on an international scale, small, nimble firms can certainly survive, but larger firms are more likely to prosper. Firms such as Hyundai, LG, and Samsung from Korea; Haier, Huawei, Legend, and ZTE from China; and Acer, Hon Hai, Mostek, Quanta, Asustek, and Vitelic from Taiwan (China) are beginning to respond to the changing global environment by seeking an international presence and aiming for the status and profitability of OBMs. Nevertheless, firms from China and Korea, even the medium-size ones, remain too highly diversified and vertically integrated compared with their rivals from the Western countries. Vertical integration is a means of containing transaction costs when market institutions are weak, and diversification is often the only means of growing when competition mounts in existing product lines. But there are limits to such strategies, and it remains to be seen whether East Asian firms can grow and compete without changing their structures and acquiring greater focus ("Samsung Way" 2003).

Government Policies

To date, the East Asian governments have initiated and supported technological advance through six broad policies:

1. Human capital deepening
2. Creation of publicly financed research institutes
3. Grants, subsidies, and tax incentives for private R&D activities and government contracts

22. On balance, the size of even the largest Chinese firms is still relatively small compared with their foreign competition, a point that Nolan and Zhang (2002) and Nolan (2002) have carefully documented. However, a few firms, such as Sinochem, Baosteel, and Shougang Steel, are approaching the size of the multinational companies ("Beijing Fails to Dent" 2003; "Steel Magnolia" 2003).

4. Technology licensing policies and technology transfer arrangements through FDI in high-tech industries
5. R&D by public sector firms
6. Incentives for information technology.

These are discussed in more detail below.

Circulation and Creation of Technology

Complementing R&D spending is the availability of knowledge workers with the requisite skills and experience. This is vital for the productivity of research endeavors and for the commercialization of findings, a point stressed by Hill (chapter 8), Nabeshima (chapter 9), and Jefferson and Zhong (chapter 10) in this volume. Governments in East Asia have emphasized investment in primary and secondary education for decades. More recently, they have increased the outlay on tertiary-level education and are taking steps to raise its quality. Governments in Japan, Korea, and Taiwan (China) have also come to recognize the desirability of orienting the leading universities toward research, promoting university-business links, and encouraging universities to patent research findings. Universities in the OECD countries account for close to half of all basic and applied research, whereas in East Asia, the share is lower, with universities in China, for example, accounting for just a quarter of the country's applied research and a third of its basic research (OECD 2002).[23] Starting with Japan in the 1950s (and earlier), overseas training has been viewed as a way of supplementing the domestic supply of trained workers, enhancing the quality, and acquiring relevant experience. Now several East Asian economies have large diasporas of highly skilled workers, many of whom have started to return as opportunities in their home economy have multiplied. Looking forward, the higher-income countries will need to sharpen their focus on the quality of graduates from tertiary-level institutions and to impart a stronger research orientation to the leading universities. Middle-income countries such as Malaysia and Thailand will also need to shift attention and resources to building those types of skills that will contribute to innovation and not just manufacturing capability.

High- and middle-income East Asian economies have already instituted tax incentives to encourage research by the private sector.[24] These

23. By giving researchers higher royalties, U.S. universities have induced an increase in investment (Lach and Schankerman 2003).

24. Wang and Tsai's (2003) research on Taiwan (China) shows that R&D has contributed substantially to the productivity of manufacturing firms.

are supplemented by grants and subsidies for specified activities and by government-financed (and some privately financed) venture capital for high-tech start-ups (Kenney 2004). Singapore, for instance, has earmarked close to $3 billion for the development of its biotech industry. Singapore and Taiwan (China) have, among others, also used government-funded research institutes to absorb new technologies and to serve as the nucleus for an industrial subsector.[25] The Industrial Technology Research Institute (ITRI) in Taiwan (China) is an oft-cited example of a public research entity that helped to launch the microelectronics industry and whose expert midwifery was responsible for the birth of both TSMC (Taiwan Semiconductor Manufacturing Corporation) and UMC (United Micro-Electronics Corporation), which are among the two leading foundries in the region (see Mathews and Cho 2000). However, public research institutes in China, Korea, and Japan have been less productive in generating both basic research and commercially viable innovations. It is likely that basic research might continue to require public funding, but whether it goes to public institutes or is channeled to universities and private research centers is a matter of debate. Commercially oriented research might be left more profitably to private industry, with the state providing suitable financial inducement. The examples of early successes such as ITRI in Taiwan (China),[26] the Korea Advanced Institute of Science and Technology (KAIST) in Korea, and the Karolinska Institute in Sweden are few in number, and the experience was not repeated in the 1990s. In fact, the trend in Europe and the United States is for the state to scale down its funding for public research institutes, which can be small and isolated, in favor of support for the private sector and the university system.

Avenues for technology transfer through licensing and trade are well marked, and firms in East Asia are savvy enough to seek out the technologies they require and to negotiate reasonable terms, whether through trade or licenses. Arguably, there is less need and also less scope for governments to involve themselves with specific technologies and with the details of contractual terms, as happened in the 1960s and 1970s. Markets are more competitive, participants are more experienced, and the rules of the game have changed.

Debate continues to swirl around the modalities of technology transfer through FDI. Most researchers would agree that FDI generally leads to technology transfer through a variety of channels, more in some

25. As Frantzen (2000) and others have shown, for non-G-7 countries, assimilating research done elsewhere initially can be more important for technological advance than domestic research.

26. According to Hill (chapter 8 of this book), ITRI has working relations with 20,000 companies.

industries than in others. But researchers differ on the degree to which FDI results in horizontal spillovers to other firms in a subsector versus vertical transmission of technology, mainly to subsidiaries and affiliates of foreign firms. Blalock and Gertler (2003) show that vertical transmission dominates, and their findings are supported by other research. However, evidence of horizontal spillovers—in particular, through the circulation of workers—is not lacking, and we have certainly not heard the last word on this.[27] What can be stated with some assurance is that FDI will remain advantageous for East Asian economies in the future, just as it was in the past, especially for Malaysia, Singapore, and Taiwan (China), because it is a vehicle for transferring a range of technologies and capital. Moreover, it offers avenues for linking firms to global production networks. Even in countries, such as China, where the aggregate supply of capital more than matches demand, financial market distortions impede the cross-provincial flow of resources and constrain the access to funding. For many firms, FDI is often the only means of overcoming the dearth of financing (Huang 2003).

Recognizing the gains from FDI—in particular, its potential contribution to move up the value chain—East Asian economies are continuing to engage in contests to attract FDI. They are offering relatively closely matched fiscal incentives, together with the infrastructure and facilities of special economic zones and increasingly liberal trading arrangements that facilitate intra-industry trade and transactions that sustain global production networks. Even Korea, which historically viewed FDI with a skeptical eye, is actively wooing foreign investors and preparing to carve out special economic zones. In exchange, East Asian economies can demand technology transfer, set thresholds for local content, and require that some fraction of the output of joint ventures or foreign-owned firms be exported.

Market Institutions and Incentives

Most East Asian countries have progressed well beyond the basics of technology policy. They have a sophisticated understanding of the relative importance of human capital, the role of R&D in private companies and universities, and the role of FDI and trade. Countries have learned from their own experiences, from that of their neighbors and

27. See Sjöholm (1997), who notes that technology transfer to local Indonesian firms tends to be greatest when markets are highly competitive.

the advanced countries, and from numerous contacts with the international business community. To a greater or lesser extent, they are disengaging from old-style industrial policies and concentrating on augmenting the supply of skills, research, foreign capital flows, technology transfers, and trade. The direction being taken is broadly appropriate given the current state of knowledge. But the early returns still lag behind expectations, suggesting that a new mix of policies might be needed.

It is far from obvious that development can be accelerated beyond a point through greater outlay of resources. Instead, East Asian economies that are eager to enhance their technological capability may need to combine investment in technological capacity with greater emphasis on reforming regulatory policies, including crafting and enforcing competition policies that will eventually be harmonized across the region; building intellectual property rights regimes that are credible and promote innovation; permitting firms to engage in mergers and acquisitions (M&A) subject to the rules defined by competition policy; and encouraging international research collaboration through links among companies, universities, and research institutes. In addition to national policies, there is a major role for municipal policies that can influence the location of firms—in particular, firms producing high-value-added goods and services, which are the largest employers of knowledge workers. Most East Asian countries have legislated such policies but have yet to put much energy into implementing them.

Familiarity with information technology and its use by businesses and for research has risen quickly throughout East Asia in response to demands of the leading electronics and automotive companies and aided by government incentives. This trend could be reinforced by actions on two fronts. First, firms need to be more proactive in acquiring infrastructure and building information technology skills. This can only enhance their competitiveness as members of global production networks; a strong base of information technology, continuously upgraded through investment, is essential for a firm with the ambition to become a global ODM or OBM. Second, government regulatory policies toward the telecommunications sector and Internet access affect the supply of services, cost, quality, technological advance, and access, as analyzed by Crandall and Litan (2002) and by Yusuf and Evenett (2002).

Competition can be a powerful spur to innovation, while an active market for corporate control through M&A not only reinforces the competitive pressures emanating from the market and trade in goods but also enables successful firms to grow and reap the benefits of size. Research on competition policy and its harmonization across a region suggests that there are significant longer-term gains to be realized from defining and policing rules governing firm behavior. Harmonization, in turn, can facilitate cross-border transactions, including M&A, and diminish transaction costs. Globalization has enhanced the attractiveness of institutional uniformity in these areas, and whether or not the trend toward global integration persists, a more limited goal of creating common regional institutions itself could be worthwhile.

Until recently, there was little demand from the business community for a competition policy and scant interest on the part of governments in introducing or pursuing such a policy. Similarly, the business culture throughout East Asia is resistant to the idea of unbridled M&A, as owners can be possessive about the companies they have set up, worry about a loss of face if their firm is taken over, and are often at pains to avoid upsetting patterns of market coexistence and imperiling settled relationships. This has influenced growth in the size of East Asian firms from middle-income countries, limited the consolidation of firms within industrial subsectors into more optimally sized units, and encouraged growth through diversification.

As the region integrates with the global economy and faces stern competition, reducing the institutional obstacles to M&A and reconsidering the resistance to foreign M&A might initiate the shakeout of industries and the emergence of a few large firms able to challenge MNCs on the global market.[28] At the same time, competition policy should seek to ensure that the barriers to entry are kept low so as to induce a steady flow of start-ups that bring promising new producers, innovations, and entrepreneurial energy to the industrial sector. In China, as Steinfeld points out in chapter 6 of this volume, provincial and county authorities routinely prevent M&A of a local firm by a firm from another province so that their administrative control over industry is not diluted. These very same authorities on occasion force quite irrational mergers among firms under their jurisdiction and also curtail the entry of firms from other parts of the country that might pose a competitive threat.

Parallel actions pertaining to intellectual property and the formation of industrial clusters in urban areas can reinforce such institutional stops. All of the middle- and high-income economies of East Asia are signatories of the WTO-TRIPS (Trade-Related Aspects of Intellectual Property Rights) agreement and are committed to the rules governing intellectual property. As these countries are now seeking to buttress their future competitive position by relying more extensively on innovation, especially in the high-tech

28. On December 2, 2002, China introduced new regulations making it easier to take over listed companies ("China, M&A Rules" 2002). Similarly, attempts to deregulate markets are under way in Japan and Korea ("Japan Talks Tough" 2003).

areas, defining and securing intellectual property take on far greater importance than was the case when they were firmly in the imitative phase of development. Research in the United States indicates that, by protecting intellectual property, governments provide incentives to innovations through a perceived patent premium (Arora, Ceccagnoli, and Cohen 2003). This is especially valuable in those areas where research and testing are costly, as in biotechnology and pharmaceuticals. Securing intellectual property rights is also of increasing importance for the "creative industries,"[29] including media, publishing, and software, as the copying and mass production of music, video games, and computer programs have become easier, putting great pressure on the companies responsible for the original content. Nevertheless, stronger intellectual property rights alone will not galvanize innovation—in fact, Korea's experience suggests that competition more than the intellectual property regime is largely responsible for the rise in innovation (Luthria and Maskus 2004). However, as East Asian countries move to strengthen the rules governing intellectual property and their legal enforcement,[30] these steps, in conjunction with measures to increase basic and applied research efforts, could deepen technological capability and permit East Asian countries to earn higher returns by differentiating manufactured products and entering new and more profitable product niches.

Clusters and Urban Policy

The innovative impulse is likely to be stimulated further if East Asian economies also redouble their efforts to create an urban environment conducive to the clustering of firms. Global production networks flourish when the affiliated firms can cluster in urban areas richly supplied with producer services (see Bresnahan and others 2001; Yusuf with others 2003; chapters 3 and 10 of this volume). Such clustering can be especially advantageous for high-tech and creative industries that derive benefits from local and specialized markets, from spillovers of tacit knowledge, from technological collaboration, from just-in-time delivery, and from the ability to access the services of financial, accounting, legal, and logistics firms among other providers.

In the developed countries, the dynamic clusters of innovative firms are found in a relatively small number of cities that are able to attract the tal-

29. The rising significance of these industries for East Asian cities is discussed in Yusuf and Nabeshima (2004).

30. The parallel institutional strengths of the legal system will be of critical significance, as emphasized by Steinfeld in chapter 6 and by Yusuf with others (2003).

ented knowledge workers and the type of firms that rely on their skills.[31] These cities offer an attractive physical environment, efficient infrastructure, convenient transport links with other countries, and reliable utilities. Beyond that, cities that are hubs of innovation have—by dint of regulation, investment in physical assets, and civic action—put in place cultural and social amenities that greatly enhance the appeal of what otherwise would be just efficient cities. As an example of what can be done to improve efficiency, Singapore's investment in port infrastructure, a skilled work force, customs organization, and information systems has made it possible for importers of electronics components to receive delivery six hours after the product arrives at a port or an airport. Similar efforts are under way in Shanghai.

As East Asian economies seek to improve their innovation capability in the interests of competitiveness, they will have to struggle to retain footloose high-tech firms. The urban dimension of policymaking will be no less critical than the rest. A credible regime of intellectual property rights will not contribute to the government's objectives if the leading cities within the country cannot attract and retain firms in the sectors with the most promise for growth. Hence, policies are needed to draw firms to specific urban locations in the hope of creating rapidly growing clusters of dynamic firms (Florida 2002). Innovation policy is thus becoming more complex. In a globalizing world, how cities evolve can determine whether and where pools of skilled workers collect and can influence the location decisions of firms that will decide tomorrow's economic performance.

OVERVIEW OF THE VOLUME

The contributors to this volume take us deep into the dynamics sketched above, starting with global production networking, then moving to the closely linked transformation sweeping the logistics sector, and then coming to technological change, which is the force now determining the competitiveness of firms, defining production relationships, and, increasingly, shaping the industrial geography of the East Asian region. As Timothy Sturgeon and Richard Lester show in chapter 2, with reference to the electronics and automobile industries, the lead firms dominating international production networks are adopting a global approach to their business operations and have begun to consolidate manufacturing in fewer locations, generally in

31. Contract manufacturers such as Flextronics, much like the auto assemblers, prefer to have their suppliers co-locate in clusters or business parks. Hill observes in chapter 8 that a quarter of Singapore's labor force is foreign and an even higher proportion is foreign born.

countries where potential market growth is greatest. The lead firms have focused their own activities on areas such as research, design system architecture, supply chain management, integration, and marketing. Such reorientation is proceeding hand in hand with the outsourcing of the production and often the design of entire modules to a small number of trusted suppliers. These suppliers, in turn, purchase their components from many second-tier producers, which are forced to compete fiercely for market share. Sturgeon and Lester observe that commodification of products and sharp competition are intensifying pressures on East Asian manufacturers and driving down profits. Moreover, the emergence of North American–based contract manufactures with globally distributed facilities and substantial economies of scale and scope threatens to squeeze the smaller regional producers even further. If East Asian firms in the electronics and auto industries are to grow and enhance their profitability, they will need to vie for the role of first-tier suppliers for lead firms, to operate on a global scale, and, in certain instances, to co-locate plants near the facilities of lead assemblers. Developing research and original design capability can further strengthen competitiveness, enable firms to take responsibility for entire modules, and eventually make a transition to original brand manufacturing on a regional or global scale. Sturgeon and Lester note that some East Asian firms have entered the ranks of first-tier electronics suppliers with ODM capacity. A select few, such as Samsung, have emerged as OBMs, but the vast majority remain second- and third-tier producers of standardized commodities.

What is required for such firms to upgrade is the central theme of Dieter Ernst in chapter 3. He discusses the increasing role of contract manufacturers in the global production networks, the integration of Chinese and Malaysian companies into these networks, and how digital information systems have facilitated the process. But Ernst also underscores the challenges faced by the contract manufacturers as well as gaps in technological capability between East Asian firms and their foreign competitors, a gap that is highlighted by the continuing heavy dependence on imported inputs. Although investment by the Malaysian authorities in information technology infrastructure can assist in the upgrading process, the principal bottleneck is the shortage of skilled workers and research personnel. Ernst concludes that, by easing skill shortages, the state can shift industry along the value chain and heighten technological capability.

Although industrial upgrading can help firms in countries such as Malaysia to remain a part of global production networks and possibly even improve their ranking as suppliers of components, the impending regional consolidation of the automobile and electronics industries will be painful for some producers. In chapter 4, Richard Doner, Gregory Nobel, and John Ravenhill describe the dispersed development of the auto industry across

East Asia behind tariff barriers that are now being dismantled. The freeing of trade is reinforcing the efforts of multinational auto companies to rationalize their production, to cut costs by concentrating on a handful of platforms to be marketed worldwide, and to rely on a few hundred first-tier suppliers for parts and modules. Doner, Nobel, and Ravenhill visualize an increasing concentration of the East Asian auto industry in three hubs. Korea's auto sector, the most advanced in East Asia after Japan, will benefit from strong domestic demand, expanding exports, and investment by multinational companies that have entered the business of assembling and producing auto parts, in the process buying up many local firms. A second hub is emerging in Thailand, which has used fiscal incentives to encourage the manufacture of light trucks and has since attracted investment in the assembly of sedans and the production of auto parts. Major auto companies are likely to use Thailand as the base from which to export automobiles and parts to other countries in Southeast Asia. This is likely to affect the development of the auto industry in Indonesia, Malaysia, and the Philippines, each of which has built up substantial production capacity. Malaysia, in particular, has aggressively used tariff and other incentives to promote a domestically owned auto industry that will have difficulty competing against imports once the tariff barriers are scaled down.

A third hub is taking shape in China, driven by the rapid expansion of the domestic market, massive investment in capacity by leading automakers as well as by local producers, and the efforts by the Chinese authorities to localize parts and production and to assimilate foreign technology. With demand in China likely to grow over the foreseeable future, automakers such as General Motors and Toyota will be relying on their sales in China and exports from their factories located there to sustain their profitability. Thus the new geography of regional production and marketing is likely to favor firms from the Northeast Asian countries and Thailand that are part of automotive global production networks.

Chapter 5 by Ken'ichi Takayasu and Minako Mori and chapter 6 by Edward Steinfeld explore the implications for Chinese and Thai firms against the backdrop of industrial changes occurring in these two countries. Takayasu and Mori note how the major Japanese and Western assemblers have been developing a base of suppliers to reduce their costs and ensure just-in-time delivery. Most of these suppliers, especially first-tier ones, are subsidiaries of foreign firms that work closely with lead assemblers in their home country. Thai firms figure among the second-tier suppliers and have been slow to acquire the technology and design skills that would enable them to undertake the manufacture of higher-value parts and to work more closely with assemblers and first-tier suppliers in R&D and in design. Thai producers, much like firms in Indonesia and Malaysia, are constrained by a shortage of workers with the technical and managerial skills that could assist with the assimilation of design and production technologies. These scarcities have been apparent for some time, and it will require concerted action by both government and industry to build a stronger base of Thai auto component suppliers that can participate in the rapid expansion of the auto sector in Thailand and—through global production networks—in other countries as well. However, skills are not the only bottleneck. Thai firms have also been slow to acquire expertise in information technology and the infrastructure that is now essential for effective participation in the finely tuned business of auto manufacturing. For all the leading assemblers, a careful orchestration of the supply chain is vital to achieve efficient and profitable production and avoid excess outlay on inventories.

In some respects, the situation in China is not dissimilar. Over the past decade, Chinese firms have engaged in trade on a vastly larger scale than their counterparts in Thailand as a result of FDI in export-oriented industry and through the initiative of local enterprises. This has multiplied their links with global production networks for a wide range of products. However, most Chinese firms remain relatively small, and typically they work with codified technologies to produce standardized commodities with the minimum of design and R&D inputs. Larger firms, such as Huawei, Legend, and ZTE, have moved up the value chain and are beginning to penetrate international markets under their own brand name. But even these firms continue to operate in the medium-tech range, producing standardized products or modules and selling these for minimal profits in highly competitive markets.

In order to grow and to increase their profits, the larger Chinese firms are still relying on vertical integration and product diversification. This has yielded decent short-term dividends; however, Steinfeld concludes that in the longer run such firms will have to imitate their foreign competitors, develop proprietary technologies that generate higher rents, and be able to market products or services straddling a number of industries. Small firms, and larger ones, will need to enhance their technological capability in the interests of competitiveness and profitability not just in products and processes, but also in logistics and supply chain management. These topics are discussed in the latter part of the volume, starting with chapter 7 on logistics.

The massive increase in East Asian trade and production networking has led to a gap in logistics capacity across the region, although more so in China and Vietnam than in Southeast Asia. As Trevor Heaver shows in chapter 7, it is not only the physical infrastructure that needs to be augmented but also the institutions, skills, procedures, and technologies. Producers and buyers order more frequently in smaller lots and expect to track their shipments so that they can synchronize deliveries with their own production schedules and with a minimum of warehousing. They also expect to conduct and settle more of their transactions electronically. Thus the logistics system in East Asia must cope with an enormous increase in volume, while at the same time it must adapt to the demand for speed in the delivery of time-sensitive items often shipped in small lots.

Economies such as Singapore have responded by expanding their ports, airports, and information technology facilities to accommodate the new traffic and to facilitate electronic clearance of goods through customs. In Hong Kong (China), Malaysia, and Singapore, the interfaces between different transport modes are becoming easier to negotiate, greatly facilitating multimodal transport. Shipping has improved as a result of efficiencies in customs clearance and freight forwarding and also the emergence of firms that monitor, consolidate, and manage freight movement for their clients. Firms such as APL Logistics and Maersk Logistics now provide integrated logistics services that are further reducing the transaction costs for firms and helping to multiply as well as cement networking arrangements within and beyond the East Asian region.

Progress has been slower in China, although port and airport capacity has been rapidly augmented in the principal gateway cities such as Shanghai and Tianjin and in the Pearl River Delta area. China's deficiencies are most serious in three areas. First, multimodal transport, especially the inland surface transport facilities, is inadequate, and the interfaces between transport modes confront shippers with numerous problems. Second, forwarding, consolidation, and other logistics-related services are also deficient, and Chinese customs clearance falls well short of the standards and technology attained by Hong Kong (China) and Singapore. Alongside these weaknesses, China's information technology infrastructure and the skills to effectively assimilate information technology with logistics are still relatively underdeveloped. Nevertheless, an amazing amount of progress has been made during the past 15 years in China and throughout the region in building physical infrastructure and modernizing logistics services to support the growth of commerce. In Southeast Asia continuing investment and stronger competition should further improve services, but competition will also put pressure on providers to pare costs and continuously upgrade technology.

In China and in Vietnam the ground to be covered is greater, but the momentum, as Heaver shows, is strong. Infrastructure—both soft and hard—is being created, and the use of information technology for production, networking, and logistics is widening steadily. Moreover, with accession to the WTO, China is increasing access to foreign logistics suppliers, and this, coupled with policy reform and the emergence of domestic firms, should bring about a progressive upgrading of the logistics system.

The centrality of innovation for industrial progress under conditions of globalization is the theme of the remaining three chapters. In chapter 8, Hal Hill reviews the building of innovation capacity in Indonesia, Korea, Malaysia, Singapore, and Taiwan (China) and draws a number of lessons. He indicates that economies, such as Singapore and Taiwan (China), that are at the forefront with respect to the development of domestic innovation systems and have integrated most tightly with global production networks have adopted similar policies. The two economies pursued macroeconomic stability, openness, and the building of a modern transport and communications infrastructure. This was combined with aggressive efforts to attract investment by MNCs, which created a base of export-oriented industries and offered access to new products and process technologies. In both economies, the assimilation and leveraging of technology to move up the value chain were promoted by the elastic supply of skills and by well-funded research institutes. These were the outcome of industrial policies that focused increasingly on innovation systems rather than on targeting manufacturing industries, as was the case in Indonesia, Korea, and, to a lesser extent, Malaysia. Hill observes that Korea sacrificed some innovation capability by discouraging FDI, a move that also limited the integration of Korean firms into global production networks. Both Indonesia and Malaysia poured substantial resources into major industrial projects that have yielded few technological gains. As the competitive pressures from Chinese exporters intensify, other East Asian economies will need to strengthen their innovation capability by building technical skills, stimulating research effort, and pulling FDI into higher-technology activities with greater value added.

The complementarity between the supply of skilled workers, FDI, and technology transfer is the principal message of Kaoru Nabeshima's wide-ranging survey in chapter 9. After carefully examining the various conduits for technology transfer such as capital investment, licensing, and FDI, Nabeshima concludes that investment in human capital and R&D by local firms—although geared more toward learning—are necessary to absorb technology from overseas, to maximize the spillover benefits from FDI, and to start the cycle of domestic innovation.

In the final chapter in the volume, Gary Jefferson and Zhong Kaifeng focus on the determinants of innovation at the level of the firm based on data from a survey of 1,826 in firms in 11 cities in seven East Asian economies. The chapter also assesses the relative innovativeness of individual cities. Jefferson and Zhong's empirical estimates validate many of the findings reported in earlier chapters. For example, they show that R&D capability and associated firm performance are explained by human capital intensity and the institutional setting, especially the level of competition and the degree of international exposure. The survey findings further show that firms benefit from clustering in cities where this improves access to information technology and physical infrastructure and increases networking relationships. Among the cities surveyed, Seoul has the highest human capital intensity, but also the lowest degree of international exposure and FDI, which is consistent with Hill's observations. Kuala Lumpur, Singapore, and Manila are ranked lower in terms of human capital but are more open and enjoy the advantages of higher levels of FDI. Among Chinese cities, Guangzhou and Shanghai lead the field in terms of firm performance and openness, with Chengdu and Tianjin trailing behind. The performance of firms in Shanghai reinforces the key theme of this volume: in an integrating world environment, firms in more open economies that benefit from FDI and are linked to global production networks have an edge over other firms. Beyond this, Jefferson and Zhong's research also reinforces the point made repeatedly in earlier chapters that the winners from international networking are the firms that rise to the uppermost tiers by harnessing research, design, and production skills and acquiring innovation capability. For firms that develop such capability, global production networks can provide the route to wider markets and larger profits.

In sum, the message of this volume is a simple one. The freeing of world trade, the commodification of many manufactured products, and the changing relations between lead firms and others participating in global production networks are forcing firms to compete more on the basis of innovation, the effective harnessing of logistics, and the assimilation of information technology. Success ultimately rests on the initiative of firms themselves, while government policies strongly influence the supply and quality of human capital, the dynamism of the institutional environment, and the efficiency of the physical infrastructure.

Export-led growth throughout much of industrializing East Asia was spearheaded by small- and medium-size firms, and these will remain a key source of industrial vigor. However, the experience of a number of economies such as Finland, Korea, Sweden, Switzerland, and Taiwan (China), for example, points insistently toward the contribution of large firms to innovation as well as to the branding and marketing of products on a global scale. An environment that is conducive to the growth of national firms that can compete against the MNCs on world markets, match their innovative capability, and vie with them in creating global production networks would be part and parcel of a development strategy pegged to technological advance.

Although competitiveness will be determined by the performance of firms, as Doner, Nobel, and Ravenhill note in chapter 4, converting "strategic options into reality [will depend] on policies, institutions, and politics." As described above, the supply of human capital, which is the main input into the production of technology, is determined in East Asia largely by government policy. Government policy also influences the expenditure on R&D by firms, universities, and research institutes. Regulatory policies of the state determine the access to information technology and its assimilation into the mainstream of business activities. Moreover, public regulatory policies—central and municipal—have a large hand in determining the growth of dynamic industrial clusters in key urban areas. Many layers of regulatory policies complement significant institutional measures. In this chapter, I have noted three sets of institutions that govern the innovation of firms: M&A activity, intellectual property rights, and venture capital. They are by no means the only ones. Indeed, one can think of a host of others that impinge on innovation, albeit more indirectly and with less force. However, these three certainly deserve prominence and are clearly at the center of ongoing efforts to develop technology in East Asia.

In the coming decades, East Asian countries will confront a far more demanding global market environment. Their ability to capitalize on the opportunities inherent in this environment will rest in no small part on acquired technological capability and on how East Asian firms enlarge the gains from participating in global production networks. The East Asian firms have shown that they are fast learners. If they retain this knack, their future could be as bright as it has been for the past three decades.

REFERENCES

The word *processed* describes informally reproduced works that may not be commonly available through libraries.

Amsden, Alice H., and Wan-Wen Chu. 2003. *Beyond Late Development.* Cambridge, Mass.: MIT Press.

Arora, Ashish, Marco Ceccagnoli, and Wesley M. Cohen. 2003. "R&D and the Patent Premium." NBER Working Paper 9431. National Bureau of Economic Research, Cambridge, Mass. Processed.

Arteta, Carlos, Barry Eichengreen, and Charles Wyplosz. 2001. "When Does Capital Account Liberalization Help More Than It Hurts?" NBER Working Paper 8414. National Bureau of Economic Research, Cambridge, Mass. Processed.

"ASEAN: Free Trade Fears." 2002. *Oxford Analytica*, August 13.

"Autos and AFTA." 2002. *Business Asia*, August 26.

"Beijing Fails to Dent Steel Industry's Rapid Expansion." 2003. *Financial Times*, October 30.

Bhagwati, Jagdish. 2004. *In Defense of Globalization.* New York: Oxford University Press.

Blalock, Garrick, and Paul Gertler. 2003. "Technology Acquisition in Indonesian Manufacturing." Paper prepared for East Asian Prospects Study. World Bank, Washington, D.C. Processed.

Bourguignon, François, and others. 2002. *Making Sense of Globalization.* CEPR Policy Paper 8. Paris: Centre for Economic Policy Research for the European Commission Group of Policy Advisors.

Bresnahan, Timothy, Alfonso Gambardella, Anna-Lee Saxenian, and Scott Wallsten. 2001. "Old Economy Inputs for New Economy Outcomes: Cluster Formation in the New Silicon Valley." SIEPR Policy Paper 00-43. Palo Alto, Calif.: Stanford Institute for Economic Policy Research.

Buchel, Bettina, and Steffen Raub. 2002. "Building Knowledge—Creating Value Networks." *European Management Journal* 20(6, December):587–96.

Cairncross, Francis. 2002. *The Company of the Future.* Boston, Mass.: Harvard Business School Press.

"China, M&A Rules to Entice Investors." 2002. *Oxford Analytica*, October 31.

"China, R&D Spending." 2003. *Oxford Analytica*, November 20.

Christensen, Clayton M., Mark W. Johnson, and Darrell K. Rigby. 2002. "Foundations for Growth: How to Identify and Build Disruptive New Businesses." *MIT Sloan Management Review* 43(3, Spring):22–31.

Crandall, Robert, and Robert Litan. 2002. "The Internet, Telecommunications, and Economic Growth in East Asia." Paper prepared for East Asian Prospects Study. World Bank, Washington, DC. Processed.

Crotty, James. 2002. "Why There Is Chronic Excess Capacity." *Challenge* 45(6):21–44.

Cutler, Harvey, David J. Berri, and Terutomo Ozawa. 2003. "Market Recycling in Labor-Intensive Goods, Flying-Geese Style: An Empirical Analysis of East Asian Exports to the U.S." *Journal of Asian Economics* 14:35–50.

Denis, David J., Diane K. Denis, and Keven Yost. 2002. "Global Diversification, Industrial Diversification, and Firm Value." *Journal of Finance* 57(5):1951–79.

Dicken, Peter. 2003. *Global Shift: Reshaping the Global Economic Map in the 21st Century.* London: Guildford Press.

"Does Science Drive the Productivity Train?" 2000. *Science*, August 25.

Dyer, Jeffrey H., Prashant Kale, and Harbir Singh. 2001. "How to Make Strategic Alliances Work." *MIT Sloan Management Review* 42(4):37–43.

Edison, Hali J., Michael W. Kleing, Luca Ricci, and Torsten Sloek. 2002. "Capital Account Liberalization and Economic Performance: Survey and Synthesis." NBER Working Paper 9100. National Bureau of Economic Research, Cambridge, Mass. Processed.

Eichengreen, Barry, and David Leblang. 2003. "Capital Account Liberalization and Growth: Was Mr. Mahathir Right?" NBER Working Paper 9427. National Bureau of Economic Research, Cambridge, Mass. Processed.

Feinberg, Susan E., and Michael P. Keane. 2003. "Accounting for the Growth of MNC-based Trade Using a Structural Model of U.S. MNCs." University of Maryland,

Robert H. Smith School of Business; Yale University, Department of Economics. Processed.

Florida, Richard. 2002. *The Rise of the Creative Class: And How It Is Affecting Work, Leisure, Community, and Everyday Life*. New York: Basic Books.

Frantzen, Dirk. 2000. "Innovation, International Technological Diffusion, and the Changing Influence of R&D on Productivity." *Cambridge Journal of Economics* 24:193–210.

Gereffi, Gary, and Olga Memedovic. 2003. "The Global Apparel Value Chain: What Prospects for Upgrading by Developing Countries?" United Nations Industrial Development Organization, Vienna. Available at www.unido.org. Processed.

Geroski, Paul A., John Van Reenen, and Hossein Samiei. 1996. *How Persistently Do Firms Innovate?* Discussion Paper 1433. London: Centre for Economic Policy Research.

Hallward-Driemeier, Mary, Giuseppe Iarossi, and Kenneth Sokoloff. 2003. "Exports and Manufacturing Productivity in East Asia: A Comparative Analysis of Firm-Level Data." NBER Working Paper 8894. National Bureau of Economic Research, Cambridge, Mass. Processed.

Hanson, Gordon H., Raymond J. Mataloni Jr., and Matthew Slaughter. 2003. "Vertical Production Networks in Multinational Firms." NBER Working Paper 9723. National Bureau of Economic Research, Cambridge, Mass. Processed.

Hatch, William. 2003. "Japanese Production Networks in Asia: Extending the Status Quo." In William W. Keller and Richard J. Samuels, eds., *Crisis and Innovation in Asian Technology*. Cambridge, U.K.: Cambridge University Press.

Held, David, Anthony McGrew, David Goldblatt, and Jonathan Perraton. 1999. *Global Transformations*. Palo Alto, Calif.: Stanford University Press.

Henry, Peter Blair. 2003. "Capital Account Liberalization, the Cost of Capital, and Economic Growth." NBER Working Paper 9488. National Bureau of Economic Research, Cambridge, Mass. Processed.

Hobday, Michael. 1998. "Product Complexity, Innovation, and Industrial Organization." *Research Policy* 26(6):689–710.

"How to Move Gently Downstream." 2001. *Financial Times*, October 16.

Huang, Yasheng. 2003. *Selling China: Foreign Direct Investment during the Reform Era*. New York: Cambridge University Press.

Humphrey, John, and Hubert Schmitz. 2002. "How Does Insertion in Global Value Chains Affect Upgrading in Industrial Clusters?" *Regional Studies* 36(9):1017–27.

"Japan Talks Tough about Closing the Gap." 2003. *Financial Times*, November 25.

Kenney, Martin. 2004. "Venture Capital Industries." In Shahid Yusuf, M. Anjum Altaf, and Kaoru Nabeshima, eds., *Global Change and East Asian Policy Initiatives*. New York: Oxford University Press.

Klein, Michael W. 2003. "Capital Account Openness and the Varieties of Growth Experience." NBER Working Paper 9500. National Bureau of Economic Research, Cambridge, Mass. Processed.

Lach, Saul, and Mark Schankerman. 2003. "Incentives and Invention in Universities." NBER Working Paper 9727. National Bureau of Economic Research, Cambridge, Mass. Processed.

Lakenan, Bill, Darren Boyd, and Ed Frey. 2001. "Why Cisco Fell: Outsourcing and Its Peril." *Strategy+Business* 24:54–65.

Lederman, Daniel, and William F. Maloney. 2003. "R&D and Development." World Bank, Washington, D.C. Processed.

"Link by Link." 2003. *Business China*, May 26.

Luthria, Manjula, and Keith Maskus. 2004. "Protecting Industrial Inventions, Author's Rights, and Traditional Knowledge: Relevance, Lessons, and Unresolved Issues." In Kathie Krumm and Homi Kharas, eds., *East Asia Integrates*, pp. 139–62. Washington, D.C.: World Bank.

Mathews, John A., and Dong-Sung Cho. 2000. *Tiger Technology: The Creation of a Semiconductor Industry in East Asia.* Cambridge, U.K.: Cambridge University Press.

Micklethwait, John, and Adrian Woolridge. 2000. *A Future Perfect.* New York: Crown Publishers.

"Multinationals and Globalization." 2001. *Oxford Analytica*, June 14.

Nolan, Peter. 2002. "China and the Global Business Revolution." *Cambridge Journal of Economics* 26(1):119–37.

Nolan, Peter, and Jin Zhang. 2002. "The Challenge of Globalization for Large Chinese Firms." *World Development* 30(12):2089–107.

OECD (Organisation for Economic Co-operation and Development). 2002. *OECD Science, Technology, and Industry Outlook 2002.* Paris.

———. 2003. *OECD Science, Technology, and Industry Scoreboard 2003.* Paris.

O'Rourke, Kevin H., and Jeffrey G. Williamson. 1999. *Globalization and History.* Cambridge, Mass.: MIT Press.

Ozawa, Terutomo. 2001. "The 'Hidden' Side of the 'Flying-Geese' Catch-up Model: Japan's *Dirigiste* Institutional Setup and a Deepening Financial Morass." *Journal of Asian Economics* 12:471–91.

———. 2003. "Pax Americana-led Macro-clustering and Flying Geese–Style Catch-up in East Asia: Mechanisms of Regionalized Endogenous Growth." *Journal of Asian Economics* 13:699–713.

"The Samsung Way." 2003. *Business Week*, June 16.

Scarpetta, Stefano, Phillip Hemmings, Thierry Tressel, and Jaejoon Woo. 2002. "The Role of Policy Institutions for Productivity and Firm Dynamics: Evidence from Micro and Industry Data." Working Paper 329. Organisation for Economic Co-operation and Development, Paris. Processed.

Sjöholm, Fredrik. 1997. "Technology Gap, Competition, and Spillovers from Direct Foreign Investment: Evidence from Established Data." Working Paper 38. European Institute of Japanese Studies, Stockholm. Processed.

"Steel Magnolia." 2003. *Economist*, November 22.

Stiglitz, Joseph E. 2002. *Globalization and Its Discontents.* New York: Norton.

Tybout, James R. 2000. "Manufacturing Firms in Developing Countries: How Well Do They Do, and Why?" *Journal of Economic Literature* 38(1, March):11–44.

Wang, Jiann-Chyuan, and Kuen-Hung Tsai. 2003. "Productivity Growth and R&D Expenditure in Taiwan's Manufacturing Firms." NBER Working Paper 9724. National Bureau of Economic Research, Cambridge, Mass. Processed.

Whittaker, D. Hugh. 2003. "Crisis and Innovation in Japan: A New Future through Techno-Entrepreneurship." In William W. Keller and Richard J. Samuels, eds., *Crisis and Innovation in Asian Technology.* Cambridge, U.K.: Cambridge University Press.

World Bank. 2000. *Entering the 21st Century: World Development Report 1999–2000.* New York: Oxford University Press.

Wyplosz, Charles. 2001. "How Risky Is Financial Liberalization in the Developing Countries?" CEPR Discussion Paper 2724. London: Centre for Economic Policy Research.

Yeung, Henry Wai-chung, and Kris Olds. 2000. "Globalizing Chinese Business Firms: Where Are They Coming from, Where Are They Heading?" In Henry Wai-chung Yeung and Kris Olds, eds., *The Globalisation of Chinese Business Firms*, pp. 1–28. London: Macmillan.

Yusuf, Shahid, with M. Anjum Altaf, Barry Eichengreen, Sudarshan Gooptu, Kaoru Nabeshima, Charles Kenny, Dwight H. Perkins, and Marc Shotten. 2003. *Innovative East Asia: The Future of Growth.* New York: Oxford University Press for the World Bank.

Yusuf, Shahid, and Simon J. Evenett. 2002. *Can East Asia Compete? Innovation for Global Markets.* New York: Oxford University Press for the World Bank.

Yusuf, Shahid, and Kaoru Nabeshima. 2004. "Creative Cities in East Asia." *Cities.* Forthcoming.

CHAPTER 2

THE NEW GLOBAL SUPPLY BASE: NEW CHALLENGES FOR LOCAL SUPPLIERS IN EAST ASIA

Timothy J. Sturgeon and Richard K. Lester

In the 1970s and 1980s the newly industrialized economies (NIEs) of East Asia moved along high-growth trajectories. A combination of prudent and stable macroeconomic policies, the targeting of specific sectors for development, and country-specific industrial and social structures enabled the absorption of key technologies and the accumulation of the skills required to achieve industrial upgrading and growth (Amsden 1989; Evans 1995; MacIntyre 1994; Wade 1990; World Bank 1993). In the Republic of Korea, development was driven by the *chaebol*, large-scale conglomerates with privileged access to state-provided capital; in Hong Kong

The preparation of this chapter benefited greatly from the ongoing work of the Globalization Study team at the Massachusetts Institute of Technology's Industrial Performance Center. The authors are grateful to Sara Jane McCaffrey for providing invaluable research assistance. Qualitative field research was conducted in Southeast Asia by Raphael Bonoan, Douglas Fuller, and Vincent Sawansawat. Suzanne Berger made many important contributions throughout. The authors would like also to thank the other participants in the Industrial Performance Center's Globalization Study, including Akintunde Akinwande, Dan Breznitz, Brian Hanson, Donald Lessard, Richard Locke, Teresa Lynch, Mike Piore, Charles Sodini, and Edward Steinfeld for valuable discussions. Shahid Yusuf and Simon Evenett of the World Bank were helpful and patient during the paper's preparation. Helpful comments were also provided by the participants in the East Asia's Future Economy Conference, organized jointly by the Development Economic Research Group of the World Bank and the Asia Pacific Policy Program of the Kennedy School of Government at Harvard University and held on October 1–2, 2001. The comments of Dennis Encarnation were especially valuable.

(China) and Taiwan (China) it was driven by small- and medium-scale enterprises that were financed largely with private or family capital; in Singapore the state played a key role in encouraging foreign direct investment (FDI) by multinational corporations (MNCs) and in fostering the development of small- and medium-scale enterprises to serve them. Industries in these economies upgraded by building links to international markets and to the necessary sources of technology, expertise, managerial experience, and capital in the advanced countries. As Guillèn (2001) has shown, country-specific links between domestic structures and national policies produced different growth trajectories among the Asian NIEs.

Less well understood is that each of these growth trajectories was also influenced by the industrial structures of the advanced countries with which the NIEs were interacting. It is commonly observed that the rapid growth of the NIEs required trade openness in the West. This is indeed a necessary condition for export-oriented development. But the development paths pursued in the Asian NIEs were also influenced and enabled by the competitive strategies of American, European, and Japa-nese firms, which involved establishing local operations in the NIEs, identifying local firms as suppliers, transferring skills and technologies to them, investing in them, and buying from them. The NIEs exploited these strategies to establish and upgrade a critical set of domestic technological and industrial capabilities. In this chapter we refer to this process as supplier-oriented industrial upgrading—developing a supply base tuned to serving advanced-economy lead firms as a key mechanism of industrial upgrading. Both firms and government policymakers tend to view supplier-oriented industrial upgrading as a stepwise learning process, beginning with manufacturing services only, perhaps in an export processing zone, then gradually progressing to manufacturing plus design, and culminating with developing-country firms having the capability to design, manufacture, and sell their own branded products on world markets.

Today, however, lead firms based in the West are reorganizing production networks in ways that still seek to engage with the newly industrialized and emerging economies of East Asia, but with different understandings about how global production networks should be organized and what roles local firms should play in them. The central argument of this chapter is that changes emanating from the advanced industrial economies, particularly the United States, have begun to alter substantially the prospects for supplier-oriented industrial upgrading in East Asia. A key finding of our research is that American and European lead firms have recently become more dependent on a set of "global suppliers" based in the West, even as they have increased their direct involvement in Asian production and Asian markets. These lead firms are increasingly relying on large suppliers and contract manufacturers from within their own societies to support their global operations. These global suppliers have, in turn, experienced rapid growth and global expansion and have become influential global actors. In the electronics sector, where the global suppliers are most highly developed, industry consultants estimate that the $90 billion of business that went to contract manufacturers in 2000 accounted for 13 percent of the total market for world circuit board assembly, product-level electronics manufacturing, and associated services. We argue that, in the key manufacturing sectors of electronics and motor vehicles, the rapid rise of global suppliers based in Europe and the United States presents an important new challenge to the supplier-oriented industrial upgrading paths that Asian economies have pursued in the past. Not only do these global suppliers offer stiff competition to established Asian suppliers and allow lead firms to bypass the painstaking process of bringing new suppliers on stream, but they also provide lead firms with a strategic alternative to relying on a group of Asian suppliers that appear to be focused on becoming, sooner or later, competitors to the lead firms in product markets.

The chapter is organized as follows. We begin by introducing the notion of supplier-oriented industrial upgrading and then document the rise of global suppliers in the electronics and motor vehicle industries. Next, we draw on evidence from our field research to argue that the threshold requirements for suppliers to participate in GPN rose sharply in the 1990s. For East Asian firms, an especially challenging requirement is for participating suppliers to expand the geographic scope of their operations. We then outline some of the risks and uncertainties introduced by the emergence of the global supplier model for both lead firms and global suppliers and briefly discuss the changing architecture of Japanese production networks in East Asia and their impact on development. Finally, we explore some policy implications of the evidence presented for the economies of East Asia.

Some of the empirical evidence presented here is drawn from more than 100 field interviews conducted during the period 1999–2003 by a team of researchers at the Industrial Performance Center (IPC) of the Massachusetts Institute of Technology (MIT) as part of its ongoing Globalization

Study.[1] Other data were collected during the course of the World Bank's Project on East Asia's Economic Future.

A NEW FOCUS ON SUPPLIER-ORIENTED DEVELOPMENT STRATEGIES FOR INDUSTRIAL UPGRADING IN EAST ASIA

By the late 1990s the debate over the relative roles of macroeconomic (that is, fiscal and monetary) policies and micro-level industrial policies in driving development in East Asia, summarized in the World Bank's 1993 volume *The East Asian Miracle*, had begun to run out of steam. There were two main reasons for this. The first was the careful work of authors such as Evans (1995) and Guillèn (2001) showing that both had been important and that macroeconomic and industrial policies had, in fact, been effectively combined and coordinated in a variety of ways in different countries. The most important lesson of this work was less about the effectiveness of one set of policies relative to the other and more about the need to build up the capacity of states to act effectively in a variety of realms through the nurturing of a competent, professionalized, politically insulated bureaucracy with effective but transparent links to a country's business elite. The second factor was the growing strength of the World Trade Organization (WTO). The traditional debate about whether development strategies such as infant industry protection and technical assistance, followed by trade liberalization and ex-

1. For the past four years (1999–2003), a team of 24 researchers from the IPC has been investigating the confluence of globalization and industry reorganization in several sectors, including electronics, motor vehicles, software, textile, and apparel. Field research for the IPC Globalization Study has consisted of semi-structured qualitative interviews with company personnel and relevant individuals from government agencies, labor unions, and academia. In-person interviews and plant tours have been conducted in Canada, mainland China, France, Germany, Indonesia, Ireland, Israel, Italy, Japan, Korea, Malaysia, Mexico, the Philippines, Singapore, Spain, Taiwan (China), Thailand, Romania, and the United States. By the end of 2002, more than 350 interviews had been conducted. Of these, 108 were conducted in the electronics industry, 72 were conducted in the textile and apparel industry, and 61 were conducted in the motor vehicle industry. Sponsors of the IPC Globalization Study include the Chinese National Federation of Industries, the Fujitsu Research Institute, the Volkswagen Foundation, and the Alfred P. Sloan Foundation. For more information about the IPC Globalization Study, including our methodology and detailed research findings, see http://globalization.mit.edu. The direct quotes included in this chapter are intended to provide a window into the attitudes and desires of industry participants that might not be otherwise available. They are not meant as statements of fact or accurate predictors of outcomes. Many of the statements are highly controversial and are meant to highlight important issues and areas of tension. These are individual opinions with which the authors may or may not agree. Manager and company names have been excluded to protect the confidentiality of the research subjects. Unless otherwise noted, all interviews were conducted by IPC researchers.

port promotion, were effective or not was becoming less relevant as more of these policies became "actionable" (illegal) under WTO rules.

Attention accordingly began to shift to the role of global production networks in stimulating and sustaining industrial upgrading in Asian societies (Dolan and Humphrey 2000; Encarnation 1999; Gereffi 1994, 1999; Kaplinsky 2000; Lee and Chen 2000; Shimokawa 1999; Tachiki 1999). The demands made by advanced-economy firms on local enterprises tended to be above and beyond what was required for the local market, and this gap stimulated the rapid advance of supply-base capabilities in East Asia (Keesing and Lall 1992). Through their roles as suppliers of parts and products and as purchasers of specialized process equipment, local firms gained access to important product and process know-how without violating WTO rules. As the case of hard disk drives in Singapore illustrates (McKendrick, Doner, and Haggard 2000), East Asian economic development has been strongly influenced by the strategies of advanced-economy firms in specific industries. Borrus, Ernst, and Haggard (2000), Dedrick and Kraemer (1998), Ernst (1997), Ravenhill (1995), and others note that production networks emanating from different advanced economies have had different characteristics and, in particular, that the "openness" of U.S. networks—that is, the willingness of lead firms from the United States to install local management and ratchet up their demands on local firms—speeded the upgrading process in East Asian locations such as Hong Kong (China), Korea, Singapore, and Taiwan (China). These insights led Dedrick and Kraemer (1998) to argue that encouraging and facilitating the participation of local firms in global production networks was the only effective policy tool left for developing countries.

The intensifying global trend toward economic liberalization and the declining appeal of autarkic development policies that sought to wall off domestic industries until "national champions" were strong enough to compete with foreign rivals have brought new attention to policies that seek to develop the capabilities of local firms as suppliers to lead firms from advanced economies. But advocates of the "supplier-oriented" approach to industrial upgrading often fail to account fully for the fundamental organizational changes taking place in the industries that have been key to the creation of cross-border production networks, the new actors that have arisen within them, and the new demands being placed on suppliers as a result.

In the conventional supplier-oriented model of economic development, domestic suppliers continuously upgrade their capabilities either by serving the needs of the local affiliates of MNCs or by supplying lead firms in advanced countries from a distance. In both cases, if the model is extended further, the expectation is that the local firms will leverage their experience by building up design competencies of their own. These design capabilities not only provide new sources of revenue, they also enable the firms eventually to develop their own branded product lines and perhaps even to emerge as direct competitors to advanced-economy lead firms. The upgrading process can proceed in stepwise fashion, beginning with simple assembly, where labor is applied to components and designs supplied by foreign buyers; followed by the supply of complete products with locally sourced components manufactured to specifications provided by foreign buyers, the so-called original equipment manufacturing (OEM) relationship; followed by the addition of post-conceptual design services to the manufacturing function, a combination known as original design manufacturing (ODM). Once design competencies are well established, the supplier can begin to conceptualize, develop, and manufacture finished products, for sale first under the brand labels of its customers and later under its own brand name. At that point, the local firm becomes what is sometimes referred to as an original brand manufacturer (OBM). In this fully blown version of the supplier-oriented upgrading path, the local firm eventually steps fully out of the supplier role to become a lead firm in its own right.

The supplier-oriented upgrading path, although straightforward in concept, is often far from smooth in practice, especially if it is to be followed to its ultimate conclusion. At each stage along the way, difficult problems have to be addressed by the companies involved. One price of entry into cross-border networks, especially for suppliers whose operations are tightly coupled with those of customers in advanced economies, has been heightened vulnerability to rapid changes in customer strategy. Thus, for example, in 1997, when Compaq adopted a new strategy of marketing a complete desktop personal computer system for under $1,000—a strategy that competitors quickly imitated—many small manufacturers of personal computers in Taiwan (China) were forced to close. Those that survived did so only by rapidly shifting production to China.

The development of in-house product design capabilities allows firms to capture more of the final price of the product and can also provide some protection against uncertainties in the business environment by enabling more rapid response to shifts in demand. In fact, many Asian firms have successfully made the shift from OEM to ODM status (although, as table 2.1 suggests, many more have not done so).

The transition from ODM to brand-name production has been more problematic. There are certainly success stories among Asian brand-name producers. Firms like Fang Brothers and VTech in Hong Kong (China), Samsung in Korea, Creative Labs in Singapore, and Acer in Taiwan (China) have developed brands for Western as well as for Asian markets. The OBM

Table 2.1 Interaction of East Asian Manufacturing Firms with Foreign Customers and Suppliers, by Sector, 2000 (percent of total)

	All sectors		Apparel		Consumer goods		Electronics		Vehicles	
Type of interaction	**ASEAN-4 (n = 58)**	**China (n = 310)**	**ASEAN-4 (n = 18)**	**China (n = 63)**	**ASEAN-4 (n = 4)**	**China (n = 40)**	**ASEAN-4 (n = 23)**	**China (n = 142)**	**ASEAN-4 (n = 12)**	**China (n = 65)**
Supply parts designed in-house to foreign customers	28	15	29	13	40	16	24	17	29	14
Supply design and R&D services to foreign customers	19	7	10	4	40	6	24	10	25	5

Source: Survey conducted by the World Bank in early 2001.

model also has been influential throughout the region, especially in Taiwan (China), where Acer, the flagship firm of the Taiwanese (Chinese) computer industry, apparently has made the transition successfully (Dedrick and Kraemer 1998). The head of a firm in Hong Kong (China) that produces both brand-name goods and OEM products for clients explained some of the challenges and attractions of brand-name production (Berger and Lester 1997, p. 39):

> With a label, you take on a series of challenges. It's a baby to continuously enhance. If you work only OEM, if you work only for others, you're taking commands from them. You're on the hand-me-down side. You're not in the decision making seat.

But in the course of our interviews we encountered numerous firms that have retreated from a brand-name strategy back to an ODM or OEM focus, and our fieldwork in Taiwan (China) and elsewhere suggests that for many firms the OEM→ODM→OBM upgrading path has stalled at the ODM phase. Even Acer struggled and ultimately failed to establish a significant presence in the U.S. market, and Fang Brothers has sold off its chain of retail outlets in the United States.

These problems have several causes. For would-be OBM manufacturers, the importing, sales, marketing, and distribution functions associated with brand-name products are entirely new and very different from their deep competencies in manufacturing and design. In our interviews, many Asian newly industrialized countries (NICs) firms mentioned the high cost of penetrating advanced-economy markets, especially at the retail level. Second, lead firms from the West continue to enjoy substantial design and marketing advantages resulting from their proximity to advanced-economy markets and lead users. While some Asian manufacturers have gained a great deal of expertise in post-architectural, detailed design, advanced-economy lead firms appear to have a strong advantage in the realm of new product creation and conceptual design, which allows them to continue to set product strategy and drive the broad trajectory of market development. Third, Asian brand-name manufacturers that have retained OEM and ODM operations have had difficulty reassuring customers that the customer's intellectual property and the manufacturer's quality of service will not be compromised as the manufacturer develops its own branded products; as a result, firms that have achieved a measure of success with their own brands sometimes have seen their status as preferred supplier slip away. Acer chairman Stan Shih recently acknowledged the problems that Acer's own-brand operations created for its contract manufacturing business: "In many cases we were in the final list [as a contract manufacturer], but . . . when they made a decision, they picked our competitors in Taiwan [China]" ("Acer Plots Path" 2001). The "pure-play" OEM-ODM suppliers that had not developed their own-brand business, such as Hon Hai and Quanta, were seen as less of a competitive threat to lead firms, and their business grew accordingly.

Like other companies that have struggled with the OBM strategy, Acer appears to be refocusing on its OEM and ODM businesses. (According to its latest corporate restructuring plan, the company intends to divest its contract manufacturing operations from its brand-name business; "Acer Plots Path" 2001.) However, the OEM-ODM strategy, important as it has been for industrial upgrading in the region, now faces new questions. Indeed, one of the biggest challenges to the supplier-oriented upgrading approach in East Asia may turn out to be the emergence of a new class of highly sophisticated pure-play suppliers, mostly U.S.-based, capable of supporting the manufacturing needs of European and U.S. lead firms in Asia and around the world.

The supplier-oriented model of economic development has generally assumed that lead firms will continue to seek out or develop distinct supply bases in multiple locations and that to achieve these objectives they will support the upgrading efforts of local firms. But our interviews with managers at dozens of lead firms and suppliers in the electronics and motor vehicle industries have cast doubt on both assumptions. This research has revealed a growing propensity on the part of lead firms to expect considerably more from their suppliers, both functionally and in terms of geographic scope. The new requirements go well beyond excellent manufacturing performance and low costs, which are perceived as being widely available and commodified. Today, suppliers must provide a capability for independent process development and an ability to perform a wide range of value adding functions associated with the manufacturing process, including help with product and component design, component sourcing, inventory management, testing, packaging, and outbound logistics.

Lead firms are also demanding that suppliers have the ability to support the lead firm's operations and market-serving activities around the world. Getting the right part or process in the right place at the right time—as well as at the right cost and with a minimum of inventories in process and in transit—is critical. Even slightly out-of-date components, products, and processes quickly lose their utility and value. As the market and operational scope of many lead firms has become truly global, suppliers are being selected on the basis of their ability to provide global support. Such suppliers have the scale and scope to coordinate component sourcing and inventory management at a global level. Finally, as noted, lead firms have become less willing to use actual or potential competitors as suppliers, especially as globally operating pure-play suppliers have appeared on the scene as an alternative.

As we show in the following sections, these new requirements pose difficult challenges for the larger, established suppliers in the Asian NIEs. The new requirements are also raising the barriers to market entry, throughout the region, for smaller and younger suppliers that are seeking to participate in GPNs. On a larger scale, the emergence of a global supply base serving lead firms in the West raises questions about the long-term viability of the supplier-oriented upgrading model and suggests a need to rethink at least some of the economic development policies that countries in the region have been pursuing. We take up these questions in the following sections.

ELECTRONICS: THE RISE OF GLOBAL CONTRACT MANUFACTURERS

Historically, most U.S. electronics firms purchased electronic components and assembled them in-house into subsystems and final products of their own design. MNCs used a mix of onshore and offshore assembly plants. Since manufacturing processes were quite labor-intensive, plants assembling high-volume, price-sensitive products were often located in areas with low labor costs, especially Mexico's northern border region and East Asia. In the 1960s Japanese firms began licensing local firms in Hong Kong (China), Korea, and Taiwan (China), initially to produce transistor radios and, later, hand-held calculators. Beginning in the 1960s American (and, later, European) semiconductor firms moved labor-intensive chip assembly processes to low-wage locations in East Asia, especially Hong Kong (China), Malaysia, Singapore, and Thailand (Scott 1987). The attraction was low labor costs and, for American firms, tariff exemptions that allowed them to pay duties only on the value that had been added through the assembly process, not on the semiconductor chips themselves. Once the semiconductors were assembled in Asia, they were shipped back to Europe and the United States, where they were sold to brand-name electronics companies and assembled into final products. For a time, nearly all semiconductor wafer fabrication, circuit board assembly, and product-level assembly stayed in Japan, Northern Europe, and the United States. For circuit board assembly, domestic contract manufacturers were used as "shock absorbers" during times of peak demand, when the internal capacity of brand-name companies was fully utilized, but not as a serious alternative to in-house manufacturing. In electronics, the typical contract manufacturing arrangement during the 1970s was for the brand-name firms to provide labor contractors—then known as "board stuffers"—with kits of components from their own inventories on a consignment basis. The contractors then supplied the labor needed to assemble the customer-supplied kits. Some overseas firms, in locations that included Hong Kong (China), Singapore, and Taiwan (China), also stepped into the role of board stuffer.

In the 1980s American brand-name electronics firms operating in high-volume, price-sensitive market segments, such as disk drives and personal computers, began following the semiconductor industry offshore as a way to tap low-cost labor in Asia, either by establishing local subsidiaries or by tapping the growing capability of Asian producers, especially in Taiwan (China). In the late 1980s as progress toward European unification continued, American electronics companies moved the assembly of some products destined for the European market to Scotland and Wales in an effort to increase value added within Europe. Scotland and Wales were chosen most often because of their relatively low labor and engineering costs, a well-trained English-speaking work force, and aggressive efforts by local economic development agencies to attract FDI in electronics. IBM, for example, set up its largest European personal computer assembly facility in Greenock, Scotland.

Over time significant numbers of local contract manufacturers assembling circuit boards for the affiliates of American brand-name electronics firms emerged, especially in Mexico, Scotland, Singapore, Taiwan (China), the United States, and Wales. Drawing on components from nearby semiconductor assembly plants, both brand-name firms and their contract manufacturers, especially those located in Asia, began sourcing the bulk of their components locally. The local availability of components allowed brand-name firms to ask their Asian contractors to begin buying components on a turnkey basis, that is, with financing provided by the supplier.

In the early 1990s some brand-name electronics firms in the United States moved beyond the tactical use of their contractors as providers of overflow capacity and began to use the most capable of them for more strategic purposes. The advantages included manufacturing close to end markets or with low-cost labor, subjecting internal operations to market forces, keeping abreast of fast-moving assembly technologies, and focusing their own activities on increasingly challenging "core competencies," such as product definition, design, sales, and marketing. As the contract manufacturers grew in size, additional cost advantages accrued from scale economies derived from the pooling of manufacturing capacity and component purchasing. Lead firms also discovered dynamic advantages to moving manufacturing further out of house: they could ramp production levels up and down more quickly and with less cost at contractor facilities than with in-house manufacturing. This proved to be an important asset in the face of a highly volatile and contentious market environment. Today, production outsourcing in electronics has become a widely accepted practice for both large and small brand-name electronics firms based in the United States and, increasingly, Europe.

More recently, globally operating lead firms have been consolidating their contract manufacturing relationships by giving a larger share of their manufacturing to a smaller group of large, technologically sophisticated contract manufacturers, nearly all of them of North American origin. Brand-name electronics firms are demanding that their contractors have a "global presence" as a way of streamlining the management of their outsourcing relationships. As a result, North American contract manufacturers have themselves been aggressively internationalizing their operations since the mid-1990s.

During the past decade, global suppliers in the electronics industry experienced rapid revenue growth, consolidation at the top levels, and geographic expansion. In the 1990s established North American electronics firms in the computer and networking sectors, such as Apple Computer, Hewlett-Packard, IBM, Lucent, Maxtor, Nortel, and 3Com, rapidly moved toward outsourcing their circuit board and product-level assembly, notably by selling off much of their domestic and offshore production facilities to the five largest contract manufacturers. Many newer North American electronics companies such as Cisco Systems, EMC, JDS Uniphase, Juniper Networks, Network Appliance, Sun Microsystems, and Sycamore Networks outsourced most of their production from the outset, and their rapid growth during the late 1990s fueled growth of the largest electronics contract manufacturers.

All of the top five contract manufacturers are based in North America. They consist of Celestica, based in Toronto, Canada; Flextronics International, incorporated in Singapore, but managed from its San Jose, California, headquarters; Jabil Circuit, based in St. Petersburg, Florida; Sanmina/SCI, based in San Jose; and Solectron, based in Milpitas, California. These five firms collectively grew at an average annual rate of 36 percent a year between 1994 and 2001 (see table 2.2).

In the latter part of the 1990s, the outsourcing trend began to spread to most of the major European suppliers of communications infrastructure as well, especially Alcatel, Ericsson, and, to a lesser degree, Nokia and Siemens. In 1997 Ericsson made a decisive series of moves, first by outsourcing production to Flextronics, SCI, and Solectron and then by selling its principal domestic production facilities in Karlskorna, Sweden, to Flextronics and a

Table 2.2 Revenue at the Top Five Electronics Contract Manufacturers, 1994 and 2001 (millions of U.S. dollars)

Company	1994	2001	Average annual growth rate (percent)
Celestica	1,989,000[a]	10,004,000	26
Flextronics	210,700	12,110,000	78
Jabil Circuit	404,056	4,331,000	40
Sanmina/SCI	2,363,581	11,248,651	25
Solectron	1,641,617	18,692,000	42
Total	6,608,954	56,385,651	36

a. In 1994 all of Celestica's revenues were from IBM.

Source: Company annual and quarterly reports.

plant in Brazil to Solectron (Dunn 1997). Solectron established a local presence in Sweden as well but shifted the bulk of Ericsson's circuit board assembly to its existing network of plants in France, Germany, and Scotland (Jonas 1997). In 2000 Ericsson shifted the remainder of its cell phone production to these American contract manufacturers and sold its U.S. production facilities to SCI ("Ericsson Shifts Phone Production" 2000).[2]

In the past two years even a few Japanese electronics firms have tested the waters. In December 2000 NEC, whose cell phone handset business was doing poorly in the midst of a fierce shakeout, announced that it was selling its cell phone production facilities in England and Mexico to Solectron, while keeping facilities in China and Japan ("NEC to Sell Cell Phone Plants" 2000). In October 2000 Sony announced that it was selling two underutilized Asian facilities to Solectron, one in Miyagi, Japan, and a second in Kaohsiung, Taiwan, China (Levine 2000). In January 2002 NEC announced the sale of two of its advanced manufacturing facilities in Miyagi and Yamanashi, Japan, to Celestica.[3] About 1,200 highly skilled NEC manufacturing specialists and related support staff became Celestica employees. As part of the deal, Celestica assumed management of the supply chain as well as responsibility for subassembly, final assembly, integration, and testing of a broad range of NEC's optical backbone and broadband access equipment. The companies expect the deal to generate revenue of approximately $2.5 billion for Celestica over a five-year period. According to Kaoru Yano, senior vice president of NEC and company deputy president of NEC Networks,

2. In 2001 SCI was acquired by Sanmina.

3. The design and development functions currently performed by NEC Miyagi will remain with NEC, and NEC Miyagi will continue as a developer of optical transmission systems. The development and manufacturing of optical devices and optical submarine cable systems at NEC Yamanashi's Otsuki plant will also remain with NEC.

> NEC's growing partnership with Celestica will allow us to improve our competitive positioning by further leveraging our leading-edge R&D [research and development], product development, and manufacturing expertise with Celestica's global manufacturing capabilities and supply chain management expertise. Through the alliance with Celestica, NEC intends to improve price competitiveness, production lead times, and supply chain flexibility to optimize overall manufacturing efficiency. NEC also chose to work with Celestica based on its reputation for providing global, advanced manufacturing capabilities and cost-effective supply chain solutions for the world's best communications and information technology companies.

As mentioned, most of the growth in electronics contract manufacturing has taken place in the very top tier of firms. Electronic Trend Publications (2000) estimated that the top five contract manufacturers had captured 38 percent of the electronics contract manufacturing market by 1999 and projected this share to grow to almost two-thirds in 2003. This rapid expansion, fueled by the acquisition of competitors and customer facilities as well as organic expansion in existing and newly established facilities, was aided by the U.S. stock market run-up in the late 1990s, which concentrated a significant share of market capitalization in contract manufacturing in the top five firms.

Each of the largest electronics contract manufacturers has established a global network of plants that consists of (a) low-product-mix, high-volume production sites, mostly in Asia, East Europe, and Mexico; (b) high-product-mix, medium- to high-volume production sites in Canada, the United States, Western Europe, and now Japan; (c) engineering-heavy "new product introduction" centers, often located near an important customer's design activities; and (d) facilities that perform final assembly and product configuration to order or that provide after-sales repair service, often located near major transportation hubs, such as Amsterdam and Memphis, Tennessee. All have large-scale investments for high-volume production in East Asia, especially in Southeast Asia and, increasingly, in China.

The rapid geographic expansion of these firms is worth noting in some detail. Celestica, which spun off from IBM in 1996, began with two production locations, a large complex near Toronto, Canada, and a small facility in upstate New York, since closed. Today, after completing 29 acquisitions, Celestica operates nearly 50 facilities in Asia, Europe, North America, and South America and generates annual revenues in excess of $11 billion (see figure 2.1). In Asia the company established or acquired facilities in Dongguan, Suzhou (two), Shanghai, Xiamen, and Hong Kong in China; Johor Baru (three), Kulim, and Parit Buntar in Malaysia; Batam and Bintan in Indonesia; and Singapore (three). Other high-volume production sites were established in Kladno and Rajecko in the Czech Republic;

Figure 2.1 Celestica's Global Footprint, 2001 Source: Celestica.

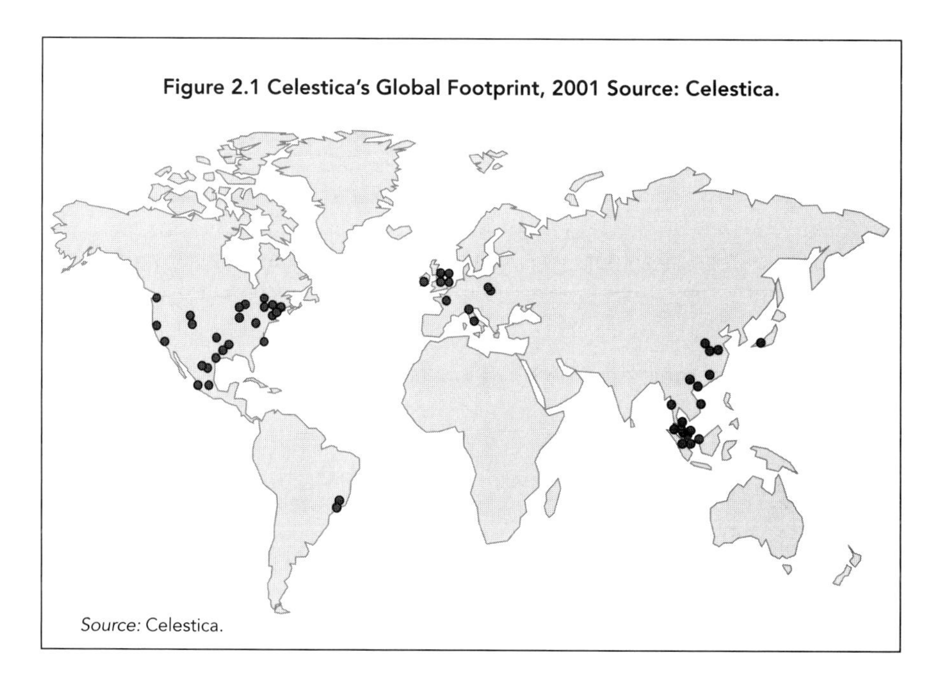

Source: Celestica.

Guadalajara, Monterrey (two), and Querétaro in Mexico; and Guarulhos and Hortolândia in Brazil.

Another striking example of rapid geographic expansion is the largest electronics contract manufacturer, Solectron, which was concentrated in a single campus in Silicon Valley until 1991, when its key customers, including Hewlett Packard, IBM, and Sun Microsystems, began to demand global manufacturing and process engineering support. Within 10 years, the company's footprint had expanded to nearly 50 facilities worldwide (see table 2.3). Today this network consists of global and regional headquarters, both high- and low-mix manufacturing facilities, materials purchasing and management centers, new product introduction centers, after-sales repair service centers for products manufactured by Solectron and others, and technology centers that develop advanced process and component packaging technologies.

According to estimates by Technology Forecasters, contract manufacturers penetrated roughly 17 percent of the total available market for circuit board and product-level electronics manufacturing in 2001. A recent Bear Stearns survey of brand-name electronics firms (Levine 2001) concluded that the rate and size of outsourcing agreements will continue to increase, with 85 percent of the firms interviewed planning on further increases in production outsourcing. As a group, the branded electronics firms in the survey expected to outsource 73 percent of total production

Table 2.3 Solectron's Global Locations and Functions, 2001

Location	Global and regional headquarters	Manufacturing facilities	Materials purchasing and management centers	New product introduction centers	After-sales repair service centers	Process technology centers
Americas						
Brazil						
São José dos Campos		•	•	•	•	
Hortolândia		•			•	
Canada						
Vaughn					•	
Calgary		•				
Mexico						
Guadalajara		•	•			
Monterrey		•	•	•		
United States						
Milpitas, Calif.	•	•	•	•	•	
Fremont, Calif.		•	•	•		
Austin, Tex.		•	•	•		•
Charlotte, N.C.		•	•	•		•
Columbia, S.C.		•	•	•		
San Jose, Calif.		•	•	•		
Atlanta, Ga.		•	•	•	•	
Westborough, Mass.		•	•	•		
Suwanee, Ga.		•	•	•	•	
Fremont, Calif.		•	•			•
Everett, Wash.		•	•			
Raleigh, N.C.		•		•		
Aguadilla, P.R.		•				
Aguada, P.R.		•				
Los Angeles, Calif.					•	
Austin, Tex.					•	
Memphis, Tenn.					•	
Louisville, Ken.					•	
Europe and Middle East						
Bordeaux, France		•	•	•		•
Germany						
Herrenberg		•	•	•		
Munich		•	•	•		•
Dublin, Ireland		•	•			

(continued)

Table 2.3 Solectron's Global Locations and Functions, 2001 (*continued*)

Location	Global and regional headquarters	Manufacturing facilities	Materials purchasing and management centers	New product introduction centers	After-sales repair service centers	Process technology centers
Carrickfergus, Northern Ireland		•	•			
Timisoara, Romania		•	•			
Scotland						
Dunfermline		•	•			
East Kilbride		•	•			
Östersund, Sweden		•	•	•		•
Istanbul, Turkey		•	•	•	•	
Reading, United Kingdom	•					
Australia						
Wangaratta		•				
Liverpool				•		
Asia						
China						
Suzhou		•	•			
Bangalore, India						•
Japan						
Tokyo	•					•
Kanagawa			•	•	•	
Malaysia						
Johor Baru		•	•	•		
Penang		•	•	•		•
Penang		•	•			
Singapore		•	•	•		
Taipei, Taiwan	•					

Source: Solectron.

needs on average, and 40 percent intended ultimately to outsource 90–100 percent of final product manufacturing. The chief executive officer of Flextronics has stated publicly that he expects annual revenues at his company to reach the $100 billion range in the next 5–10 years.

In some instances, the expansion of North American contract manufacturers in East Asia has come about through the acquisition of established regional firms, such as Solectron's acquisition of NatSteel Electronics (Singapore) and Ocean Electronics (Hong Kong, China) in the late 1990s. In the future, the North American contract manufacturers may compete directly with the largest indigenous electronics contract manufacturers in East Asia, and this trend may already be constraining the growth of Asian-owned firms. Given the importance of the electronics industry in East Asia's industrial upgrading, the question of how much business indigenous East Asian electronics suppliers will capture in the ongoing shift from in-house to outsourced production is critical for the future of economic development in the region. Chapter 3 explores this question further and, in particular, examines not only the strengths of the contract manufacturers but also the risks they face.

MOTOR VEHICLES: THE RISE OF THE MEGA SUPPLIERS

The interest in producing motor vehicles close to final markets has always been strong, although the motivations have changed over time. Prior to the advent of mass production in 1910, automobiles were luxury products, and the need for customization required proximity to pools of well-heeled customers. When mass production lowered the cost of the automobile to the point where mass markets developed, final assembly moved close to final markets to reduce transport costs. In the 1930s overcoming trade barriers erected by national governments become the main motivation for offshore production. Automakers were forced to establish local production or to forgo participation in the most promising emerging markets of the day. By 1928 Ford and General Motors were assembling vehicles in 24 countries, including Brazil, India, Japan, and Malaysia. A decade later both companies were operating large-scale integrated "transplant" facilities in Europe. When trade barriers were extended to automotive parts, automakers moved to integrate offshore production and source parts locally to the extent possible. Both tariff and nontariff barriers to trade in finished vehicles—or the threat thereof—continue to be a key motivation for the growth of offshore production today.

The number of automakers willing and able to compete in the arena of international production increased markedly in the 1980s, but American automakers had begun to face competition in the developing world in the late 1950s, when European producers had recovered sufficiently from World War II to begin investing in Australia, Latin America, and South Africa. During the 1960s and 1970s a regional pattern emerged. Most new offshore assembly plants established by American and European automakers were located in Latin America, and most offshore plants established by Japanese firms were located in Asia. There were exceptions to this pattern, namely the investments of Ford and General Motors in Taiwan (China) and a few small Japanese investments in Brazil, Ecuador, and Peru.[4] In the 1980s the remarkable successes of the Japanese automakers' export strategy in the United States at the direct expense of the American Big Three sparked a political backlash, which resulted in the setting of "voluntary" limits to continued expansion of market share via exports. In response to these quotas, Japanese automakers embarked on a wave of plant construction in the United States (Kenney and Florida 1993). By 1995 Japanese automakers were locally manufacturing two-thirds of the passenger vehicles they sold in the United States. A similar dynamic led to a wave of Japanese transplants in Europe, beginning with Nissan's plant in the United Kingdom in 1986. By 1995 Japanese automakers were locally manufacturing nearly one-third of the passenger vehicles they sold in Europe.

In the 1980s American and European automakers, under increasing pressure from the sudden appearance of fierce new competition from Asia, began importing finished vehicles into their home markets from operations in lower-cost peripheral locations (for example, Canada, Eastern Europe, Mexico, and Spain) within the context of regional trade agreements such as the European Union and the North American Free Trade Agreement (NAFTA).

During the 1990s there was a wave of new assembly and supplier plant construction in emerging markets such as Brazil, China, Eastern Europe, India, Mexico, Thailand, and Vietnam. These new investments were driven by increased competition and market saturation at home, the opening of new spaces for investment following the end of the cold war, host-country requirements for local content and production, and an effort by automakers to cut costs within the context of regional trade arrangements such as the European Union and NAFTA.

The idea that emerging markets, particularly in Asia, would be the locus of rapid economic growth in the medium term stimulated a huge wave of new investment in Asia during the 1990s, particularly in China and the Association of South East Asian Nations (ASEAN). Industry consultants projected that Asia and Eastern Europe would have the highest rates of pro-

4. Investments by Japanese automakers, however, tended to be of a very different character than those of American and European firms. American and European firms tended to build larger, more integrated plants, whereas Japanese firms built plants that relied heavily on the assembly of vehicle "kits" sourced from home factories. Japanese investments were highly conservative, in that assembly plant investments remained scaled to the actual, not potential, size of the local market—something that is still true today. Still, in places where Japanese automakers faced no competition from more aggressive investors, such as the ASEAN countries of Indonesia, Malaysia, and Thailand, they were able to capture the lion's share of these markets, especially in countries where local content rules became more stringent over time (Doner 1991).

duction growth, with Asia outside of Japan generating 55 percent of the world's new production. Developing countries, taken together, were projected to account for 80 percent of all new production (Sturgeon and Florida 2003). The Asian financial crisis brought the investment boom to an abrupt end, and many of the motor vehicle investment projects in Asia were scaled back or put on hold. Still, the long-term projections for the development of the motor vehicle market in large Asian countries such as China and India remain very positive. In an interview conducted in 2000, the manager of a German automotive parts supplier in China asserted that 50 million cars would be sold annually in China by 2050, 3.5 times the size of the current U.S. market.

As the number of production locations multiplied during the 1990s, automakers sought to streamline operations on a global scale, particularly in the area of vehicle design and component sourcing. Most automakers today are seeking to place a greater number of car models on fewer underbody platforms, allowing for greater commonalization and reusability of parts, while retaining the ability to adapt specific models to local tastes and driving conditions. Such strategies call for global sourcing, tighter coordination of worldwide design efforts, and in cases where platform design activities have become geographically dispersed over time (that is, in American firms), consolidation of project management in core locations and the formation of international design teams. At the same time, the need to respond to unique market requirements has created pressure to localize body design, prompting highly centralized automakers (that is, Japanese firms) to set up regional design studios to cater to local tastes. Since the benefits of global platforms can only be reaped when they are used and reused across a broad product line, there has been a wave of consolidation in the industry, as large players acquire small, specialty producers. Efforts to create global platforms—often thought to have begun with the "world car" strategies of the 1980s—in fact long predate this and can be traced as far back as Ford's failed "1928 Plan," which aimed to supply Model As to the world from three giant River Rouge–style plants in Canada, Detroit, and England. The current trend, however, includes an entirely new feature: the formation of a global supply base.

The recent round of globalization in the motor vehicle industry has helped to change the nature of relationships between automakers and their largest suppliers. First-tier suppliers are moving to module design, second-tier component sourcing, and the provision of local content in the context of emerging markets. The growing need to provide automakers with modules on a worldwide basis is driving a wave of consolidation and geographic expansion among first-tier suppliers, just as it has in the electronics industry. For suppliers that serve multiple automakers, the geographic scale of operations can surpass that of any single customer. In the long run it may well be suppliers, not automakers, that generate the vast majority of the industry's future FDI—and the associated economic and social benefits (such as employment).

As in electronics production, the outsourcing trend in the motor vehicle sector has been strongest among American firms. The structure of the motor vehicle industry in the United States—and the characteristics of the jobs within it—have changed radically since 1986, when long-time rough employment parity between the assembly and parts sectors began to diverge (see figure 2.2). Since the 1980s the supply sector has been the main source of job growth in the U.S. automotive industry, adding 220,900 jobs since 1982, compared with only 25,300 jobs in assembly.

In the realm of vehicle manufacturing, automakers are performing far fewer functions within their assembly facilities than they have in the past. Vehicle assembly lines have been streamlined; integrated "feeder" lines that build up subassemblies such as seats, cockpits, and climate control systems within vehicle assembly facilities have all but disappeared. Assembly workers now bolt together a greater number of large subassemblies of individual components, known as "modules," that have been preassembled off-site by

Figure 2.2 Employment in Assembly and Parts in the U.S. Motor Vehicle Industry, 1958–2000

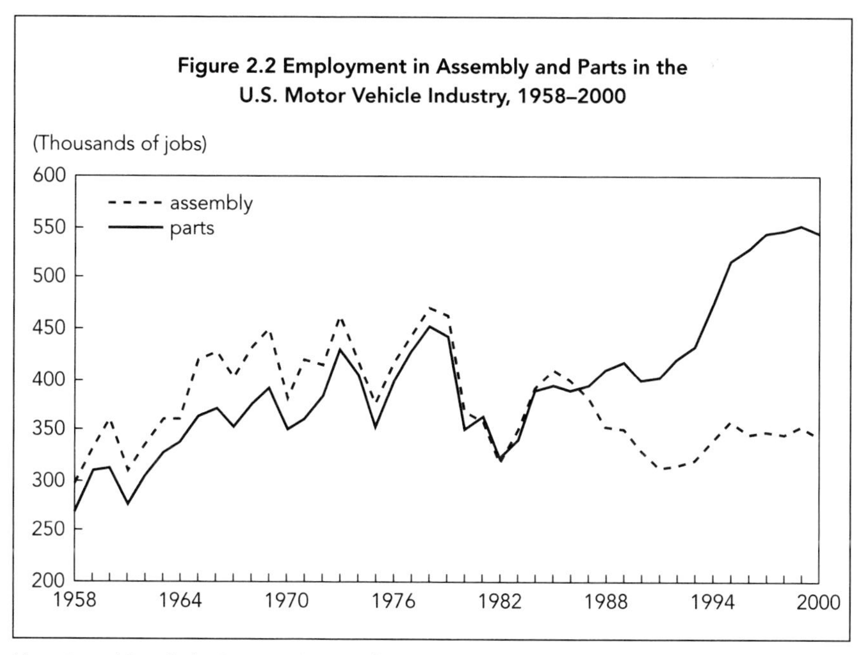

Note: Assembly includes SIC 3711 (motor vehicles and car bodies), and parts includes SIC 3714 (motor vehicle parts and accessories).

Source: U.S. Bureau of Labor Statistics (various years).

suppliers. Modules arrive fully assembled on the loading docks of final assembly plants, ready to be attached to vehicles as they move down the line. The result is more production workers in supplier plants and fewer in final assembly plants.

Modularity

Modularity has already been mentioned in the context of final assembly, but it has important consequences for the scope of supplier activities as well. As automakers do less within their assembly plants, suppliers have an opportunity to do more. For example, vehicle doors can be delivered with the glass, fabric, interior panels, handles, and mirrors preassembled. Dashboards can be delivered complete with polymers, wood, displays, lights, and switches. One manager interviewed during our research estimated that 75 percent of vehicle value can be accounted for by only 15 modules. Important modules are suspension (supplied as "corners"); doors; headliners (which can come with grip handles, lighting, wiring, sunroof, sun visors, and trim preassembled); heating, ventilation, and air conditioning (HVAC) units; seats; dashboards; and drive trains (that is, engines, transmissions, and axles). A continuation of the trend toward modules would mean that suppliers will provide automakers with groups of related modules, in what might be called "module systems." For example, seats, interior trim, the headliner, dashboard, and cockpit module could be supplied as a complete "interior system." Figure 2.3 provides a graphic representation of the trend from discrete parts to modules and then module systems.[5]

The drive toward modularity is associated with supply-base consolidation, as first-tier suppliers buy second-tier suppliers to gain the broader range of capabilities needed to supply modules and systems. TRW's recent

5. Some modules comprise contiguous subassemblies, while others do not. For example, seats and HVAC units comprise physically contiguous subassemblies, while vehicle electronics or occupant safety can consist of a variety of physically discrete components that work together to make up a functional system. Contiguous subassemblies provide the key benefit of assembly-line simplification, while noncontiguous systems do not. Sourcing noncontiguous modules from a single supplier is a way for automakers to pass the responsibility for system integration to suppliers. For example, an electronics supplier such as Bosch, Delphi, Siemens Automotive, or Visteon can take responsibility for ensuring that engine controls work properly with temperature, pressure, revolutions per minute, and other sensors that provide information to the control unit. In other instances, sourcing noncontiguous modules is a way for automakers to pass warranty responsibility for entire aspects of vehicle quality—such as engine and transmission sealing or occupant restraint—on to suppliers. This is not to imply that industry nomenclature in the area of modules and systems has been standardized; some automakers refer to contiguous subassemblies as "modules" and functionally related noncontiguous parts as "systems."

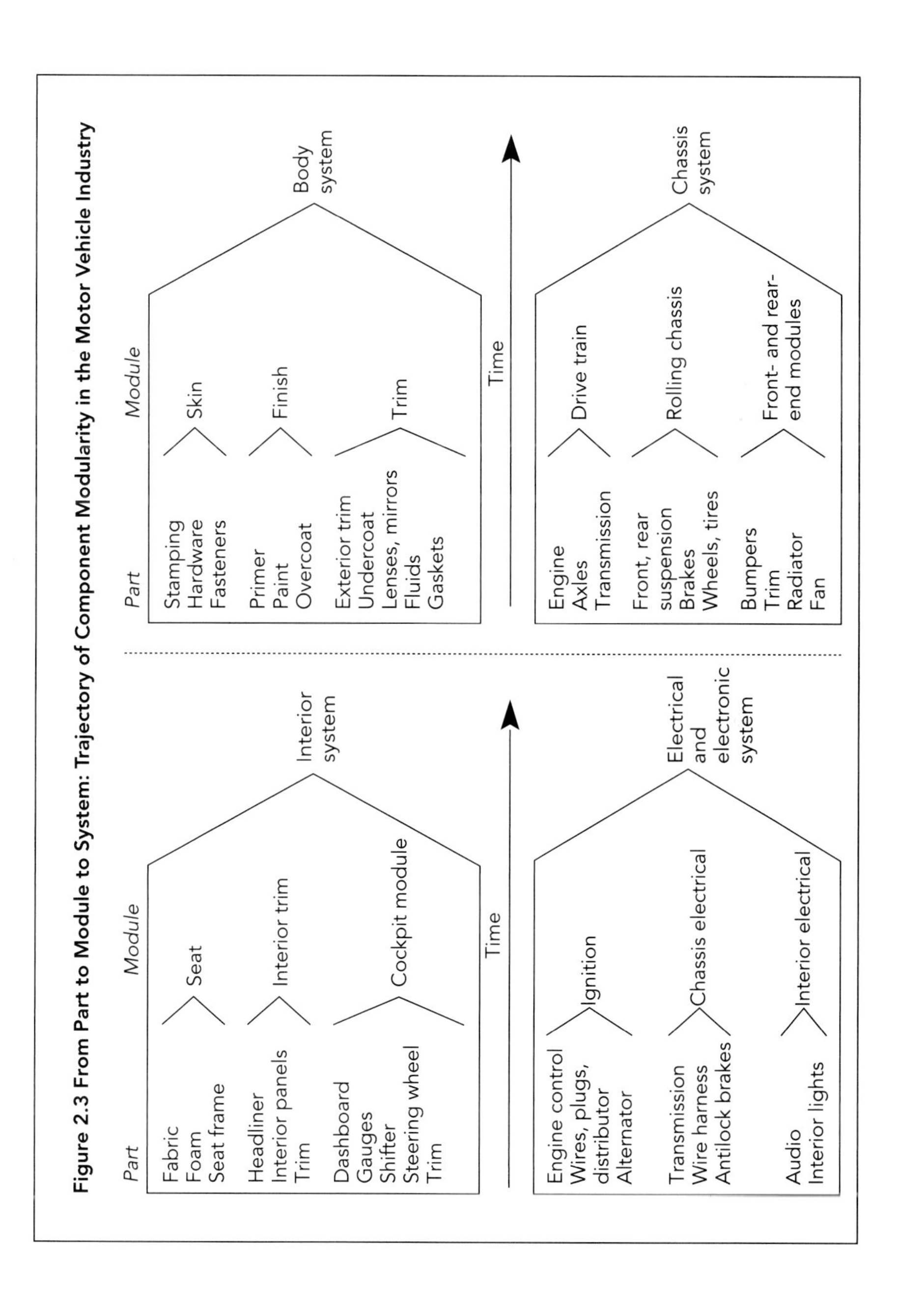

Figure 2.3 From Part to Module to System: Trajectory of Component Modularity in the Motor Vehicle Industry

acquisitions, for example, have given the company the capability to deliver all aspects of occupant restraint systems. Modularity creates natural breakpoints in the value chain and makes the outsourcing or relocation of module design and production more feasible. But modularity can also be pursued largely as an in-house strategy, as at Volkswagen, where internal subsidiaries have moved beyond the manufacture of parts to the assembly of modules and systems that are delivered to streamlined assembly plants.

Because larger modules are more difficult and expensive to ship over long distances and are more likely to be sequenced (that is, coordinated tightly with the final assembly process), the adoption of modular assembly processes is associated with the co-location of assembly plants and supply plants. In-line sequencing has accelerated the adoption of "just-in time" parts delivery, where modules are delivered according to the sequence of cars moving down the assembly line. A key motivation for in-line-sequencing is color matching. Mirrors, interior panels, seats, dashboards, carpets, door handles, and bumpers all have to match or accent the body color and thus must be tightly sequenced with the colors of vehicles on the final assembly line.

Globalization is occurring at the same time as increased outsourcing and the move to sourcing modules and systems, and so many suppliers are taking a larger role in the globalization process. Companies like Bosch, Johnson Controls, Lear, Magna, Siemens Automotive, TRW, Yazaki, and others have become the preferred suppliers for automakers around the world. Many first-tier suppliers have responded by embarking on a wave of vertical integration (through mergers, acquisitions, and joint ventures) and geographic expansion to gain the ability to deliver parts and modules on a global basis. The entry of Ford's and General Motors' former component divisions into the merchant market for vehicle components, modules, and systems has, almost overnight, created the world's two largest, most diversified, and geographically extensive automotive suppliers.

Unlike the global contract manufacturers in electronics, global suppliers in the motor vehicle industry have emerged from Europe and Japan as well as North America.[6] Still, the trends toward rapid growth, geographic expansion, and consolidation are most pronounced among suppliers based in North America. Bosch and Siemens Automotive, both based in Germany, have tended to remain more focused on their core activities, but since their

6. The integral nature of product architecture in motor vehicles makes it harder for lead firms in the industry to switch suppliers, so they tend to bring their suppliers with them when they set up international operations, while the more standard interfaces and production processes in electronics have allowed contract manufacturers from North America to surge ahead and more easily win business from lead firms of many nationalities.

focus has long been electrical and electronic systems for vehicles, their growth has been in part due to the increased electronic content in vehicles and in part to increased sales to Japanese and especially American automakers. Major European suppliers are experimenting with a modular approach, however, especially with their American customers. In 2000 Siemens Automotive acquired another German firm, VDO, which added cockpit instrumentation capability to Siemens' climate control and interior plastics capability. This has allowed the firm to bid on completely built-up dashboard modules.

To illustrate the radical pace of change in the motor vehicle supply base, we consider the case the German tire manufacturer Continental AG in some detail. Continental has long specialized in tires for the retail market and had already established a global manufacturing presence in tires by the early 1990s.[7] As late as 1995, Continental was still concentrated in the retail tires market and ranked 52nd in the world in direct sales to automakers. That year, the company established Continental Automotive Systems Group and began acquiring automotive suppliers with a wide variety of competencies and geographic attributes, such as TBA Belting (United Kingdom) and ITT Brake and Chassis (United States). The latter acquisition, which was valued at nearly $2 billion, added 23 plants and 10,000 employees. To round out the company's global footprint, Continental made a series of additional acquisitions in Argentina, Brazil, Mexico, Romania, Slovakia, and South Africa. In 2001 Continental acquired Temic Microelectronic GmbH, a medium-size ($900 million in revenues) German automotive electronics firm with 3,000 employees and nine manufacturing facilities in Germany and also a small, newly established global footprint, including 2,800 workers at factories in Brazil, China, Hungary, Mexico, and the Philippines, and two technical centers, one at its headquarters in Germany and a new center in Auburn Hills, Michigan, just north of Detroit. Continental's product strategy is threefold: to leverage competence in synthetic rubber by entering markets for power transmission belts and other rubber parts for motor vehicles; to develop integrated modules from the tire

7. Continental established a global manufacturing presence in tires largely through acquisition. In 1979 the company acquired the European assets of Uniroyal (United States), a deal that included plants in Belgium, France, Germany, and the United Kingdom. In 1985 Semperit Reifen AG (Germany) was acquired, adding plants in Austria and Ireland (since closed). In 1987 Continental acquired General Tire (United States), including four plants in the United States, two in Mexico (since sold), and a series of joint venture operations in Asia, Africa, and South America. In 1991 a joint venture agreement was signed with Yokohama (Japan) and Toyo (Japan) to make commercial tires for the U.S. market (http://www.conti-online.com/). In 1992 the company acquired the Swedish tire producer Nivis Tyre.

inward, including assembled wheels, brakes, and suspension parts; and to enter the high-growth area of vehicle electronics. Today Continental's automotive divisions operate 140 facilities in 36 countries and employ 64,000 workers. Development centers are located in Detroit and Germany. By 2000 Continental had jumped to number 12 in the global ranking of sales to automakers (see table 2.4.)

As tables 2.4 and 2.5 show, most of the largest and most rapidly growing suppliers providing auto parts and modules are based in North America. Consider the example of Lear. The company's focus is on interior modules and systems, which are used in vehicles bearing the nameplates of Audi, BMW, Chrysler, Daewoo, Fiat, Ford, General Motors, Honda, Hyundai, Isuzu, Jaguar, Mazda, Mercedes, Mitsubishi, Nissan, Opel, Peugeot, Porsche, Renault, Rover, Saab, Subaru, Suzuki, Toyota, Volkswagen, and Volvo,

Table 2.4 Top Fifteen Motor Vehicle Parts Suppliers, by World Rank, Sales, and Compound Annual Growth Rate, 1995–2000

Home country and region	Company	World rank		World OEM sales (millions of U.S. dollars)		Compound annual growth rate (percent)
		1995	2000	1995	2000	1995–2000
North America						
United States	Delphi	1	1	26,400	29,100	2
United States	Visteon	6	3	9,200	19,500	16
United States	Lear	13	5	4,707	14,100	25
United States	Johnson Controls	15	6	4,420	11,869	22
United States	TRW	7	7	6,100	11,000	13
Canada	Magna	19	8	3,223	10,099	26
United States	Arvin Industries	32	13	1,792	5,153	24
United States	DuPont Auto	18	14	3,500	5,100	8
Average				7,418	13,240	17
Japan						
Japan	Denso	2	4	15,000	16,392	2
Japan	Aisin World	5	9	11,587	8,301	–6
Japan	Yazaki	10	11	5,000	6,000	4
Average				10,529	10,231	0
Europe						
Germany	Robert Bosch	3	2	14,200	20,550	8
Germany	Continental	52	12	800	5,500	47
France	Valeo	11	10	5,000	8,200	10
Average				6,667	11,417	22
Average for the top 15				7,924	12,205	14

Source: *Automotive News* (1996); Crain's Detroit Business (2001).

Table 2.5 Top 15 Motor Vehicle Parts Suppliers, by Share of Sales in North America, 1995 and 2000

		OEM sales in North America (millions of U.S. dollars)		Share of sales in North American market (percent)	
Home country	**Company**	**1995**	**2000**	**1995**	**2000**
North America					
United States	Delphi	21,800	23,600	83	81
United States	Visteon	8,140	14,400	88	74
United States	Lear	3,373	8,600	72	61
United States	Johnson Controls	3,257	7,596	74	64
United States	TRW	3,300	5,610	54	51
Canada	Magna	2,579	6,111	80	61
United States	Arvin Industries	892	3,252	50	63
United States	DuPont Auto	2,500	2,550	71	50
Average		5,730	8,965	72	63
Japan					
Japan	Denso	2,300	3,803	15	23
Japan	Aisin World	563	664	5	8
Japan	Yazaki	1,600	2,400	32	40
Average		1,488	2,289	17	24
Europe					
Germany	Robert Bosch	1,576	6,200	11	30
Germany	Continental	350	1,650	44	30
France	Valeo	600	2,246	12	27
Average		842	3,365	22	29

Source: Automotive News (1996); Crain's Detroit Business (2001).

among others. Headquartered in Southfield, Michigan, Lear has grown to 120,000 employees working at more than 200 locations in 33 countries. Lear rose from the world's 13th largest automotive supplier in 1995 to the 5th largest in 2000, with record sales of $14.1 billion.

As mentioned, the spin-off of the internal parts divisions of Ford and General Motors in the late 1990s created the world's two largest and most diversified automotive parts suppliers, with capabilities to supply complete modules and with global operations from the outset. For example, Visteon has system and module capabilities in chassis, climate, electronics, glass and lighting, interior, exterior trim, and power train. The company currently operates 42 facilities in Canada and the United States; 29 in western Europe; 22 in Asia, 9 in Mexico, 6 in eastern Europe, and 4 in South America. In East Asia Visteon operates plants in Changchun, Nanchang, and Shanghai (3) in China; Japan (5); the Philippines; Republic of Korea (5);

Taiwan (China); and Thailand (2). (Three other plants are located in India.) System and module engineering work is carried out in plants in England (3), Germany (3), Japan, and the United States (4).

Although outsourcing is an industry-wide phenomenon, our research has also identified significant variations in the speed, extent, and nature of deverticalization among automakers. Ford and General Motors, long among the most vertically integrated automakers, have been aggressively outsourcing to cut costs and reduce overhead, both by increasing their use of outside suppliers and, as noted, by moving to spin off their internal parts subsidiaries as independent "merchant" firms. Even so, sourcing is still fairly traditional at Ford and General Motors, which have globally centralized and notoriously predatory purchasing organizations. In the resulting atmosphere of price pressure and mistrust, suppliers are only slowly and irregularly gaining influence over design. There is some experimentation with preselection of suppliers and involvement prior to project approval where suppliers are asked to bid on the parts they would like to design and produce, but the drive toward lowest-cost sourcing and ongoing cost reduction is still very strong. As a result, there is tension between the purchasing organization, which pushes for lower costs, and manufacturing, which pushes for modularity, local content, and co-location. DaimlerChrysler's Chrysler Division, by contrast, has long sourced as much as 70 percent of the value of its vehicles from outside suppliers. The relationship of Chrysler with suppliers is far more consultative than that of Ford or General Motors, and the company has asked suppliers to perform a significant amount of module design and engineering work.

Japanese automakers are well known for their extensive reliance on multitiered supplier networks and high levels of outsourcing. The nature of Japanese supplier networks tends to be more "captive" than those that have been developed by American and European firms—that is, Japanese suppliers tend to be more dominated by their largest customer. For example, Japan's largest supplier, Denso, a Toyota Group company, generated half of its revenues from Toyota in 1997 and none from Toyota's arch rival, Nissan. On the one hand, such captive relationships inhibit the build-up of external scale economies and engender financial and technological dependence of suppliers on their largest customers. On the other hand, the integral nature of product architecture in motor vehicles may well favor this kind of tight integration between lead firms and their suppliers, as the continued superior performance of firms such as Toyota may attest. In this hierarchical system, it is not surprising that the trends toward modularity and the outsourcing of component design and engineering are much weaker among Japanese automakers. As mentioned, Volkswagen has pursued modularity and final assembly plant simplification largely as an in-house strategy, although its plants outside of Germany appear to be making much greater use of external suppliers. Premium European brands such as BMW and Mercedes have outsourced very little of their component design and engineering.

Prospects for Local Suppliers in East Asia

Global suppliers in the motor vehicle industry most often establish production facilities in developing countries at the behest of their customers who have set up final assembly plants and are trying to meet local content requirements. While suppliers are hard pressed to establish plants in all of the locations where their customers operate, they are often willing to invest in certain countries, especially those where multiple automakers are present and where lower operating costs raise the possibility of large-scale exporting. China is seen as such a location, as are smaller countries with dynamic vehicle markets and the possibility of raising quality to world standards without too much trouble, such as Thailand. What is clear is that having a plant in each East Asian country makes no economic sense to suppliers and that trade liberalization will very likely lead to dramatic consolidation.

The rise of suppliers with global reach and the technical sophistication necessary to design and produce complex modules and subsystems is making supplier-oriented industrial upgrading in the motor vehicle industry increasingly difficult. To paraphrase a manager of a Taiwanese (Chinese) automotive parts producer,

> Despite lower operating costs than Western suppliers, we cannot win business in ASEAN outside of our home economy because we are not present in the industry's design centers. In China, we suffer from the additional problem of not being part of the local supply base. We have not been asked to follow our American customers to new locations such as India. They prefer to work with American suppliers worldwide. In general, it is difficult for Taiwanese [Chinese] firms to get into the American supplier networks because all Big Three design is done in Detroit. To supply Ford, a company needs to have at least an office in Detroit and, in some cases, manufacturing facilities there too. If we were in Detroit, we could get assembly drawings and make bids on parts. We have more or less given up on building business with the Big Three. This is a critical problem because the local market is slow growing. Only by expanding regionally, especially into China, will we be able to attain better scale economies. But the [vehicle] platforms must be the same.

The ongoing globalization of the motor vehicle industry and the new role of suppliers in supporting the increasingly integrated global operations of automakers have important implications for supplier-oriented development in East Asia. The potential for local sourcing in the motor vehicle sector is high because of the large number, size, and weight of components and materials. With tariffs on imported parts, locally operating suppliers have a huge advantage. For example, a manager at a global automotive parts supplier in China stated a willingness to pay local suppliers a price equivalent to the worldwide price plus tariffs and transportation costs. What is clear is that local content, when it is provided, will come from the affiliates of global suppliers; a lack of technical competence has generally confined locally owned suppliers to simple, standardized, and slow-changing components, such as bearings, where there is a wider market that supports adequate scale economies.

Local suppliers in Asian NIEs have gained little responsibility for design. As a manager of a German automotive parts producer in China put it, "Design responsibility remains in Germany." According to a local supplier interviewed in Taiwan (China), "Ford does not allow changes in the 'hot' or critical points of cars, such as gear boxes, engine, chassis but only 'soft point' changes, for example, front- and back-end styling." The result is a somewhat loose technological relationship between actors in the offshore supply chain, while tight linkages are forged and maintained in the advanced economies. This puts a ceiling on the engineering capabilities that can be developed at suppliers based in emerging economies. According to a manager at an American-owned auto parts firm in China, "The auto industry in China is a 'veneered industry'—while it appears that Chinese firms are major suppliers to assemblers, all are propped up by huge amounts of imports." Another manager at a foreign-invested manufacturer of fuel system components and wire harnesses in China stated,

> We use only five or six local suppliers, and local content is only around 2 percent. Even for relatively simple products like wire harnesses 85 percent of materials is imported from places like Japan and Korea. Because of customs duties, transportation costs, "harassment" at customs, etc., these materials are very expensive. Automakers do not even buy all their steel in China because they cannot get the necessary quantities at the right quality. [The lack of] true local sourcing is the biggest impediment to growth of the industry in China. China hasn't done enough to allow its basic industries to survive. If the raw materials sectors do not develop, Chinese firms that do not have access to a global supply system will die after WTO is implemented. First-tier suppliers are pressuring their material suppliers to move to China. We have given many of our suppliers an ultimatum: by a certain year, they must either establish facilities in China or provide the company with imported material at a "domestic" price—the price they could achieve if they were located in China and not paying customs duties, transport costs, and so forth. Material suppliers, though, are reluctant to move to China because they are not sure they can reach critical economies of scale and because there is a shortage of basic commodity in-

puts in China. Material suppliers are particularly sensitive to commodity prices and quality because as one moves further up the supply chain in wire harnesses, materials become far more important than labor in determining total costs.

The result of these tensions is that the supply chains that are emerging in developing countries are increasingly foreign owned and very "thin." Core design activities remain concentrated in advanced economies, and many parts and materials continue to be imported. Trade, market, and investment liberalization will cause a further thinning as specific activities consolidate in particular places and export either regionally or globally. Some of the respondents interviewed in the Chinese motor vehicle industry believed that foreign automakers have no intention of manufacturing locally when China joins the WTO and drops its import restrictions. This view sees current operations as loss leaders that provide foreign automakers with a means to develop brand recognition and distribution and service networks for vehicles that they will later import.

Thun (2001) suggests two alternative strategies for local motor vehicle parts suppliers facing these difficulties. The first calls for local suppliers to join forces with global suppliers, either through mergers, joint ventures, or alliances. Local suppliers can offer global suppliers additional production locations and help to create "complementarity schemes," where vehicle components are exchanged within a region to offset tariffs and concentrate assembly so as to enhance scale economies. The second strategy requires local suppliers to serve global suppliers at a lower tier, supplying local content. But as local content rules are phased out, many parts will be produced on a regional basis to increase scale economies, undermining the rationale for regional complementarity schemes and putting further pressure on local firms. This pressure may force local suppliers from smaller Asian countries to invest in regional production centers, where they will either come into direct competition with global suppliers or, if the second-tier option is followed, with local suppliers following the same strategy. A third option, of course, is to exit the business entirely.

OLD AND NEW REQUIREMENTS FOR SUPPLIERS IN GLOBAL PRODUCTION NETWORKS: DRIVERS, RESPONSES, AND OUTCOMES

The evidence presented in the previous two sections points to the emergence of a new organizational structure underpinning the GPN in two key manufacturing industries. The rise of this global supply base can best be understood in terms of drivers, responses, and outcomes.

First, the "deverticalization" of value chains has been driven by increased levels of international competition, rising competence in the supply base, and a belief among lead firm managers in the West in the doctrine of "core competence" (Prahalad and Hamel 1990). Second, the ascendancy of the WTO and the strengthening of regional trading blocs in Europe and North America is driving high-volume production to low-cost sites within the blocs (that is, Eastern Europe and Mexico) and to the largest countries of Asia (that is, China and India). This trend toward the regional organization of production has been reinforced by the rising demand in end-user markets for rapid-response, build-to-order, and configure-to-order performance on the supply side. At the same time, deverticalization has been enabled by the increased use of information and communications technologies throughout the supply chain, but particularly in the design process and in supply chain management.

These drivers have led to the emergence of lead firms with little if any in-house manufacturing and the rapid expansion, growing financial strength, and increasing competence of the largest external suppliers of core manufacturing services, which in a self-reinforcing dynamic of "industry co-evolution" has fueled further outsourcing by lead firms and more consolidation within the supply sector (Sturgeon and Lee 2001). An important outcome of this process has been the growing requirement for greater involvement by suppliers early in the design process, which has strengthened the competitive position of suppliers that are able to co-locate and coordinate engineering and production activities with those of their customers at a global level. These changes are raising the threshold of performance for supplier firms in several areas, including design and engineering, sourcing, the effective use of information and communications technology, and the ability to operate in and coordinate between multiple locations. In the following paragraphs we discuss several of these developments in more detail, with an emphasis on the new requirements they raise for suppliers, paying particular attention to the implications for supplier firms in East Asia.

Deverticalization

The deverticalization of firm structures is at least partly a response to the increased volatility of the markets in which the firms are operating. Shorter product life cycles, the complexity and high cost of introducing new products, and unforgiving end-user and capital markets are combining to impose new pressures on all areas of the firm, from research and development to manufacturing to marketing and sales. In response to these pressures, and to offload risk, firms in a wide range of sectors and locations have sought to outsource non-core functions, especially those, like manufacturing, that are associated with large-scale fixed capital. Outsourcing has been especially prominent in competitive and industries like electronics, motor vehicles, and textiles and apparel (Fine 1999; Quinn and Hilmer 1994).

Lead firms have focused on the areas and functions that they believe to be essential to the creation and maintenance of competitive advantage, especially product innovation, marketing, and other activities related to brand development, and have increasingly come to rely on specialized suppliers to provide non-core functions. The belief is that by divesting non-core functions, lead firms can reap value more quickly from innovations while spreading risk in volatile markets (Venkatesan 1992). Firms that develop, market, and sell electronic hardware, cars, and clothing have turned to suppliers for production and, increasingly, post-architectural (that is, detailed) design services. By tapping the competencies of suppliers, lead firms are able to maintain substantial market presence without the fixed costs and risks of building and supporting a vertically integrated corporate organization (Sturgeon 2000, 2002). Among the advantages claimed for these production networks are that they are more adaptable than an integrated firm would be and are capable of providing better economic performance in highly competitive and volatile markets (Cooke and Morgan 1993; Powell 1990).

Design and Engineering

Lead firms in the advanced economies are asking their suppliers to take on more responsibility for the design and engineering of products and subsystems. Suppliers are increasingly being chosen and brought into the development process before products are fully designed. By doing so, the lead firm is able to spread risk and reduce costs. When suppliers participate in the development of prototypes, moreover, they typically improve their performance in design for manufacturability and in the implementation of subsequent engineering change orders. Product redesign for different markets is also easier and quicker if suppliers are actively involved from the outset.

One of the keys to achieving heavy supplier involvement in early development activity is the promise of future business. Small numbers of suppliers or even sole sourcing has become common for specific product models and generic product platforms. Conversely, situations in which a new supplier is brought on board after volume production has begun are becoming increasingly rare. In industries such as motor vehicles, where models remain in production for two to five years, business is won and lost by suppliers in large blocks and for long periods. The lead firms try to reduce supplier market power in various ways (by using different suppliers for different models and types, for example), but a supplier for a successful model can gain a great deal of leverage.

Many Asian suppliers have aggressively pursued design and engineering competencies (although, as table 2.1 shows, many more remain at the OEM stage). Leading Asian suppliers in the electronics, motor vehicle, and textile-apparel industries now offer a full range of design and engineering services. Advanced Taiwanese (Chinese) notebook computer ODMs and Hong Kong (Chinese) garment producers, for example, provide their customers with completely designed finished products as well as products that are made-to-specification and co-designed.

However, few Asian suppliers are yet able to provide design services that are co-located with those of their customers. One manager at an American-based global motor vehicle parts supplier observed, "Co-location of design with our customers' vehicle development is important because the design interface is very complex and technology changes very quickly. Face-to-face meetings are still needed to resolve design issues." Continuing advances in broadband communications and information technologies may make long-distance concurrent engineering between lead firms and suppliers a more viable alternative to co-location, but to the extent that co-location remains important or becomes more so, suppliers with design facilities located near their customers will have an advantage.

Information and Communications Technology

The increasing use of advanced information and communications technology (ICT) has accelerated the possibilities for effective integration in geographically and organizationally dispersed value chains. Firms carrying out functions at different points along a value chain have a greater ability to exchange data so as to achieve high degrees of conformance with specifications and tight coordination of productive activities. In the past, achieving such conformance and coordination required firms to locate the relevant functions within their own vertically integrated organizations or within tightly controlled networks of subordinate suppliers. The promise of ICT is that codifiable specifications and standard interfaces will make it possible to coordinate activities through market-based exchanges among autonomous organizations.

So far, the most important contribution of ICT has been in management of the supply chain and in the design and manufacturing of products and components. Specific applications in the area of supply chain management include enterprise resource planning systems, business-to-business e-commerce marketplaces, and electronic data interchange. The application of digital technology to the product and component design process involves tools such as electronic design automation, computer-aided engineering, and computer-aided design. These technologies, when combined with computer-aided manufacturing systems embedded in numerically controlled and robotic production equipment, allow complex product specifications to be handed off to outside suppliers. Using these tools, suppliers can create fully computer-integrated manufacturing environments to track product quality and inventory and shorten design and production cycles.

New ICTs enable lead firms to ask much more of established suppliers in terms of rapid response, design collaboration, lower costs, and close monitoring. It is not unusual for lead firms to exert great pressure on their suppliers to adopt the latest ICT to improve quality, facilitate the tracking of in-process inventory, and streamline the order and reorder process.

Whether new ICT systems raise or lower barriers to entry is an open question. Leading-edge ICT systems are expensive and often must be adapted to the specific requirements of the firms that use them. Successful adoption of ICT requires competent vendors and specialized personnel to build and operate the new systems and frequently also requires user firms to adapt their organizational routines to accommodate them. ICT systems typically have short life cycles and must be continually upgraded to remain compatible with customer systems and changing industry standards. Staying on this treadmill of ongoing capital spending, learning, and organizational change is only possible at considerable effort and expense. Leading Asian merchant manufacturers and suppliers such as Taiwan Semiconductor and Li and Fung are among the most sophisticated users of ICT to be found anywhere, and in our field research we came across many firms that had adopted the latest information technologies. Examples include semiconductor foundries, producers of personal computers and peripherals, and textile-apparel suppliers based in Taiwan (China); electronic component suppliers based in Singapore; and textile-apparel suppliers in Hong Kong (China). But many smaller firms in East Asia, especially from the developing economies in the region, are struggling to keep up. One indication of the general situation with respect to the adoption of ICT is provided by table 2.6, which reports the (very low) rates of Internet usage by manufacturers in China and the four ASEAN economies of Indonesia, Malaysia, the Philippines, and Thailand.[8]

8. About 24 percent of the ASEAN firms also reported that they participate in an electronic data interchange network.

Table 2.6 Internet Use by ASEAN-4 and Chinese Manufacturing Firms (percent)

Type of use	ASEAN-4	China
Communicate with clients via Internet	24	6
Communicate with suppliers via Internet	15	5
Place orders via Internet or e-mail		
1999	10	2
2000	14	4
2005 (estimated)	23	6

Source: Survey conducted by the World Bank in early 2001.

ICT systems, especially for supply chain management, are relatively new and still rapidly evolving. Vendors have emerged from various fields (manufacturing automation, enterprise computing, and so forth), business models are often untested and experimental, battles over standards are in full swing, and merger and acquisition activity among information technology vendors is proceeding at a rapid pace—and, since the bursting of the dot-com bubble in the United States, some have gone out of business. Moreover, the vendors and standards used in different industries are often very different. Our interviews suggest that, given the rapidly changing technology and standards, user firms may derive advantage from being located near the nexus of information technology innovation. In places such as Silicon Valley, users and vendors have set up informal and formal working groups intended to set standards and develop better applications (Sturgeon 2003). Although all have sales offices around the world, the leading vendors of information technology in the field of supply chain management are based in California (Ariba, Oracle, and PeopleSoft), Colorado (J. D. Edwards), Germany (SAP), and the Netherlands (BAAN). None of the major vendors is based in Asia, which may be a disadvantage for Asian suppliers.

Global Suppliers

As lead firms have outsourced more of the manufacturing, sourcing, and logistics functions that were previously carried out in-house, their preference for dealing with suppliers with international reach has grown. The reasons are several. First, the lead firms are in many cases marketing their products globally and require engineering, manufacturing, and logistics support in multiple locations. Second, the lead firms often seek to economize on development costs by creating global product platforms that share and reuse many common parts, modules, and subsystems. Partnering with a small number of suppliers, or even a single supplier, enables lead firms to exploit these economies of scope more fully, while avoiding the cost of requalifying new suppliers for each new market. Third, cost pressures require purchasing organizations to scan the world for low-cost, high-quality parts, and, to the degree that suppliers are taking on these responsibilities, they too must have global sourcing capabilities. Fourth, suppliers based in protected final markets can combine global sourcing with local sourcing and subassembly to help lead firms meet local content requirements. Fifth, the preference for key suppliers to take on a more active role early in the development process requires these suppliers to be able to co-locate at least some of their own design activities with the design facilities of their customers. Some lead firms have given their key suppliers an ultimatum: provide support on a global basis or lose the business entirely. Managers at three global automotive suppliers made essentially the same point during separate interviews with IPC researchers:

> The industry began to change 5–10 years ago. Today it is a requirement to serve platforms—it is part of the bid. If a supplier doesn't have a global strategy, it can't bid. New projects are no longer seen as an opportunity to expand globally—instead, a supplier must have a global base in place to even make a bid. This forces suppliers to have a global supply system in place.
>
> Suppliers must support assemblers as a sole source for global product lines to support commonalization. We must supply the same part, with the same quality and price, in every location. If [the automaker] says to go to Argentina, we must go or lose existing, not just potential, business. Logistics are becoming a key competitive advantage; we must have the ability to move production to where customer's facilities are.
>
> We want our plants to be present where vehicles are produced. Sometimes customers ask us to locate near offshore assembly plants to provide local content . . . . We will follow our customer's strategy by establishing local engineering operations in large emerging markets only, such as Brazil, Korea, and Mexico.

Providing this kind of support involves coordinating flows of components, subassemblies, and products across production networks that often span several countries or even continents. It also requires setting up design operations close to the design centers of the lead firms. As a result of the popularity of outsourcing among American and European lead firms, the great majority of the suppliers that have risen to this challenge so far have originated from Europe and North America, where the lead firms have the bulk of their design activities, where there is a deep pool of management talent with long experience in international operations, and where capital is available to finance global expansion. The pressure to expand has been met partly by internal growth but even more by aggressive merger and acquisition activity.

Acquisitions of competitors in similar lines of business have yielded sudden jumps in geographic coverage. Acquisition of firms with upstream or downstream capabilities has broadened the range of products and services on offer (frequently, acquisitions have served both purposes simultaneously). Finally, acquisitions of customer facilities have also helped suppliers to win new business while expanding their geographic and functional scope.[9]

The rise of regional production systems under trade arrangements such as the European Union and NAFTA has strengthened the hand of global suppliers in several ways. First, the drive to serve advanced economies from proximate low-labor-cost locations such as Eastern Europe and Mexico, for both tariff and speed-of-response reasons, has reduced the competitiveness of suppliers serving advanced-economy markets solely from Asian locations. Some Asian suppliers in the electronics and textile-apparel sector have responded by opening plants in places like Central America and Mexico, but most have not. For global lead firms and suppliers, fragmented national production systems are being replaced by regional systems that allow increases in scale economies in plants serving regional markets. This regional strategy is nested within a global strategy that seeks to commonalize and reuse as many design elements as possible among regions. Suppliers that do not or cannot put a global or regional system in place to match the strategy of the lead firms in their industry, because either they are too small or are concentrated in a single region, may find long-term success elusive. For Asia the increased popularity of the regional or global model has meant the consolidation of production in China, and more global suppliers are scaling back existing investments in Southeast Asia and building up very large-scale operations in China. In the words of the president of Manufacturing Services Limited, a medium-size electronics contract manufacturer based in Concord, Massachusetts, "We want to consolidate [the Malaysia and Singapore operations] in China . . . we want to supersize our China operation. They have a great pool of available labor" (Serant 2001).

9. For example, Sanmina, an electronics contract manufacturer based in San Jose acquired Nortel's Wireless Electro-Mechanical Subsystem Assembly in August 1999. This deal included inventory, production equipment, and 230 employees in manufacturing facilities in Calgary (Canada) and Châteaudun (France) as well as a small engineering design group located in Guyancourt (France) outside Paris. Both the Calgary and Chateaudun facilities were located adjacent to Nortel's wireless system design houses, allowing the production facilities, now owned by Sanmina, to remain poised to bring new products quickly into production. The acquisition provided Sanmina with new locations in Canada and France, locations that can support not only Nortel but its other customers, as well as new expertise in radio-frequency electronics manufacturing.

For established East Asian manufacturing firms, the requirement for global reach may be the most challenging of all the performance requirements associated with the emerging supply base. According to a manager at one global motor vehicle parts supplier, "Supplier logistics need to be synchronized with [lead firms], and many local suppliers do not have the capability to do this." Except for the Japanese, few East Asian suppliers have expanded geographically outside East Asia. Most have lacked access to the capital that would enable the aggressive merger and acquisition strategies needed to build an international position. Many East Asian firms are still controlled by their founders or the founders' family members. Other, publicly held, firms are listed on thinly capitalized local exchanges that lack the liquidity of U.S. and European public equity markets. Most of these firms, too, have lacked access to the managerial expertise needed to operate successfully outside Asia. Of 100 ASEAN-4 manufacturing firms responding to the World Bank's 2001 survey, only 16 percent reported employing managers with any foreign work experience.

RISKS AND UNCERTAINTIES OF THE GLOBAL SUPPLIER MODEL

The organizational and technological changes we have described in the previous sections are still emerging, and many questions have been raised about their sustainability. Some of the strongest doubts about the global supplier strategy were expressed during our interviews with Japanese lead firms in the electronics sector, but our research at lead firms in Europe and the United States also revealed conflicts and confusion over the desirability of increased outsourcing. Some of these concerns are of long standing but have grown more acute as the pace of outsourcing and supply base consolidation has accelerated. There are fears of the loss of innovative capacity as production-related activities migrate to global suppliers, of the leakage of proprietary information to competitors through shared suppliers, of the creation of competitors if suppliers choose to move into the production of final products, and, more generally, of the increased market power of the largest suppliers. There are also concerns about the feasibility of product design modularization in cases where unanticipated or complex interdependences between the components of the product lead to much higher coordination and transaction costs. There are also examples of outsourcing leading to additional effort. In some lead firms we observed the emergence of "shadow" engineering organizations, whose ostensible role in monitoring cost and quality and maintaining the ability to switch suppliers has expanded into large-scale duplication of supplier engineering work.

The reliance of lead firms on one or just a few external suppliers, or sometimes even a single plant, when coupled with the reduced inventories associated with just-in-time delivery practices, has significantly increased the vulnerability of their operations to disruption from external work stoppages, accidents, or natural disasters. When key supply plants go off-line, the whole system can go down.

Other questions have been raised about the sustainability of the suppliers' business model itself. Some global suppliers have encountered difficulties in finding a way to get lead firms to pay for the additional engineering, component purchasing, and logistics support they are providing, leading to very low or negative profit margins. Moreover, the rapid international expansion of supplier firms, often driven by acquisition of companies in different institutional settings with dissimilar corporate governance systems, industrial relations practices, product and process technologies, and ICT systems, may be re-creating, or even exacerbating, the operational inefficiencies that outsourcing was (in part) intended to remedy in the first place.

The separation of product strategy and design from manufacturing also raises concerns—especially at Japanese electronics firms, where there is a strong tradition of using what is learned on the shop floor to inform and improve product design—that outsourcing will undermine innovation in the long run. Whether product innovation can be sustained in the face of high levels of outsourcing remains to be seen. Similarly, there are concerns that process innovation will suffer when so much production is in supplier firms with fewer capabilities in and funding for research and development.

The recent global economic downturn presents one more challenge to these global supply networks, which owe much of their growth to the long, uninterrupted economic expansion of the 1990s and whose stability in recessionary conditions is only now being tested. It also remains to be seen whether this pattern of deverticalization and global-scale organization will be affected by the September 11 terrorist attacks and their aftermath. The pattern is real and has considerable momentum. Even so, a caution against assumptions of irreversibility and inevitability is appropriate, especially at the present juncture. Research on these questions is continuing at the MIT Industrial Performance Center.

JAPANESE SUPPLY CHAINS IN EAST ASIA

Japanese firms have had a major impact on industrial upgrading throughout East Asia. The flows of technology, investment, and training resources from Japan to other East Asian economies have been substantial, and today Japanese firms are sourcing throughout the region on a large scale. Supplier-oriented upgrading in East Asia based on linkages with Japanese firms has taken two distinct, though sometimes overlapping, forms.

The first has occurred through the acquisition by developing-country suppliers, often through licensing or joint ventures, of process technologies for the manufacture of inputs and components that have become highly competitive, standardized, and unprofitable for Japanese producers. Japanese firms, pursuing the "flying geese" strategy of letting older, less profitable industries migrate to their developing neighbors, have systematically developed suppliers in East Asia for inputs that are becoming commodities or are subject to great demand volatility. The logic of the flying geese strategy is to create a "shock absorbing" supply base outside Japan capable of meeting the variable portions of world demand. This is achieved by off-loading process technologies for product categories where commoditization results in intense price competition and low profit margins. With an alternative source of supply for low-value items in place, higher-value components and inputs, and many of the final products using these components, can continue to be produced in Japan.[10]

In the 1970s and 1980s Japanese firms began to transfer synthetic textile and steel production technology to firms in Korea and Taiwan (China). In the 1980s process technology for the production of computer memory chips, or DRAMs, was transferred to Korean firms.[11] In the mid-1990s Japanese firms began to transfer process technology for flat panel displays to Korean and Taiwanese (Chinese) firms (Akinwande, Fuller, and Sodini 2001). In this way, an important path to industrial upgrading in Korea and Taiwan (China) served simultaneously (and, from the Japanese perspective, more importantly) as a strategy for promoting industrial transformation in Japan. Technical progress in developing countries based on the flying geese model has been widely documented (see, for example, Encarnation 1999). Amsden (1989) refers to it as "apprenticeship," the acquisition of foreign technology through licensing and technical assistance programs, and differentiates it from "imitation," the copying of foreign technology through

10. American and European firms have also licensed process technology to firms in East Asia, but these moves have often been motivated by short-term tactical considerations (usually the generation of cash in times of trouble) rather than strategic ones. In some cases, however, Western firms have taken a more strategic approach. The development of a supply base in Taiwan (China) for low-cost desktop and notebook personal computers by American computer firms such as Compaq and Dell is a case in point.

11. Taiwanese (Chinese) firms acquired DRAM technology from the Japanese and also from European producers but have met with very limited success.

tactics such as reverse engineering, a method that has also been widely used in the Asian NIEs.[12]

For developing economies, the risk of upgrading via the flying geese model comes down to one of low profitability due to technological lag. Since the technology that has been transferred has invariably been one or more generations behind the leading edge, countries that have relied on this strategy for development find themselves on the "bleeding edge" of many markets, stuck in low-profit and volatile sectors such as low-cost personal computers, computer monitors, scanners, power suppliers, batteries, keyboards, DRAMs, and mass market apparel and footwear products. More profitable product segments, such as high-end computers and servers, communications equipment, software, logic semiconductors, and high-fashion apparel and footwear, continue to be pursued primarily by American, European, and Japanese firms working closely with advanced users, who are concentrated in the advanced economies.

The second variant of industrial upgrading that has been based on linkages with Japanese firms is that of FDI. Japanese firms have a long history of FDI in East Asia, and Japanese investment has in the past been widely dispersed throughout the region—much more so than U.S.-originated FDI (Mason and Encarnation 1995). Until the mid-1980s these investments, which were concentrated in consumer and white goods, consumer electronics, and automobiles, were established mainly to serve protected local markets. Over time, Japanese firms gained a strong foothold in all but the most protected markets in East Asia. In East Asian countries other than Korea, for example, Japanese motor vehicle firms have a market share that exceeds 90 percent (Doner 1991; Sturgeon and Florida 1999). The appreciation of the yen in the mid-1980s accelerated Japanese investment in Southeast Asia, and some production platforms were upgraded and expanded to serve ex-

12. Amsden (1989, p. 110) argues that the acquisition of advanced production equipment has been a major avenue for industrial upgrading in East Asia. Productivity increases are drawn from the world "technology shelf" through imports of foreign technology embedded in advanced production equipment. Operating advanced production equipment at scale economies sufficient to minimize unit costs, and learning to use it more efficiently than foreign rivals, enhances competitiveness on world markets. As long as profits from increased output are invested in new equipment that embodies the latest technology, growth can be maintained. Increased output results in greater scale economies and expands the opportunities for learning by doing, improving efficiency and increasing wages, and driving growth in the domestic market. Although the state can do a lot to initiate this process, effective application of new technology depends on what happens on the shop floor, which according to Amsden helps to explain the importance of managers over entrepreneurs in countries, such as Korea, that have pursued this manufacturing-led approach to industrial upgrading.

port markets in the West.[13] But the cross-border networks led by Japanese firms have tended to rely on internal subsidiaries and on Japanese suppliers, and, given the scale of investment, the opportunities for local firms to upgrade their competencies via supply relationships have consequently been very limited.[14] To the extent that local firms were included, it was usually in the smallest and least promising markets such as the Philippines and Vietnam, where they were used as final assemblers and retailers who would work under license, source all components from Japanese firms, and absorb the risk of distributing and selling finished products in small and uncertain markets. In a few cases where the network relationships were of long standing and the local supplier's capabilities were high, local firms were tapped to serve the export market. But such firms were never part of the core of the production network and were particularly vulnerable to changes in the lead firm's strategy.

The experience of a Philippines-based independent consumer electronics contract manufacturer is a case in point. The firm started out in 1965 as a television distributor for a major Japanese consumer electronics company, and, following the introduction of tariff barriers for fully assembled sets, began assembling televisions for the local market under license in 1971. By 1994 production at this dedicated facility had increased to 12,000 units per month. In 1977 the company began assembling televisions under license for a second Japanese consumer electronics firm, initially for the local market and, then, beginning in 1993, for export. In 1994 the company built a 1 million unit per year plant in an export-processing zone. Production at this plant quickly rose to 40,000 units a month. The license agreements required the company to source components from its Japanese partner's subsidiaries, some of which were located in Japan and Singapore, and also allowed the company to provide after-sales service, marketing, and distribution in the Philippines.

As the company grew, it added a plastic injection molding division, which was spun off as an independent company in 1994. Its strategy was eventually to develop its own original brand of products in addition to its contract

13. Japanese export platforms in Southeast Asia were not as extensive as they might have been, because Japanese firms began at the same time to make very large investments within or adjacent to Western markets (Abo 1994; Curry 2000).

14. Some analysts have contrasted the closed Japanese production networks in Asia with the relatively open networks led by North American—and to a lesser extent European—firms, which they argue have created more opportunities for East Asian firms to upgrade their capabilities (see Borrus, Ernst, and Haggard 2000; Ernst 1997; Ravenhill 1995). On the captive, hierarchical character of Japanese-led production networks, see Aoki (1987), Dore (1986), Gilson and Roe (1993), Sako (1989), Schonberger (1982), Sturgeon (1999), and Womack, Jones, and Roos (1990).

manufacturing and distribution business. After the Asian financial crisis, however, the first Japanese electronics customer began to shift its export production to its plant in Malaysia, and production at the dedicated plant declined from 12,000 to 2,000 units a month; orders ceased completely at the end of 2001. By 1998 the second customer had also withdrawn all its orders. The plant survived by serving a new customer, a Taiwanese (Chinese) producer of monitors.

Management believes that its television business is unlikely to recover because the global market is saturated, and television manufacturers are focusing on high-value television sets whose large size and weight leads to a preference for assembly close to final markets. The company is trying to diversify into broadband communications services, real estate development, corporate financial services, and food and agriculture businesses, but the future is uncertain. The injection molding unit did develop new business in beer crates and motorcycle, refrigerator, and water cooler parts, but management has found that many of the Japanese firms producing these products locally prefer to rely on their own suppliers, which have followed them to the Philippines.

The recent experience of this Philippine manufacturer is far from unique. Japanese production networks throughout East Asia have been undergoing a process of rationalization. As a local supplier to both the Ford and Nissan joint ventures in Taiwan (China) noted, "Japanese firms have slowed investment in new capacity, rationalized production in Southeast Asia, increased investment in China, scaled back internal capacity, and outsourced more activities, including engineering." The combination of the protracted economic downturn in Japan, the continuing weakness of other Asian economies, the lowering of tariff barriers for finished products in the region, and the saturation of markets for some products in smaller East Asian countries is leading to the centralization of production at a smaller number of locations in order to eliminate excess capacity and to exploit economies of scale. In many cases this has meant a new focus on production in China at the expense of other East Asian locations. For local suppliers elsewhere in East Asia, therefore, the prospects for upgrading through links with Japanese lead firms do not seem as promising as in the past.

In a sense, the new preference of Western lead firms to use Western suppliers on a global basis simply brings these production networks more in line with those led by Japanese firms. While FDI of this type can do much to create jobs and train workers in advanced technical and business processes, its contribution to the development of a vibrant locally owned supply base, and thus to the strategy of supplier-oriented industrial upgrading, is less evident.

POLICY IMPLICATIONS OF THE NEW GLOBAL SUPPLY BASE FOR EAST ASIA'S FUTURE DEVELOPMENT

Throughout East Asia, many manufacturing firms have moved along the industrial learning curve by mastering contract manufacturing under the tutelage of some of the world's most advanced lead firms, which have, in turn, marketed the products under their own brands. In the automobile and electronics sectors on which our research has focused, East Asian managers described learning how to make products well with help from their early customers and then improving their products by selling to ever more demanding buyers. A company's capabilities as a manufacturer are proven, in effect, by its clientele. The fact remains, however, that in this common scenario, the manufacturer captures only part (and often relatively little) of the final price of the good, remains at the beck and call of the buyer, and is vulnerable to competition from lower-cost competitors that threaten to take the business away. Although this is true for many suppliers in all economies, the problem is made acute in places such as East Asia, where a great many suppliers are not co-located with their customers or with their final markets and so have little opportunity to move beyond this highly subordinate status.

The concentration of lead firms in advanced economies and the increased demands they are imposing on their suppliers mean that many Asian firms are dealing with fewer and stronger customers than they faced in the past. This is not a situation in which the Asian suppliers are inevitably the losers. If the evidence in table 2.7 is anything to go by, some larger Asian suppliers still appear to hold their own in negotiations with their customers. Although it is difficult to conceptualize and to measure power relations in supply chains, the levels of profitability and return on capital of the suppliers compared with those of the lead firms—often their customers—show the suppliers capturing a comparable share of the rewards in the system. The suppliers compared in this table stand out, however, not only for producing at low cost but also for recruiting and nurturing well-educated managers, engineers, and technicians who can work in partnership with customers. These firms have innovative design capabilities that allow them to suggest new model designs to customers (ODM) as well as work to their clients' specifications (OEM). They have implemented major changes in their business structures and adopted advanced ICT systems that allow them to coordinate widely dispersed activities in different countries. Like the chip foundries and electronics contract manufacturers in Taiwan (China), some East Asian firms have aggressively exploited the possibilities of deverticalization and so have achieved some success in capturing the business that is being spun off by formerly vertically integrated firms.

Table 2.7 Profitability of Contractor and Lead Firms (percent)

Firm	Economy	Sector	Profit growth, five-year average	Return on shares	Sales growth, four-year average
Johnson Electric	Hong Kong (China)	Micromotors	19	15	10
Black and Decker	United States	Tools	10	13[a]	–1
Quanta	Taiwan (China)	Laptop maker	11[b]	57	73
Dell	United States	Laptop lead firm	9	58	45
Techtronics	Hong Kong (China)	Tools	7	23	23
Home Depot	United States	Retail	8	18	25
Hon Hai	Taiwan (China)	Cables	11	33	57
Intel	United States	Microprocessors	34	32	24
TSMC	Taiwan (China)	Microprocessors	0	38	28
Philips	Holland	Microprocessors	44	10	6
Hung Hing	Hong Kong (China)	Printing	16	21	16
Bertelsmann	Germany	Press	7	25	6
Li and Fung	Hong Kong (China)	Trading	3	49	25
The Limited	United States	Retail	8	35	6

a. Median value.

b. Four-year average.

Source: Boillot and Michelon (2001, p. 142).

The suppliers in table 2.7 are exceptional. The resources that they have deployed for each of these advances are well beyond the reach of most small Asian suppliers. In the early days of the Asian miracle, even rather small firms might have been linked up to large international customers, who in turn provided many of the services and inputs the firms needed in order to produce to specifications and gradually upgrade their capabilities. Today, however, there are fewer such supports for the small firm. The international customers are looking for suppliers that are already able to make the products—not for firms that can be brought up to the needed level of performance.

For Asian manufacturers today, moving away from old-style OEM is a precondition for enhancing the value they create, for protecting their share of the market against competitors—both rivals elsewhere in the region and global suppliers based in the West—and, more broadly, for contributing to the continued economic development of their home societies. We do not reject supplier-oriented approaches to industrial upgrading, but new actors and new requirements are posing challenges to any notion of an unproblematic or inevitable OEM→ODM→OBM upgrading path. There are a variety of ways in which firms can balance or replace OEM production with new higher-value activities, and the scope for industrial upgrading in the Malaysian context is explored in chapter 3. One route is to acquire or create a brand and sell some or all of the firm's production under the brand name. We have already discussed the pitfalls of competing directly with customers in end markets, but there are approaches, such as selling products or selling in markets that customers are not interested in, that can avoid conflict. A second approach is to continue to make products or components that will be sold under another label, but to develop internal capabilities sufficient to invest these products with design and functional features allowing them to command higher prices and to raise the barriers to the entry of competitors, the classic ODM route. A third is to specialize in process-specific technologies that enable the firm to provide high-quality, low-priced manufacturing and manufacturing-related services for a number of (increasingly demanding) brand-name customers, some of whom may themselves have shed those functions entirely; this is essentially the OEM route with an emphasis on continuous improvement. A fourth is to move to the second tier of the supply base, providing domestic content for the local affiliates of global suppliers. The fifth and perhaps most radical option for Asian firms is to follow the same strategy of those lead firms in the West that have shed the manufacturing function entirely and to tap the resources of the increasingly ubiquitous global supply base, which has in many ways commodified the manufacturing process. This will prove challenging in many Asian societies, which have for so long based their industrial upgrading and industrial policies on the initial establishment of excellence in manufacturing. As we have stressed in this chapter, each of these routes to upgrading poses formidable technological and managerial challenges, but the key point is that industrial upgrading is an unending process that requires constant attention and investment in time, money, and human effort.

What concrete steps can East Asian governments take to strengthen the capabilities of their domestic suppliers to participate successfully in emerging global production networks? The most important policies are those whose outcomes benefit all sectors of the economy: macroeconomic stability; transparent and efficient capital markets; confidence in public institutions; protection of individual rights under rule of law; and a strong foundation of education and training. Beyond these general measures, however, additional steps can be taken to address the specific capabilities of the supply base. The details will depend on the stage of development of the economy, the strength of its public institutions, the particular capabilities of both government and industry, and the prevailing attitudes toward the role of government in the economy. But while the particular prescriptions will vary from one country to another, the empirical evidence on supplier-oriented industrial upgrading presented in this chapter points toward several important general lessons.

First, there is more than one upgrading path. The best approach will vary from one industry to another; moreover, even in the same industry different firms may choose different paths depending on their internal capabilities, the regional environment in which they operate, and the particular overseas market or markets they are serving. Government policies that do not acknowledge the range of upgrading possibilities for domestic suppliers, that presume that all firms will follow the same path, or, worse, that attempt to dictate such a path are likely to fail.

Equally important, however, is to recognize that remaining at the same level of competence is not a viable option. This is the second key lesson for policymakers. The process of upgrading is unending; there is no threshold of adequacy. Taken on its own, the strategy of pursuing low-cost labor wherever it is to be found, a strategy that many Asian manufacturers have pursued, is not a viable approach in anything other than the short term. Cost and quality will remain important to customers, but performance in these dimensions is increasingly becoming a commodity and will not be enough to yield attractive margins. Providing innovative design content, production flexibility, and the ability to dependably deliver the right product at the right time and in the right place are gaining in importance, as are the international sourcing and logistic capabilities on which these service enhancements are based. Even as the competitiveness of lead firms in Western markets is increasingly determined by the efficiency of their supply chains, these lead firms are electing to outsource more of the key supply chain functions, and this continued deverticalization will certainly put ongoing pressure on suppliers—from East Asia and elsewhere—to continuously upgrade their capabilities and their geographic reach.

Third, the innovations that fuel new generations of products and processes grow both out of new ideas that a society develops on its own and those it finds in other countries and adapts and develops for its own purposes. Even an advanced economy like the United States, with its vast domestic research and development infrastructure, still needs mechanisms to monitor science and technology in leading centers abroad and to pull the most promising new developments into its own industries. East Asian nations need to enhance those institutions that enable them to tap into the most promising technologies developed abroad. In new areas of technology in which firms wish to advance, there may be much to be gained by recruiting international experts with experiences in leading-edge firms. Immigration and housing policies can often be effective in aiding the recruitment efforts of domestic corporations. Governments should also look for opportunities to strengthen links with leading research universities overseas, including universities elsewhere in the region.

The importance of information technology to the effort to upgrade supplier performance has been a pervasive theme of this chapter and is a fourth key lesson for policymakers. We have commented on the central role of information technology in the design, manufacture, and delivery of the service-enhanced products demanded by customers in advanced economies. For the emerging economies of East Asia, participating successfully in GPNs will require stronger domestic capabilities to exploit advanced technologies of information generation, storage, and communication. Notwithstanding the deflation of the information technology bubble in the United States and elsewhere, electronic commerce is rapidly becoming a *sine qua* non for small- and medium-size manufacturing enterprises throughout East Asia to compete effectively in global markets. The technologies of electronic commerce, combined with the computerization and codification of both the design and manufacturing processes, are accelerating the possibilities for effective integration in geographically and organizationally dispersed value chains. Although great strides have been made, the process of codification that supports this sort of "value chain modularity" is, in many ways, in its infancy.[15] Of course, vast areas of knowledge and information remain uncodified, and some are extremely difficult to codify for technical reasons. Furthermore, the ongoing processes of technological change and innovation can render existing standards obsolete, restarting the clock on the process of codification (David 1995; Storper 1995). Indeed, internalizing such realms of tacit knowledge, and packaging them as services for a wide range of customers, remains one of the most promising routes for industrial upgrading.

Related to this is a fifth important lesson. The cultivation of vertically integrated "national champions" can no longer serve as a primary goal of industrial development policies. Rather, the development of key supply chain capabilities should be the objective of these policies. The presumption that such capabilities should be combined within vertically integrated corporate structures is not consistent with the emerging pattern of GPNs and may actually inhibit the acquisition of knowledge and technology on which both domestic upgrading and international competitiveness depend.

15. For more on the concept of value chain modularity, see Baldwin and Clark (2000); Dolan and Humphrey (forthcoming); Gereffi, Humphrey, and Sturgeon (forthcoming); Schilling and Steensma (2001); Sturgeon (2002, 2003).

REFERENCES

The word *processed* describes informally reproduced works that may not be commonly available through libraries.

Abo, Tetsuo. 1994. *Hybrid Factory: The Japanese Production System in the United States.* New York: Oxford University Press.

"Acer Plots Path to Unbeatable Service." 2001. *Financial Times,* March 30.

Akinwande, Akintunde, Douglas Fuller, and Charles G. Sodini. 2001. "Leading, Following, or Cooked Goose: Explaining Innovation Successes and Failures in Taiwan's Electronics Industry." IPC Globalization Working Paper 01-001. Massachusetts Institute of Technology, Cambridge, Mass. Processed.

Amsden, Alice H. 1989. *Asia's Next Giant: South Korea and Late Industrialization.* New York: Oxford University Press.

Aoki, Masahiko. 1987. "The Japanese Firm in Transition." In K. Yamamura and Y. Yasuba, eds., *The Political Economy of Japan.* Palo Alto, Calif.: Stanford University Press.

Automotive News. 1996. Available at http://www.autonews.com/ Top 50 automotive suppliers, 1995.

Baldwin, Carliss Y., and Kim B. Clark. 2000. *Design Rules.* Cambridge, Mass.: MIT Press.

Berger, Suzanne, and Richard Lester, eds. 1997. *Made by Hong Kong.* New York: Oxford University Press.

Boillot, Jean-Joseph, and Nicolas Michelon. 2001. *Chine, Hong Kong, Taiwan.* Paris: Documentation Française.

Borrus, Michael, Deiter Ernst, and Stephan Haggard, eds. 2000. *International Production Networks in Asia.* London and New York: Routledge.

Cooke, Phillip, and Kevin Morgan. 1993. "The Network Paradigm: New Departures in Corporate and Regional Development." *Environment and Planning D: Society and Space* 11(5):543–64.

Crain's Detroit Business. 2001. "Largest OEM Parts Suppliers Ranked by 2001 Sales of Original-Equipment Manufacturers' Parts." *Crain's List,* July 8. Available at http://www.crainsdetroit.com/.

Curry, James. 2000. "San Diego/Tijuana Manufacturing in the Information Age." Paper prepared for San Diego Dialogue: The Global Engagement of San Diego/Baja California. Available at http://www.sddialogue.org/pdfs/brpaper percent20mfg.pdf. Processed.

David, Paul A. 1995. "Standardization Policies for Network Technologies: The Flux between Freedom and Order Revisited." In Richard Hawkins, Robin Mansell, and Jim Skea, eds., *Standards, Innovation, and Competitiveness: The Politics and Economics of Standards in National and Technical Environments,* pp. 15–35. Aldershot, U.K.: Edward Elgar.

Dedrick, Jason, and Kenneth Kraemer. 1998. *Asia's Computer Challenge: Threat or Opportunity for the United States and the World?* New York: Oxford University Press.

Dolan, Catherine, and John Humphrey. 2000. "The Governance of the Trade in Fresh Vegetables: The Impact of U.K. Supermarkets on the African Horticulture Industry." *Journal of Development Studies* 37(2):147–76.

———. Forthcoming. "Changing Governance Patterns in the Trade in Fresh Vegetables between Africa and the United Kingdom." *Environment and Planning A.*

Doner, Richard. 1991. *Driving a Bargain: Automobile Industrialization and Japanese Firms in Southeast Asia.* Berkeley: University of California Press.

Dore, Ronald. 1986. *Flexible Rigidities: Industrial Policy and Structural Adjustment in the Japanese Economy 1970–1980.* Palo Alto, Calif.: Stanford University Press.

Dunn, Darrell. 1997. "Ericsson Telecom Signs Agreement with Solectron." *Electronic Business News*, March 31(1051): Service and Support section.

Electronic Trend Publications. 2000. "The Worldwide Contract Electronics Manufacturing Services Market." In *A Comprehensive Study on the Trends, Issues, and Leading Companies in the Worldwide Market for Contract Electronics Manufacturing Services*, 8th ed. San Jose, Calif. Available at http://www.electronictrendpubs.com/cems8bro.pdf.

Encarnation, Dennis J. 1999. "Asia and the Global Operations of Multinational Corporations." In Dennis J. Encarnation, ed., *Japanese Multinationals in Asia: Regional Operations in Comparative Perspective.* Japan Business and Economics Series. Oxford: Oxford University Press.

"Ericsson Shifts Phone Production to Contractors." 2000. *Electronics Weekly*, October 20.

Ernst, Dieter. 1997. "From Partial to Systemic Globalization: International Production Networks in the Electronics Industry." BRIE Working Paper 98. Berkeley Roundtable on the International Economy.

Evans, Peter. 1995. *Embedded Autonomy: States and Industrial Transformation.* Princeton, N.J.: Princeton University Press.

Fine, Charles H. 1999. *Clockspeed: Winning Industry Control in the Age of Temporary Advantage.* London: Little, Brown and Company.

Gereffi, Gary. 1994. "The Organization of Buyer-Driven Global Commodity Chains: How U.S. Retailers Shape Overseas Production Networks." In Gary Gereffi and Miguel Korzeniewicz, eds., *Commodity Chains and Global Capitalism.* Westport, Conn.: Praeger Publishers.

———. 1999. "International Trade and Industrial Upgrading in the Apparel Commodity Chain." *Journal of International Economics* 48(1):37–70.

Gereffi, Gary, John Humphrey, and Timothy J. Sturgeon. Forthcoming. "The Governance of Global Value Chains." *Review of International Political Economy.*

Gilson, Ronald J., and Mark J. Roe. 1993. "Understanding the Japanese Keiretsu: Overlaps between Corporate Governance and Industrial Organization." Stanford University, Center for Economic Policy Research, Palo Alto, Calif. Processed.

Guillèn, Mauro. 2001. *The Limits of Convergence: Globalization and Organizational Change in Argentina, South Korea, and Spain.* Princeton, N.J.: Princeton University Press.

Jonas, Gabrielle. 1997. "Ericsson, CEMs Sign Délas." *Electronic Business News*, July 14(1066): News section.

Kaplinsky, Raphael. 2000. "Globalization and Unequalization: What Can Be Learned from Value Chain Analysis?" *Journal of Development Studies* 37(2):117–46.

Keesing, Donald B., and Sanjaya Lall. 1992. "Marketing Manufactured Exports from Developing Countries: Learning Sequences and Public Support." In Gerald Helleiner, ed., *Trade Policy, Industrialisation, and Development*, pp. 176–93. Oxford: Oxford University Press.

Kenney, Martin, and Richard L. Florida. 1993. *Beyond Mass Production: The Japanese System and Its Transfer to the United States.* New York: Oxford University Press.

Lee, Ji-Ren, and Jen-Shyang Chen. 2000. "Dynamic Synergy Creation with Multiple Business Activities: Toward a Competence-Based Growth Model for Contract Manufacturers." In Ron Sanchez and Aime Heene, eds., *Advances in Applied Business Strategy*. Vol. C: *Research in Competence-Based Research*, pp. 209–28. Stamford, Conn.: JAI Press.

Leonard-Barton, Dorothy. 1994. "Core Capabilities and Core Rigidities." In H. Kent Bowen, Kim B. Clark, Charles A. Holloway, and Steven C. Wheelwright, eds., *The Perpetual Enterprise Machine: Seven Keys to Corporate Renewal Through Successful Product and Process Development*. New York: Oxford University Press.

Levine, Bernard. 2000. "Sony, Solectron Set EMS Deal." *Electronic News*, October 18.

———. 2001. "Bear Stearns Upgrades Contractors." *Electronic News*, May 7.

MacIntyre, Andrew, ed. 1994. *Business and Government in Industrializing Asia*. Ithaca, N.Y.: Cornell University Press.

Mason, Mark, and Dennis Encarnation, eds. 1995. *Does Ownership Matter? Japanese Multinationals in Europe*. New York: Oxford University Press.

McKendrick, David, Richard Doner, and Stephan Haggard. 2000. *From Silicon Valley to Singapore: Location and Competitive Advantage in the Hard Disk Drive Industry*. Palo Alto, Calif.: Stanford University Press.

"NEC to Sell Cell Phone Plants in Mexico." 2000. Bloomberg News [Britain], December 12.

Ojo, Bolaji, and Claire Serant. 2002. "IBM Deal Points to More PC Outsourcing Opportunities." *Electronic Business News*, January 14.

Powell, W. 1990. "Neither Market Nor Hierarchy: Network Forms of Organization." *Research in Organizational Behavior* 12:295–336.

Prahalad, C. K., and Gary Hamel 1990. "The Core Competence of the Corporation." *Harvard Business Review* 3(May–June):79–91.

Quinn, James Brian, and Frederick G. Hilmer. 1994. "Strategic Outsourcing." *Sloan Management Review* 35(4):43–56.

Ravenhill, John. 1995. "National Champions, Multinationals, and Networks in Asia: Why Ownership Makes a Difference in the Electronics Industry." Paper presented at the conference Does Ownership Matter? MIT Japan Program. Processed.

Sako, Mari. 1989. "Competitive Cooperation: How the Japanese Manage Inter-firm Relations." London School of Economics, Industrial Relations Department. Processed.

Schilling, Melissa A., and Harvey K. Steensma. 2001. "Industry Determinants of the Adoption of Modular Organizational Forms: An Empirical Test." *Academy of Management Journal* 44(6):1149–69.

Schonberger, Richard. 1982. *Japanese Manufacturing Techniques*. New York: Free Press.

Scott, Allen J. 1987. "The Semiconductor Industry in Southeast Asia." *Regional Studies* 21(2):143–60.

Serant, Claire. 2001. "Singapore No Longer a Magnet for EMS Companies." *Electronics Business News*, April 4.

Shimokawa, Koichi. 1999. "New Trend for Component Modules in Japan." *Japanese Automotive News*, April 1.

Storper, Michael. 1995. "The Resurgence of Regional Economies, Ten Years Later." *European Urban and Regional Studies* 2(3):191–221.

Sturgeon, Timothy J. 1999. *Turnkey Production Networks: Industrial Organization, Economic Development, and the Globalization of the Electronics Manufacturing Supply Base*. Ph.D. diss. University of California, Berkeley, Department of Geography.

———. 2000. "Turnkey Production Networks: The Organizational Delinking of Production from Innovation." In Ulrich Juergens, ed., *New Product Development and Production Networks: Global Industrial Experience.* Berlin: Springer Verlag.

———. 2002. "Modular Production Networks: A New American Model of Industrial Organization." *Industrial and Corporate Change* 11(3):451–96.

———. 2003. "What Really Goes on in Silicon Valley? Spatial Clustering Dispersal in Modular Production Networks." *Journal of Economic Geography* 3(April):199–225.

Sturgeon, Timothy J., and Richard Florida. 1999. "Globalization and Jobs in the Automotive Industry: A Study by Carnegie Mellon University and the Massachusetts Institute of Technology." Final report to the Alfred P. Sloan Foundation. IPC Globalization Working Paper 01-003. Massachusetts Institute of Technology, Cambridge, Mass. Processed.

———. 2003. "Globalization, Deverticalization, and Employment in the Motor Vehicle Industry." In Martin Kenney and Richard Florida, eds., *Locating Global Advantage: Industry Dynamics in a Globalizing Economy.* Palo Alto, Calif.: Stanford University Press.

Sturgeon, Timothy J., and Ji-Ren Lee. 2001. "Industry Co-Evolution and the Rise of a Shared Supply Base for Electronics Manufacturing." IPC Working Paper 01-003. Massachusetts Institute of Technology, Industrial Performance Center, Cambridge, Mass. Processed.

Tachiki, Dennis S. 1999. "Exploiting Asia to Beat Japan: Production Networks and the Comeback of U.S. Electronics." In Dennis J. Encarnation, ed., *Japanese Multinationals in Asia.* New York: Oxford University Press.

Thun, Eric. 2001. "Growing up and Moving out: The Globalization of Traditional Industries in Taiwan [China]." Globalization Study Working Paper 00-004. Massachusetts Institute of Technology, Industrial Performance Center, Cambridge, Mass. Processed.

U.S. Bureau of Labor Statistics. Various years. "Non-Farm Payroll Statistics." In *Current Employment Statistics.* Washington, D.C.

Venkatesan, Ravi. 1992. "Strategic Sourcing: To Make or Not to Make." *Harvard Business Review* 70(6):98–108.

Wade, Robert. 1990. *Governing the Market: Economic Theory and the Role of Government in East Asian Industrialization.* Princeton, N.J.: Princeton University Press.

Wang, Abel S. H. 2001. "Taiwan's Hardware Industry in 2000 and Emerging Trends." MIC Special Release. Taipei, Taiwan, China: Market Intelligence Center, March 23. Available at mic.iii.org.tw/. Processed.

Womack, James P., Daniel T. Jones, and Daniel Roos. 1990. *The Machine That Changed the World.* New York: Rawson Associates.

World Bank. 1993. *The East Asian Miracle: Economic Growth and Public Policy.* New York: Oxford University Press.

CHAPTER 3

GLOBAL PRODUCTION NETWORKS IN EAST ASIA'S ELECTRONICS INDUSTRY AND UPGRADING PROSPECTS IN MALAYSIA

Dieter Ernst

East Asia's catch-up in the electronics industry during the late twentieth century provides a fascinating example of the catalytic role that linkages with foreign firms can play in industrial development (for example, Borrus, Ernst, and Haggard 2000; Ernst 1997b; Ernst and Guerrieri 1998): an early integration into global production networks provided Asian producers with access to the industry's main growth markets, helping to compensate for the initially small size of their domestic markets. Network participation also provided new opportunities, pressures, and incentives for Asian network suppliers to upgrade their technological and managerial capabilities (Ernst and Kim 2002). As a result, East Asia emerged as the dominant base of global manufacturing for the electronics industry, especially for assembly and component manufacturing.

This pattern of growth survived the 1997 financial crisis (Ernst 2001b). The "new-economy" boom in the United States provided an additional boost, increasing demand for Asian electronics exports.[1] But there are limits to export-led growth, and recent transformations of global production networks force us to reconsider the region's prospects for industrial upgrading within these networks. The downturn in the global electronics industry beginning in late 2000 brutally exposed the downside of export-led

1. The size of this investment-led boost in demand for Asian exports can be gathered from the following data. Almost half of the U.S. capital investment since 1997 has gone into information technology, up from less than 24 percent during the early 1990s, and roughly 40 percent of U.S. consumption of computers and semiconductors is imported, largely from East Asia (data are courtesy of U.S. Department of Commerce, December 2001).

industrialization: a country is more vulnerable, the higher the share of electronics in its exports, the greater its integration into global production networks, and the greater its dependence on exports to the United States.

There is a broad consensus that East Asia's electronics industry is in need of upgrading (for example, Ariffin 2000 on Malaysia; Chen 2002 on Taiwan, China; Ministry of Information and Communication 2002 on the Republic of Korea; Simon 2001 on China; Toh 2002 on Penang and Malaysia). Defined as a shift to higher-value-added products, services, and production stages through increasing specialization and efficient domestic and international linkages, industrial upgrading necessitates a strong base of domestic knowledge (see, for example, Ernst forthcoming a). Building on existing strengths in volume manufacturing, industrial upgrading requires the development of complementary skills and capabilities in design and development (including the introduction of new products) as well as in "embedded" software, SoC (system-on-chip) design, intellectual property trade, system integration, and the management of resources, supply chains, and customer relations. Of critical importance is the capacity to bring in at short notice specialized experts from overseas who can bridge gaps in knowledge and catalyze necessary changes in organization and procedures required to develop these capabilities locally. In short, successful upgrading raises daunting challenges; chief among them are substantial investments in long-term assets, such as specialized skills and capabilities.

This chapter examines how three fundamental transformations in global production networks are providing new opportunities for upgrading in East Asia's electronics industry: (1) *vertical specialization:* the emergence of increasingly complex "networks of networks" that juxtapose original equipment manufacturers (OEMs) and global, U.S.-based contract manufacturers; (2) *coordination and content:* the increasing use of digital information systems to manage these networks and to build global information service networks that complement networks centered on manufacturing; and (3) *location:* the emergence of China as a priority investment target for global production networks in the electronics industry.

This analysis is based on the conceptual framework developed in Ernst (2002a, 2003a). It complements the discussion in chapter 2 but places greater emphasis on new opportunities that arise from vertical specialization within global production networks (Ernst and Kim 2002) and from the increasing use of digital information systems to manage these networks (Ernst 2003b). I argue that the reintegration of geographically dispersed specialized production and innovation sites into multilayered global production networks and the increasing use of information technology systems to manage these networks are *gradually* reducing the constraints on international diffusion of knowledge. Global production networks expand interfirm linkages across national boundaries, increasing the need for knowledge diffusion, while information systems enhance not only the exchange of information but also the sharing and joint creation of knowledge. In turn, this new mobility of knowledge creates new opportunities for using network participation as a catalyst for further upgrading of Asia's electronics industries.

This chapter also documents the pressures that rising requirements for participation in these networks are putting on mid-size countries and especially on local small- and medium-size Asian suppliers. These pressures result not only from the strengths but also from the limitations of the U.S.-style model of contract manufacturing that have been brutally exposed by the downturn in the global electronics industry. Based on this analysis, the chapter then explores the difficulties of devising realistic upgrading strategies and discusses policies and support institutions that could improve the implementation of these strategies.

It is difficult in one chapter to consider the entire range of upgrading possibilities that face the economies of East Asia. Therefore, I focus on Malaysia, a country that faces a particularly demanding challenge, due to four peculiar characteristics of its electronics industry.[2] First, Malaysia exceeds most other Asian electronics producers (with the exception of Singapore) in its vulnerability to the vicissitudes of export-led growth: electronics constitute around 60 percent of its exports, its electronics industry is heavily exposed to global production networks, and the U.S. market absorbs 25 percent of its total exports (an estimated 40 percent of electronics exports). Second, with its focus on low-end, assembly-type volume manufacturing and a weak domestic base of supply, this mid-size country is especially vulnerable to the emergence of China as a new competitor. Third, with the Penang Development Centre, with its two industrial master plans (Ministry of International Trade and Industry 1986, 1996), and with the Bill of Guarantees (drawn up for its Multimedia Super Corridor), Malaysia has developed one of the most aggressive sets of upgrading incentives for private companies (both foreign and domestic). And fourth, despite such policies, Malaysia has yet to develop a sufficiently diversified and deep industrial structure to induce a critical mass of corporate investment in specialized skills and capabilities.

A focus on Malaysia helps to highlight three important propositions that should inform the study of upgrading prospects in East Asia's electronics industry. First, as long as peculiar characteristics of industry structure

2. For related studies on the prospects for industrial upgrading in the electronics industry in Korea and Taiwan (China), see Ernst (1994b, 2000b, 2001a).

constrain the incentives for firms to invest in long-term assets (such as specialized skills), the prospects for upgrading will remain limited. Second, while investment incentives and infrastructure matter, the key to success is the development of specialized skills and innovative capabilities, *ahead* of what the market would provide. Of critical importance are incentives that encourage university professors, researchers, and students to interact closely with the private sector (through, for example, company internships and sabbaticals). Equally important are training institutions, jointly run by the private and public sector, like the Penang Skills Development Centre.

Third, in countries where the structure of domestic industry constrains upgrading efforts, international linkages through participation in global production networks can play an important catalytic role. In such a situation, it is critical to understand whether and how the current transformations of global production networks can help to bypass the constraints on upgrading. This chapter argues that there is now greater scope for diversifying international network linkages, beyond the erstwhile exclusive linkages with OEMs, and that this could facilitate upgrading into more knowledge-intensive production and services.

The chapter begins by sketching key characteristics of global production networks and introducing an operational definition of industrial upgrading. It then outlines the transformations of global production networks in East Asia's electronics industry. Finally, it asks how the upgrading prospects of Malaysian electronics firms are affected by these transformations. Specifically, it explores whether these transformations provide new opportunities for relieving the constraints on domestic upgrading, outlines feasible responses, and identifies options for further analysis.

CONCEPTUAL FRAMEWORK: GLOBAL PRODUCTION NETWORKS AND INDUSTRIAL UPGRADING

Trade economists have recently discovered the importance of changes in the organization of international production as a determinant of trade patterns (for example, Feenstra 1998; Jones and Kierzkowski 2000; Navaretti, Haaland, and Venables 2002).[3] Their work demonstrates that (a) production is increasingly fragmented, with parts of the production process being scattered across a number of countries, hence increasing the share of trade in parts and components, and (b) countries and regions that have become part of the global production network have industrialized the fastest. And

3. For details on the characteristics of global production networks, see Borrus, Ernst, and Haggard (2000) and Ernst (1994a, 1997a, 2002b, 2003a, 2003b).

leading growth economists (for example, Grossman and Helpman 2002) are basing their models on a systematic analysis of global sourcing strategies.

This chapter builds on their work but uses a broader concept of global production networks that emphasizes three characteristics that are important for industrial upgrading: (1) *scope:* global production networks encompass all stages of the value chain, not just production; (2) *asymmetry:* flagships dominate control over network resources and decisionmaking; and (3) *knowledge diffusion:* knowledge must be shared to keep these networks growing.

Characteristics of Global Production Networks

A global production network covers both intrafirm and inter-firm transactions and forms of coordination: it links together the flagship's own subsidiaries, affiliates, and joint ventures with its subcontractors, suppliers, service providers, as well as partners in strategic alliances. Although equity ownership is not essential, network governance is distinctively asymmetric. A network flagship like IBM or Intel breaks down the value chain into a variety of discrete functions and locates them wherever they can be carried out most effectively, where they improve the firm's access to resources and capabilities, and where they are needed to facilitate the penetration of important growth markets. The main purpose of these networks is to provide the flagship with quick and low-cost access to resources, capabilities, and knowledge that complement its core competencies. As the flagship integrates geographically dispersed production, customer, and knowledge bases into global production networks, this may produce transaction cost savings. Yet the real benefits result from the dissemination, exchange, and outsourcing of knowledge and complementary capabilities (Ernst 2003a).

Global production networks typically combine rapid geographic dispersion with spatial concentration on a growing, but still limited, number of specialized clusters. Two types of clusters can be distinguished (Ernst 2003b): centers of excellence that combine unique resources, such as research and development (R&D) and precision mechanical engineering, and cost- and time-reduction centers that thrive on the timely provision of lower-cost services.[4] Different clusters face different prospects for industrial upgrading, depending on their specialization and on the product composition of their global production network. The dispersion of clusters differs across the value

4. Cost- and time-reduction centers include the usual suspects in Asia (China, Korea, Malaysia, Taiwan [China], Thailand, and now also India for software engineering and Web services) but also exist in once-peripheral locations in Europe (for example, Central and Eastern Europe, Ireland, and Russia), in Latin America (Brazil and Mexico), in some Caribbean locations (like Costa Rica), and in a few spots elsewhere in the rest of the world.

chain: it increases, the closer one gets to the final product, while dispersion remains concentrated especially for high-precision and design-intensive components.[5] In short, agglomeration economies continue to matter, giving rise to the path-dependent nature of upgrading trajectories for individual specialized clusters.

Flagships. Global production networks typically consist of various hierarchical layers, ranging from flagships that dominate such networks, because of their capacity for system integration (Pavitt 2003), down to a variety of usually smaller, local specialized network suppliers. The flagship is at the heart of a network: it provides strategic and organizational leadership beyond the resources that, from an accounting perspective, lie directly under the control of management (Rugman 1997, p. 182). The strategy of the flagship company thus directly affects the growth, strategic direction, and network position of lower-end participants, like specialized suppliers and subcontractors. The latter, in turn, "have no reciprocal influence over the flagship strategy" (Rugman and D'Cruz 2000, p. 84). The flagship derives strength from its control over critical resources and capabilities that facilitate innovation and from its capacity to coordinate transactions and the exchange of knowledge between the different network nodes.

Flagships retain in-house activities in which they have a particular strategic advantage; they outsource those in which they do not. It is important to emphasize the diversity of such outsourcing patterns (Ernst 1997a). Some flagships focus on design, product development, and marketing, while outsourcing volume manufacturing and related support services. Other flagships outsource a variety of high-end, knowledge-intensive support services.

To move this model a bit closer to reality, we distinguish two types of global flagships: (a) original equipment manufacturers that derive their mar-

5. On one end of the spectrum is final assembly of personal computers, which is widely dispersed to major growth markets in Asia, Europe, and the United States. Dispersion is still quite extended for standard, commodity-type components, but less so than for final assembly. For instance, flagships can source keyboards, computer mouse devices, and power switch supplies from many different sources, including Asia, Mexico, and the European periphery, with Taiwanese (Chinese) firms playing an important role as intermediate supply chain coordinators. The same is true for printed circuit boards. Concentration of dispersion increases the more we move toward more complex, capital-intensive precision components: memory devices and displays are sourced primarily from "centers of excellence" in Japan, Korea, Singapore, and Taiwan (China); and hard disk drives are sourced from a Singapore-centered triangle of locations in Southeast Asia. Finally, dispersion becomes most concentrated for high-precision, design-intensive components that pose the most demanding requirements on the mix of capabilities that a firm and its cluster needs to master: microprocessors, for instance, are sourced from a few globally dispersed affiliates of Intel, two American suppliers, and one recent entrant from Taiwan, China (Via Technologies).

ket power from selling global brands, regardless of whether design and production are done in-house or outsourced, and (b) U.S.-based global contract manufacturers that establish their own global production networks to provide integrated manufacturing and global supply chain services (often including design) to the OEMs.

Asian suppliers. To determine whether a local company in Asia is integrated into a global production network, this chapter uses a broad set of indicators that includes (a) use of dedicated parts supplied by a foreign firm, (b) contract manufacturing of parts or final products to the specifications of a foreign firm, (c) contract manufacturing of parts or final products based on its own design, or (d) the provision of knowledge support services to a foreign firm.

It is necessary to open the black box of Asian suppliers. First, some of these suppliers have been around for quite a while. Since the 1960s, various groups of Asian suppliers have emerged, first in consumer electronics, then in contract chip assembly (Korea's Anam is the most prominent example), and, more recently, in contract wafer fabrication (so-called silicon foundries), in the own-design-manufacture (ODM) supply of computers and hand-held and wireless devices, and in the design of integrated circuits (Ernst forthcoming b). Second, Asian suppliers differ considerably in their capabilities, network position, and market power. Substantial differences also exist with regard to their capacity for component sourcing, design, development, and engineering, the capacity to provide global support services, and the use of digital information systems.

Greatly simplifying, I distinguish two types of Asian suppliers: higher-tier and lower-tier suppliers. Higher-tier suppliers, such as the Acer in Taiwan, China (Ernst 2000c), play an intermediary role between global flagships and local suppliers. They deal directly with global flagships (both brand leaders and global U.S.-based contract manufacturers), they possess valuable proprietary assets (including technology), and they have developed their own mini global production networks (Chen 2002). Building on their strengths in volume manufacturing and the provision of detailed design services (the so-called ODM services), these higher-tier suppliers are now under pressure to develop complementary skills and capabilities in the introduction of new products, process reengineering, as well as embedded software, System-on-chip (SoC) design, intellectual property trade, system integration, and the management of network resources, supply chains, and customer relations. With the exception of hard-core R&D and strategic marketing, which remain under the control of the global brand leader, Asian higher-tier suppliers must be able to shoulder all steps in the value chain. They must even take on the coordination functions necessary to manage the global supply chain.[6]

Lower-tier Asian suppliers are the weakest link in global production networks. Their main competitive advantages are low cost, speed, and flexibility of delivery. They are typically used as price breakers and capacity buffers and can be dropped at short notice. This second group of local suppliers rarely deals directly with the global flagships; they interact primarily with local higher-tier suppliers. Lower-tier suppliers normally lack proprietary assets; their financial resources are inadequate to invest in training and R&D; and they are highly vulnerable to abrupt changes in markets and technology and to financial crises.

Industrial Upgrading

An appropriate long-term development strategy for Asian electronics industries must focus on improvements in specialization, productivity, and linkages (as defined by Hirschman 1958, ch. 6), all of which necessitate a broad base of skills and capabilities (this section is based on Ernst forthcoming a). All four elements are essential prerequisites for improving a country's capacity to raise long-term capital for facility investment, R&D, and human resource development. The concept of industrial upgrading ties these four elements together in a cohesive framework for unlocking new sources of economic growth.[7] Critical prerequisites for successful upgrading are a sufficiently large pool of specialized and retrainable skills, a strong base of domestic knowledge, forms of corporate governance that facilitate innovation, sophisticated information management, and strong international knowledge linkages.

My definition emphasizes the importance of international linkages. I do not assume that industrial upgrading ends at the national border or that it occurs only if improved specialization generates pressures to create dense forward and backward linkages *within* the district or the national economy. A "closed economy" assumption is unrealistic, as globalization and infor-

6. As emphasized in chapter 2, this is a demanding agenda, but some ODM suppliers, especially Hon Hai and Mitac, have been able to mount a credible challenge to U.S.-based contract manufacturers. (Author's interviews with U.S. contract manufacturers in China, October 2002).

7. By focusing on knowledge and innovation as major sources of economic growth, my approach is consistent with leading-edge economic thinking, such as endogenous growth theories (Grossman and Helpman 1991; Romer 1990), Lipsey's structuralist growth theory (for example, Lipsey 2001), evolutionary economics (for example, Nelson and Winter 1982; Penrose 1995 [1959]; Richardson 1990 [1960]), and attempts to reunite economic growth and innovation theory and business history (for example, Lazonick 2000). A focus on knowledge and innovation also reflects a recent shift in policy debates within important international institutions, such as the European Commission, the Organisation for Economic Co-operation and Development, and the World Bank.

mation technology have drastically increased the international mobility of trade, investment, and even knowledge (Ernst 2003b). This increases the scope for cross-border forward and backward linkages, in the manner in which improved specialization generates pressures to create dense forward and backward linkages within the economy (Ernst 2002b).

Equally important, most countries are constrained by a narrow domestic base of knowledge and limited linkages. Both constraints are particularly important for small developing economies, which tend to have a narrow and incomplete set of domestic linkages (for example, Ernst, Ganiatsos, and Mytelka 1998; Lall 1997). The result is an inverted production pyramid: a growing sector of final products rests on a weak and much smaller domestic base of mostly inefficient support industries. Rapid growth in the final products sector necessitates considerable imports of intermediates and production equipment. In addition, highly heterogeneous economic structures constrain agglomeration economies, weak and unstable economic institutions obstruct learning efficiency, and a high vulnerability to volatile global currency and financial markets constrains patient capital, which is necessary for the development of a broad base of domestic knowledge. As a result of this vicious circle, very limited sharing and pooling of resources and knowledge occur *within* the country and often even within the export-oriented cluster. This implies that the model of industrial upgrading needs to integrate international knowledge linkages. To compensate for their narrow base of domestic knowledge and limited linkages, Asian developing economies have to rely on foreign sources of knowledge to catalyze the formation of domestic capabilities. International linkages need to prepare the way for an upgrading of East Asia's electronics industries.

I focus on two aspects of industrial upgrading found in the literature: *firm-level upgrading* from low-end to higher-end products and value chains and *industry-level linkages* with support industries, universities, and research institutes. Without industry-level linkages, firm-level upgrading will soon reach its limits.[8] Two additional features distinguish my concept of industrial

8. The other three forms of industrial upgrading are (a) inter-industry upgrading proceeding from low-value-added industries (for example, light industries) to higher-value-added industries (for example, heavy and higher-tech industries); (b) inter-factor upgrading proceeding from endowed assets (that is, natural resources and unskilled labor) to created assets (physical capital, skilled labor, social capital), and (c) upgrading of demand within a hierarchy of consumption, proceeding from necessities to conveniences to luxury goods. See Ozawa (2000) for discussion of upgrading taxonomies. Most research has focused on a combination of the first two forms of industrial upgrading, based on a distinction between low-wage, low-skill "sunset" industries and high-wage, high-skill "sunrise" industries. Such simple dichotomies, however, have failed to produce convincing results, for two reasons. First, there are low-wage, low-skill value stages in even the most high-tech industry, and high-wage, high-skill activities exist even in so-called traditional industries like textiles. And, second, both the capability requirements and the boundaries of a particular "industry" keep changing over time, which makes an analytical focus on the industry level even more problematic.

upgrading. First, firm behavior is a key dimension, allowing for the co-evolution of industry structure and firm behavior in response to actions of key participants and also to the policy environment. And second, a broad definition of innovation expands the focus beyond R&D and patenting.[9] There is now a widespread consensus that "innovation efforts" should be defined broadly to include engineering, technology purchases, expenditures on licensing and consulting services, and technology search as well as the accumulation of tacit knowledge required to absorb imported technology (for example, Nelson 1990). That broader focus is necessary to capture the proliferation of knowledge-intensive professional services, made possible by information and communication technology.

NETWORKS OF NETWORKS: OUTSOURCING BASED ON CONTRACT MANUFACTURING

The new-economy boom in the United States accelerated a long-standing trend toward vertical specialization in the electronics industry: outsourcing based on contract manufacturing became the "panacea of the '90s" (Lakenan, Boyd, and Frey 2001, p. 3), a "new American model of industrial organization" (Sturgeon 2002). Two inter-related transformations need to be distinguished: supply contracts and mergers and acquisitions (M&A). Global brand leaders, like Dell—the original equipment manufacturers—increasingly subcontract manufacturing and related services to U.S.-based global contract manufacturers, like Flextronics. Equally important is that the very same contract manufacturers have acquired the existing facilities of OEMs, as the latter are divesting their internal manufacturing capacity, seeking to allocate capital to other activities that are expected to generate higher profit margins, such as sales, marketing, and product development.

Argument

Sturgeon and Lester (chapter 2 in this volume) emphasize that the rise of U.S. contract manufacturers with global reach may pose a serious competitive

9. Most empirical work on industrial upgrading has explored the expansion of R&D-intensive industries. For most developing countries, that narrow focus is of very limited value. The (usually) implicit notion is that potential rates of productivity growth are higher in "emergent" R&D-intensive industries (Globerman 1997, pp. 98, 99). Hence, "specializing in the "right" technological activities directly contributes to faster growth rates of real income." A related notion is that, for R&D-intensive industries, economic rents can be extracted, in part, from foreign consumers. Specializing in the "right" technological activities contributes to higher levels of national income by promoting more favorable international terms of trade.

threat to Asian suppliers. Their analysis highlights rising threshold requirements for the suppliers to participate in global production networks. Complementing their analysis, this chapter highlights the other side of the coin and explores how Asian suppliers can exploit linkages with U.S.-based contract manufacturers, which now complement the original linkages with OEMs, for their upgrading purposes. More specifically, I ask what new opportunities for upgrading may open up for Asian suppliers, as outsourcing based on contract manufacturing has created increasingly complex, multi-tier "networks of networks" that juxtapose global ties among the two large global players (the OEMs and contract manufacturers) as well as intense regional ties with smaller firms (as argued, for instance, in Almeida and Kogut 1997).

A focus on complex, multi-tier networks of networks distinguishes this analysis from Sturgeon's modular production network model (Sturgeon 2002). That model focuses on two actors only: global OEMs and contract manufacturers, most of them of American origin. OEMs and contract manufacturers are perceived to interact in a virtuous circle where each of them can only win. In that model, nothing can stop continuous outsourcing through contract manufacturing: "Turnkey suppliers and lead firms co-evolve in a recursive cycle of outsourcing and increasing supply-base capability and scale, which makes the prospects for additional outsourcing more attractive" (Sturgeon 2002, p. 6). If that scenario were to materialize, Asian suppliers in the global electronics industry might face a considerable backlash. Specifically, Asian suppliers might be unable to compete against the vastly superior capabilities of U.S.-based contract manufacturers in four areas: (1) component sourcing; (2) design, development, and engineering; (3) global reach, the provision of support services across multiple locations in all major macro regions; and (4) network coordination, improved network efficiencies through the use of sophisticated digital information systems.

The analysis in this chapter yields a less gloomy perspective. Asian suppliers already play an important role as global contract manufacturers, and peculiar features of the U.S.-style model of contract manufacturers suggest possible limitations of that model. The U.S. model is just one possible approach, and Asian electronics firms may have a role to play, based on their accumulated experience with contract manufacturing, before it was given that name. Furthermore, there are ample opportunities for the emergence of a variety of new specialized Asian suppliers, provided necessary changes are put in place in policies and support institutions.

Three peculiar features of the U.S.-style model are important: the critical role played by financial considerations, the limited share of contract manufacturing in worldwide electronics hardware production, and the limited presence of American contract manufacturers in Asia relative to their presence in the Americas and Europe. The downturn in the global electronics industry exposed serious limitations of these arrangements, forcing both OEMs and contract manufacturers to adjust and rationalize the organization of their networks. All of this has important implications for upgrading in Malaysia's electronics industry.

Drivers

Outsourcing through subcontracting has a long history in the electronics industry (Boswell 1993). Yet, during the 1990s, outsourcing gained a new dimension, spreading across borders: global brand leaders (OEMs) have put up for sale a growing number of their overseas facilities and, in some cases, whole chunks of their global production networks. OEMs from North America like Compaq, Dell, Hewlett-Packard, IBM, Intel, Lucent, Motorola, and Nortel were first in pursuing such divestment strategies, but European OEMs (for example, Ericcson, Philips, Siemens) and, more recently, Japanese ones (for example, NEC, Fujitsu, Sony) have followed suit. The main driver is financial considerations: getting rid of low-margin manufacturing helps OEMs to increase shareholder returns.[10] Other expected benefits include hedging against losses due to volatile markets and periodic excess capacity, scale economies (surface-mount technology requires large production runs, reflecting its growing capital and knowledge intensity), and improved capacity to combine cost reduction, product differentiation, and time-to-market.

Growth and Market Share

Contract manufacturers aggressively seized this opportunity through acquisitions and capacity expansion. Within a few years, they developed their own global production networks, which now complement the networks established by the OEMs. For instance, Flextronics has 62 plants worldwide, Solectron has factories in 70 countries, and the recently merged Sanmina/SCI has 100 factories around the world. This expansion gave rise to an extremely rapid growth of the contract manufacturing industry (figure 3.1). From 1996 to 2000, capital expenditures grew elevenfold (50 percent compound annual growth rate), and revenues increased almost 400 percent (81 percent compound annual growth rate). The industry's rapid growth was driven primarily by M&A (figure 3.2).

10. In response to pressures from institutional investors and financial analysts, OEM firms are eager to "slash their balance sheets by placing the low-margin operations with hungry contract manufacturers" (Lakenan, Boyd, and Frey 2001, p. 4).

Figure 3.1 Growth of the Contract Manufacturing Industry, 1996–2000

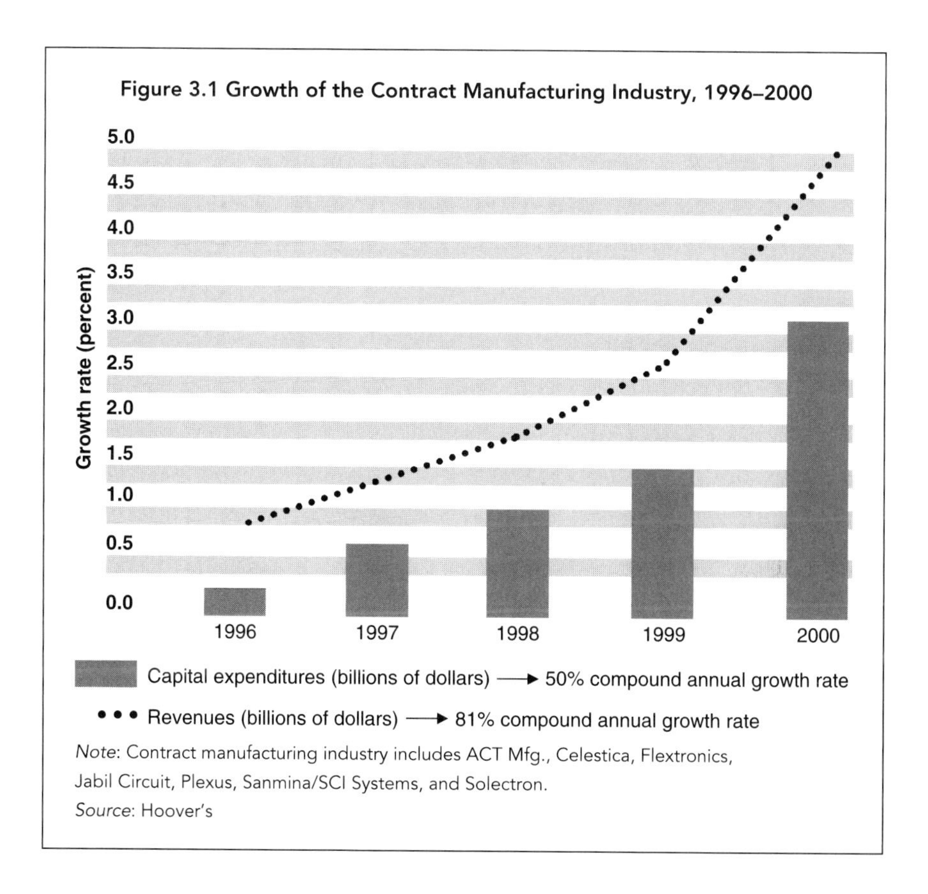

Note: Contract manufacturing industry includes ACT Mfg., Celestica, Flextronics, Jabil Circuit, Plexus, Sanmina/SCI Systems, and Solectron.

Source: Hoover's

Figure 3.2 Mergers and Acquisitions in the Contract Manufacturing Industry, 1997–2000

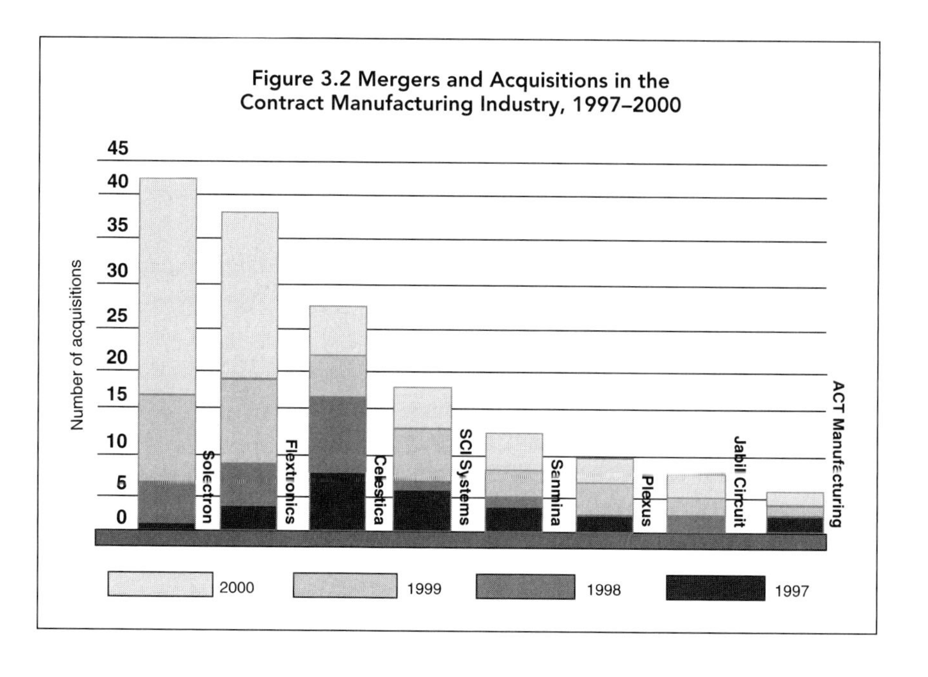

It is important, however, to emphasize the still limited share of U.S. contract manufacturers in worldwide production of electronics hardware. In 2001 this share was estimated to be around 13.7 percent (up from 13.0 percent in 2000); for 2002 it increased to 16.3 percent (e-mail from Eric Miscoll, chief executive officer, Technology Forecasters, April 15, 2002).

Late Move to East Asia

The presence of American contract manufacturers in East Asia pales relative to their presence in the Americas (Brazil, Canada, Mexico, United States) and Europe (including Eastern Europe and Israel). The move to East Asia came relatively late. During the 1990s American contract manufacturers spent most of their money acquiring global flagship facilities in the Americas and Europe. During the new-economy boom in the United States, speed-to-market due to close proximity was much more important than cost considerations. With the slowdown in the electronics industry, cost reduction again has become a central concern. Arguably, this may create new incentives for contract manufacturers to expand their East Asian networks.

Table 3.1 offers a few examples. Flextronics, which has its headquarters in Singapore, has the strongest presence in Asia, with 12 facilities in six nations: China, India, Malaysia, Singapore, Taiwan (China), and Thailand. Yet this compares with a total of 62 plants worldwide, of which 18 are in the Americas and 27 in Europe (including two in Israel).[11]

Solectron, the long-time industry leader, has factories in 70 economies, but only five of these (from the table, including Japan) are in Asia.[12] Solectron began to increase its presence in Asia only after 2001. This is primarily in response to the company's projections that, by 2005, 60 percent of its turnover will come from Asia (including Japan), up from about 30 percent in 2001. Traditionally focused on Penang (since 1991), during 1996 Solectron added facilities in Johor (Malaysia) and in Suzhou (Jiangsu Province, China). During 2000 Solectron acquired two Sony factories, one in Japan and one in Taiwan (China), as well as NEC's Ibaraki production facilities for servers, workstations, and system file products. The latter acquisition provides Solectron with 500 highly skilled Japanese employees who are well trained in build-to-order manufacturing and final test and fulfillment services.

The presence in Asia is even more limited for the remaining three major global contract manufacturers. The recently merged Sanmina-SCI has

11. Reflecting the growing importance of the Asia-Pacific market, Flextronics decided in December 2001 to make Malaysia the manufacturing and logistics hub for its operations in the Asia-Pacific region (excluding China, where Flextronics has five facilities).

12. During the 2001 recession Flextronics took over the leadership position.

Table 3.1 Contract Manufacturing Clusters in East Asia, 2002

Company	Southeast Asia[a]	Greater China[b]	Other countries
Flextronics	Singapore: headquarters Malaysia: Johore, Johor Baru, Melaka Thailand: Samutprakarn	Taiwan (China): Taipei China: Beijing, Changzhou-Jiangsu, Doumen-Zhuhai, Shanghai Wai Gai Free Trade Zone, Xixiang-Baoan Shenzhen	India: Bangalore
Solectron	Singapore Malaysia: Penang, Johor (1991)	China: Suzhou-Jiangsu Taiwan (China): Taipei	Japan: Kanagawa, Tokyo
Sanmina-SCI	Singapore Malaysia: Penang, Kuching-Sarawak, Sana-Jaya FTZ Thailand	Taiwan (China): Taipei China: Shenzhen, Qingdao, Kushan-Jiangsu	
Celestica	Singapore Malaysia: Kulim Hi-Tech Park Thailand Indonesia	Hong Kong (China) China: Dongguan-Guangdong	
Jabil Circuit	Malaysia: Penang	Hong Kong (China): Asia headquarters China: Guangzhou-Guangdong, Panyu-Guangdong, Shenzhen-Guangdong	

a. Indonesia, Malaysia, Philippines, Singapore, Thailand, Vietnam.
b. China, Hong Kong (China), Taiwan (China).

100 factories around the world, of which only seven facilities are in Asia—China, Malaysia, Singapore, Taiwan (China), and Thailand. Celestica, a spin-off from IBM Canada in 1994, has acquired 36 plants around the world. Until mid-2001, it had four plants in Asia: one in Malaysia's Kulim Hi-Tech Park (part of the northern cluster), two in China, and one in Thailand. Since then, Celestica has substantially expanded its Asian presence to meet the growing demand of Japanese OEMs for outsourcing.[13]

13. The starting point was an $890 million acquisition of Omni, one of Singapore's leading contract manufacturers, which added facilities in Indonesia, Malaysia, Singapore, and Thailand. And in January 2002 Celestica acquired two optical and broadband equipment factories in Japan from NEC as part of a five-year, $2.5 billion supply agreement.

Finally, Jabil Circuit, the smallest of the leading global contract manufacturers, has 21 facilities worldwide. As with the other leaders, Jabil's presence in Asia (three facilities) lags behind its presence in the Americas (11) and Europe (seven). Its involvement in Asia started in 1995, with its factory in Penang. In 1998 it established its Asian regional headquarters in Hong Kong (China) and a large low-cost manufacturing plant in China's Guangdong Province, in the YiXing Industrial Estate in Panyu.[14]

Limitations of the U.S.-Style Model of Contract Manufacturers

The downturn in the global electronics industry exposed serious limitations of the U.S. model of contract manufacturing, forcing both OEMs and contract manufacturers to adjust and rationalize the organization of their networks. That model was based on the assumption of uninterrupted growth in demand. In reality, however, demand and supply only rarely match. This simple truth was all but forgotten during the heydey of the new economy.

Upgrading strategies in Asian electronics industries should take note of six important limitations of the U.S.-style model.[15] First, global contract manufacturing is a highly volatile industry. While powerful forces push for outsourcing, this process is by no means irreversible. Major OEMs retain substantial internal manufacturing operations; they are continuously evaluating the merits of manufacturing products or providing services internally versus the advantages of outsourcing. Second, global contract manufacturers are now in a much weaker bargaining position than OEMs, whose number has been reduced by the current downturn and which are now much more demanding. In principle, important long-term customer contracts permit quarterly or other periodic adjustment to pricing based on decreases or increases in component prices. In reality, however, contract manufacturers "typically bear the risk of component price increases that occur between any such re-pricings or, if such re-pricing is not permitted, during the balance of the term of the particular customer contract" (Jabil Circuit 2001, p. 49).

A third important limitation represents tradeoffs between the advantages of specialization and rapid inorganic growth through M&A. In economic

14. Jabil Circuit recently expanded its long-established operations in Penang through the acquisition of Xircom, a wholly owned subsidiary of Intel that supplies personal computer and network cards (*New Straits Times*, August 23, 2001).

15. This section is based on a recent study by Booz Allen Hamilton (Lakenan, Boyd, and Frey 2001); e-mail correspondence with the study's lead author, Bill Lakenan; recent 10K reports of the leading U.S. global contract manufacturers; and author's interviews at affiliates of global contract manufacturers in Malaysia.

theory, vertical specialization is supposed to increase efficiency, that is, to reduce the wastage of scarce resources. It is not clear whether the recent rapid growth of contract manufacturers has produced this result. The excessive growth and diversification that we saw during the new-economy boom may well truncate the specialization and efficiency advantages of the contract manufacturer model. The leading contract manufacturers have aggressively used M&A to pursue four objectives that do not easily match: rapid growth, a broadening of the portfolio of services that they can provide, a diversification into new product markets (especially telecommunications equipment), as well as an expansion of their own production networks, establishing a global presence at record speed. Yet this forced pace of global expansion may well create an increasingly cumbersome organization that could undermine the primary advantage of the contract manufacturing model: a capacity to scale up and down rapidly, in line with the requirements of OEMs.

Fourth, rapid growth, based on the use of stock as a currency for M&A is extremely risky and contains the seed of future problems. It stretches the limited financial resources of contract manufacturers, which typically have to cope with very low margins. The downturn of the global electronics industry further increased these financial pressures on leading U.S.-based contract manufacturers.[16] This, of course, raises the question of whether this will lead firms to employ off-balance-sheet financing techniques to hide accumulated debt.

Fifth, in contrast to the original expectation that outsourcing based on contract manufacturing would improve inventory and capacity planning, global brand leaders in the electronics industry that rely heavily on outsourcing have experienced very serious periodic mismatches between supply and demand. On the one hand, when a product unexpectedly becomes a hit, outsourcing provides OEMs with only a limited capacity for scaling up. During a recession, on the other hand, OEMs cannot abruptly reduce orders that they had previously placed with contract manufacturers.[17]

Lastly, there seems to be a conflict of interest between OEMs, which are looking for flexibility, and contract manufacturers, which are looking for

16. Ironically, these pressures are particularly severe for those contract manufacturers, like Solectron, that have aggressively diversified beyond the personal computer sector into telecommunications and networking equipment, the high-growth sectors of the new-economy boom.

17. Take Cisco. During the peak of the new-economy boom, from 1999 to 2000, demand for its products grew 50 percent. Reliance on contract manufacturers initially produced severe component shortages and a massive backlog in customer orders. When demand fell abruptly, starting in the fall of 2000, Cisco was saddled with excess inventory worth $2.25 billion which it had acquired to meet expected growth in demand. In time-sensitive industries like electronics, such inventory depreciates at alarming speed.

predictability and scale. For instance, OEMs focus on early market penetration and rapid growth of market share to sustain comfortable margins. OEMs thus need flexible outsourcing arrangements that allow them to divert resources at short notice to a given product as it becomes a hit. This contrasts sharply with the situation of contract manufacturers: with razor-thin margins, they need to focus ruthlessly on cost cutting. Contract manufacturers need predictability: "They want to make commitments in advance to reap benefits like big-lot purchases and decreased overtime" (Lakenan, Boyd, and Frey 2001, p. 60).

These conflicting interests complicate the coordination of contract manufacturer–based outsourcing arrangements. They also require substantial fundamental changes in the organization of both OEMs and contract manufacturers as well as an alignment of incentives through contract terms and agreements. Effective outsourcing requires that both flagships and contract manufacturers acknowledge their conflicting interests. Further, with complexity comes uncertainty. In industries with rapidly shifting technologies and markets, OEMs have no way to predict with any accuracy the specifications of what they will need in terms of capacity, design features, and configuration and in terms of the specific mix of performance requirements. In the electronics industry, all of these variables can change quite drastically and at short notice.

Such high uncertainty has important implications for the reorganization of contract manufacturer–based outsourcing arrangements (table 3.2). Flexibility now becomes the key to success. Proceeding by conjecture (stochastically) takes over from a deterministic approach. Flagships need adjustable networks to "satisfy a range of possible demand profiles with a portfolio of customizable capacity." They "need access to—and the ability to turn off—big chunks of production more quickly than ever contemplated in order to capture profitability" (Lakenan, Boyd, and Frey 2001, pp. 61, 62). This has important implications for East Asia's upgrading prospects. The transition to stochastic and fluid outsourcing arrangements will substantially increase the required capabilities that local network suppliers in East Asia will have to master. However, this also opens up new possibilities for upgrading, provided necessary changes in policies and support institutions are put in place.

COORDINATION AND CONTENTS: INFORMATION SYSTEMS AND SERVICES

A second important transformation of global production networks results from the increasing use of digital information systems to manage these net-

Table 3.2 Changes in Contract Manufacturer–Based Outsourcing Arrangements

Activity	Traditional approach	Emerging changes
Capacity planning	Deterministic • Fix capacity, service levels, and performance requirements at contract closure	Stochastic • Proceed by conjecture • Create reserve capacity that covers base load • Scale up if product becomes a hit
Production planning	Precise commitment • Determine product mix and linkages at contract closure	Rolling commitment • Reserve aggregate capacity for contract period • Have OEM reserve the right to decide how to use that capacity just before actual production starts • Require contract manufacturers to provide flexible production systems that allow alternative uses
Product design	Frozen • Allow minimal changes in configuration	Flexible • Allow for variations in availability of parts and components • Allow for variations in capacity • Enable faster turnover: extend successful lines with derivates
Network governance	Strategy shapes structure • Develop network organization in line with given strategy • Control every aspect of value chain	Iterative learning • Start outsourcing arrangement without seeking perfect solution at the beginning • Correct as you go along • Accelerate the speed of iterative adjustment, which is key to profitability
Performance expectations	Focused • Reduce unit cost over a small range of production volumes • Entail limited scalability • Seek to achieve local, not global, optimum	Systemic • Combine unit cost reduction with extended scalability • Allow flexible use of reserved capacity • Focus on high-margin products and services • Achieve "90 percent of the global optimum—fast"

works (coordination) and to build global information service networks that complement networks centered on manufacturing (contents).

Digital Information Systems and Knowledge Diffusion

Digital information systems—electronics systems that integrate software and hardware to enable communication and collaborative work—are increasingly used to manage global production networks. While still at an early stage of trial and error, these systems gradually enhance the scope for knowledge sharing among multiple network participants at distant locations. Equally important, they reduce (but do not eliminate) the problems associated with the need to coordinate product design and manufacture rapidly over long distance.[18] This new mobility of knowledge may provide new opportunities for Asian suppliers to upgrade their capabilities, provided appropriate policies and support institutions are in place.

Digital information systems provide new opportunities for improving communication routines within global production networks: new combinations become feasible between old and new forms of communication. The most commonly used technologies today facilitate *asynchronous* interaction, such as e-mail or non-real-time database sharing. But as data transfer capacity (bandwidth) increases, new opportunities are created for using technologies that facilitate *synchronous* interaction. This involves video conferencing and real-time data exchange for financial control, engineering, and R&D.

Digitization implies that knowledge can be delivered as a service and built around open standards. This has fostered the specialization of knowledge creation, giving rise to a process of modularization, very much like earlier modularization processes in hardware manufacturing. Under the heading of "e-business," a new generation of networking software provides a greater variety of tools for representing knowledge, including low-cost audiovisual representations. Those programs also provide flexible information systems that support not only the exchange of information among dispersed network nodes but also the sharing, use, and creation of knowledge among multiple network participants at remote locations. New forms of remote control are emerging for manufacturing processes, quality, supply chains, and customer relations.

Digital information systems, and especially the open-ended structure of the Internet, substantially broaden the scope for outsourcing. They have allowed OEMs to shift from *partial* outsourcing, covering the nuts and

18. I am grateful to the late Keith Pavitt for this suggestion.

bolts of manufacturing, to *systemic* outsourcing, which includes knowledge-intensive support services. This has sharpened the competition among the providers of contract manufacturing services: competition now focuses on the capacity to provide integrated manufacturing, design, and supply chain management services wherever required.

In turn, this has heightened the competition among specialized clusters in the electronics industry. For lower-cost outsourcing, OEMs and contract manufacturers can now choose between alternative locations in Asia, Latin America, the former Soviet bloc, and the European periphery. For higher-end outsourcing, they can choose from among specialized clusters in Nordic countries, France and Germany, the United States, as well as in China, Hungary, Ireland, Israel, Korea, Singapore, and Taiwan (China).

Global Information Service Networks

Global information service networks complement the existing production networks with their primary focus on manufacturing.[19] They cover a variety of knowledge support services, such as software engineering and development, development of information technology applications, business process outsourcing, maintenance and support of information systems, as well as skill transfer and training. While much of this service outsourcing involves low-cost "sweatshop" activities,[20] it also provides considerable opportunities for Asian network suppliers to upgrade their capabilities.

Market pull. The growth of these service networks in Asia's electronics industry is due to a combination of market pull and government policies. With the drastic slowdown in major information technology markets, especially in the United States, the center of activity has shifted to Asia. During 2001 the region's information services market significantly outpaced that of other regions, with a growth rate double the world average and nearly three times that of North America (Gartner Dataquest data, quoted in CMPnet.Asia, December 4, 2001). The main drivers of demand are attempts by both global flagships and local suppliers to improve the efficiency and security of existing global production networks. Asian lead markets are Korea, Singapore, Taiwan (China), major export platform clusters in Malaysia and elsewhere

19. This section is based on phone interviews, company websites, and the following sources: various issues of CMPnet.asia and Asia Computer Weekly.com; Aberdeen Group (2001).

20. A typical recent example is a call center set up by General Electric in the city of Dalian (Liaoning Province). Staffed by Chinese people fluent in Japanese, the center handles inquiries from GE Consumer Credit's Japanese customers (*Nikkei Weekly*, November 12, 2001).

in Southeast Asia, China's electronics clusters, especially in the south and Shanghai, as well as India's software clusters.

Government policies. Other important drivers of global information service networks are support policies and incentives to foster the establishment of higher-level software and service development centers, especially in China, Korea, Malaysia, Singapore, and Taiwan (China). Singapore, for instance, has amended its highly successful policy to develop local manufacturing support industries (especially for the electronics industry) with a policy to promote local suppliers of information services. The former policy was called the Local Industry Upgrading Programme (LIUP), and the new service-oriented program is called Infocomm Local Industry Upgrading Programme (iLIUP). A typical example of the efforts to build a support system is the iLIUP partnership agreements that link Software AG, the German vendor of XML (extensible markup language) software, with specialized Singaporean solution providers[21] to develop customized XML-based business solutions for foreign affiliates and Asian companies. To make this network operate, Software AG has brought in a global supplier of training programs for XML and related technologies. That company (Genovate Solutions) is responsible for establishing and running an XML Academy in Singapore that serves the Asia-Pacific region, training enterprises and individuals on XML, SAP, Java, Oracle, Linux, Web Logic, and other enterprise software programs.

Of particular interest are policies, pursued in China, to develop software and information service capabilities. A core element of these policies is the development of 10 major software bases. Three of the important ones include the Qilu Software Park in Jinan (Shandong Province), Shanghai's Pudong Software Park, and the Yangtze River Software Belt. Most of these projects are quite ambitious. For instance, the Jiangsu Software Park, the center of the Yangtze River Software Belt, takes up 120,000 square meters and contains 165 software companies that focus on applications in telecommunications, network security, and e-business. This park is expected to become a major export platform for software, way beyond Jiangsu Province's current software export revenues of 2.3 billion yuan (about $277 million), one-tenth of China's total software exports.

The Qilu Software Park in Jinan, Shandong Province is expected to develop into China's largest software park. The park covers a vast area

21. These Singaporean solution suppliers focus on enterprise resource planning, system management, network security, network knowledge management, and a variety of Internet-based application services.

of 6.5 square kilometers, and sales revenues (mostly exports) are expected to reach $1.2 billion in 2005, up from sales revenues of $233 million in 2000. A very different approach has been chosen for Shanghai's Pudong Software Park. Space constraints (only 9,000 square meters are available for this park) and the high price of land have led to a very selective focus on the dominant global flagships in the information industry. The dominant flagship is IBM, which is responsible for roughly one-third ($300 million) of total current foreign investment in the Pudong park. The park's specialization is shaped by two important recent developments. First, Shanghai is rapidly developing into a regional R&D and engineering center for leading global network flagships (so far, around 40), especially in the electronics industry. Second, Shanghai is about to become one of Asia's most important clusters for the design and manufacturing of semiconductors. The Pudong Microelectronic Industrial Belt aims to build around 10 chip production lines by 2005, with a projected production value of 100 billion yuan (approximately $12.05 billion). Proximity to this cluster implies that much of the Pudong Software Park's activities are geared to circuit design and related activities.

Another example of joint cross-border software development is a new research center, established by Ericsson, the Swedish telecommunications equipment producer, in China's Southern Software Park in Zhuhai, China. This center serves as a focal point for interaction with Ericcson's local partners, enabling them to co-develop and test 2G, 2.5G, and 3G applications for the Chinese market. The center is jointly run with Zhongshan University of Guangdong Province, which provides top-notch graduates to pursue joint research projects with Ericsson related to wireless Internet protocol data network compression and encryption technology, multimedia services, mobile electronics business, bluetooth technology, embedded software, and 3G systems.

Skill development and training. Skill development and training are essential elements of these global information service networks and may open up new opportunities for industrial upgrading. Often training and service provision are closely intertwined. A first example is the development of Asian networks for wireless Java applications. This is based on the joint initiative of two global flagships: Sun Microsystems, the developer of the Java language, and Nokia, the leading supplier of mobile handsets. This initiative brings together the individual networks, established by both flagships: the Sun Developer Network and Forum Nokia. The Asia-Pacific Sun and the Nokia wireless Java developers networks have a twofold purpose: to develop the Asian market for wireless Java applications and, at the same time, to create an Asian low-cost base of developers of such systems. A key component of these networks is the Developer Training Program, which is based on tools, knowledge, and resources provided by the two network flagships. Although programs and Java tools will be available at no charge, Asian developers will have to pay for hardware, training, and technical support. The objective is to "train up to 30,000 [Asian] developers to develop wireless Java applications . . . and to bring to market 1,000 Java content providers by end-2002" (AMPnet.asia, November 29, 2001). For Nokia, the objective is to create an Asian consumer market for more than 50 million mobile terminals supporting the Java platform.

A second example of emerging global information service networks involves the transfer of skills and the outsourcing of support services for storage area networks (SAN), a technology where Asian markets are expected to play an important role. With the exception of Singapore, Asian economies are latecomers in the use of digital information systems. Asian companies and government agencies thus have the opportunity to base their storage infrastructure on the new SAN model, which is more flexible and cheaper than traditional direct-attached storage models. In response, all major providers of infrastructure for storage area networks are rushing to establish global information service networks that, in addition to providing training for potential users, seek to develop a robust supply base for software development and SAN-related support services. An example is the emerging network of Brocade Communications Systems, which has nodes in Beijing, Hong Kong (China), Seoul, Singapore, Sydney, and Tokyo. Essential elements of Brocade's network are partnerships with leading Asian universities, like Beijing's Qinghua University, to establish joint technology labs and scholarship programs.

3Com Asia-Pacific, the network equipment and software supplier, provides a third example of global information service networks that are centered on information technology–enabled training (so-called e-training). Its 3Com University network provides on-line training and certification programs in simplified Chinese as well as web-based support services to customers and suppliers in 11 major cities in China. 3Com has established similar networks in Japan and Korea and plans to expand its efforts into other Asian countries. An important motivation is the need to create new markets for 3Com's products and software. Equally important, however, 3Com needs to have access to lower-cost local supply bases for service modules.

Our last example concerns Sybase, a global vendor of database technology and e-business applications. The company has strong links with China and Korea. In Korea it has partnered with Samsung to create new e-business software designed to facilitate the management of Asian suppliers' multiple linkages with global production networks. In China it has developed strong links with leading telecommunications carriers, supplying database management software to support the billing systems for these carriers.[22] Sybase's entry into the e-training market is motivated by two goals: to develop the Chinese market for database management software and to reduce the growing deficit of information technology skills in this country. An equally important objective is to develop a robust base of low-cost human resources that Sybase can tap into at a later stage. A major component of this training network is a joint venture between Sybase and the Beijing University of Aeronautics and Astronautics to offer on-line information technology training courses that are customized to the requirements of specific industries. Topics covered include database technology, e-business applications, Java applications, and mobile and wireless applications.

CHINA: A SHIFT IN NETWORK LOCATION

A third important transformation within East Asia is the emergence of China as a priority investment target for the leading global electronics flagships (whether from Europe, Japan, or the United States), their global suppliers from Korea and Taiwan (China), and, more recently, the leading U.S. contract manufacturers. As a result, China poses a serious challenge for mid-size countries (like Malaysia) with a focus on volume manufacturing. But the new challenge from China could also be a blessing in disguise, by catalyzing serious industrial upgrading efforts. Furthermore, China's huge potential market for electronics products and services provides new trade and investment opportunities for Asian firms. Equally important, Asian electronics firms may consider tapping into China's huge pool of low-cost engineers and scientists.

China's New Role in the Electronics Industry

There is no doubt that the center of gravity of FDI in the electronics industry is moving toward China, transforming the geography of global production networks within the region. China is no longer only a provider

22. Sybase's main partners include China Unicom, Heilongjiang Telecom, Shaanxi Mobile, Shandong Telecom, Sichuan Mobile, and Zhejiang Telecom (CMPnet.Asia, October 24, 2001).

of cheap labor. China's new attractiveness results from a combination of five developments: a booming market for information technology products and services, when the rest of the world is in recession; an unlimited supply of low-cost information technology skills; abundant land and a rapidly improving infrastructure; a massive rush of capital flows into China; and, catching this opportunity, support policies pursued by the central government as well as regional and local authorities to rely on FDI as an accelerator of industrial upgrading.

The move toward China is particularly pronounced in three sectors: electronic components (especially semiconductors), computers, and telecommunications. For U.S. foreign direct investment in semiconductors, China has become the second most important recipient, after Singapore, overtaking Malaysia, which was the main recipient in 1996. Even the new incentives provided by Malaysia's Second Industrial Master Plan (IMP2) have not prevented this move toward China. A similar trend can be discerned for FDI by Japanese electronics firms: China has moved into first place ahead of Malaysia, which was its biggest recipient during the early 1990s (Malaysia Ministry of Finance data, quoted in Takeuchi 2001).

Taiwanese (Chinese) computer companies that supply leading U.S. computer OEMs have played an important pioneering role in integrating China into global production networks. Since the early 1990s, they have continuously moved production from Taiwan (China) to China. The result is that roughly 40 percent of China's electronics exports today are shipped from Taiwanese (Chinese) factories in China (courtesy of Market Intelligence Center, Institute for Information Industry, Taipei, December 2001). Taiwanese (Chinese) suppliers are serving a large share, around 33 percent, of their export orders from production lines in China.

From the United States, OEMs like AMD, Cisco, Compaq, Hewlett Packard, Intel, Microsoft, Motorola, and Sun Microsystems have all initiated significant new investment projects in China. Motorola, for instance, considers China to be of critical importance. At present, Motorola has 12 affiliates in the Asia-Pacific region. While the company's involvement started out with Korea, later followed by Singapore and Taiwan (China), China has gained substantially in importance since the early 1990s. Currently, six of Motorola's Asian affiliates are based in China, with two in Singapore, including the regional headquarters, and one each in India, Korea, Taiwan (China), and Thailand. Motorola is counting on the continuous rapid growth of the Chinese market to reduce the negative impact of the current recession.

But China is also expected to play an important role in Motorola's long-term strategy. By providing access to the world's largest pool of relatively lower-cost information technology skills, the company expects to enhance its innovation capabilities. For instance, out of the 13,000 people that Motorola currently employs in China, 1,000—that is, almost 8 percent—are active in R&D. Reflecting China's growing importance, Motorola held its annual board meeting for 2001 in China, its first board meeting outside the United States.

China also has attracted major new investments from leading European electronics flagships, like Alcatel, Ericsson, Nokia, Philips, and Siemens. Philips, for instance, has moved its entire cell phone manufacturing operations to its Shenzhen joint venture with Beijing-based China Electronics. And Nokia has committed itself to establish a large integrated cluster, the Xingwang (International Industrial Park) in Beijing, which incorporates Nokia's own mobile phone plant and brings together 15 of its long-standing and trusted suppliers of international components. This Nokia-centered cluster involves an initial investment of about $1.2 billion and is expected to create 15,000 jobs, with a projected annual sales volume of $6 billion.

In addition, all leading Japanese electronics flagships are considering major new investments in China as they struggle under a depressed domestic economy and high manufacturing costs. Toshiba is building a plant for computer hardware and components in Nanjing, while Matsushita, Mitsubishi Electric, and NEC all are expanding their phone production in China. Finally, a large surge in investments in China is also reported by the Korean electronics *chaebol* that are all gripped by a severe "China fever."

Semiconductors

Since 2000, the semiconductor industry has provided a telling example of the speed of China's integration into global production networks.[23] In terms of policies and support institutions, China's experience in the semiconductor industry also provides a useful reference point for a case study of Malaysia.

Until 1999, investment in China's semiconductor industry lagged woefully behind similar investments in Korea and Taiwan (China). The turning point came in 2000. During that year, the Chinese government apparently made a strategic decision to rely on FDI to accelerate the development of this industry. The underlying expectation was that FDI can generate a

23. If not stated otherwise, the following is based on Simon (2001) and relevant industry newsletters, such as the *Semiconductor Reporter* and *Interfax China*.

critical mass, establishing new global dynamic clusters for semiconductor manufacturing, primarily in Beijing and Shanghai. The hope is that, if it engages leading U.S. flagships (both semiconductor manufacturers and equipment producers), this strategy may create enough pressure within the United States to dismantle the remaining U.S. restrictions on exports of technology. This may be an unrealistic expectation, however, in light of the current resurgence of defense and security concerns in the United States.

China's new pragmatic policies toward FDI have induced global flagships to announce several large investment projects worth around $7 billion. During 2000, China's semiconductor output grew 42 percent. Exports during 2000 grew almost 35 percent in unit terms and 30 percent in value terms, reaching $2.1 billion. The role of global production networks has been critical. Foreign-invested enterprises dominate China's semiconductor exports, with a share of about 94 percent, while state-owned enterprises are responsible for a meager 5.3 percent.

After 2001 China emerged as the main growth market for semiconductor production equipment manufacturers. Leading equipment makers are all scrambling to expand their sales in the largely untapped China market.[24] Table 3.3 presents information on major investment projects by global flagships, Taiwanese (Chinese) contract manufacturers (so-called silicon foundries), as well as domestic firms.

The prospective boost in FDI has given rise to optimistic projections. In-Stat Group, a semiconductor industry consulting firm, argues that China has the potential to become the second-largest market for semiconductors after the United States by 2005. The Chinese government certainly leaves no doubt that it has very ambitious objectives. By 2005 it hopes to increase semiconductor sales revenues to $9.7 billion, or 2 percent of the global market. By then, domestic production is expected to meet about 30 percent of China's demand. By 2010 China expects to be where Korea is now, that is, producing roughly 5 percent of the global semiconductor market.

If these projections materialize, this would obviously pose a major threat to existing semiconductor industries in Asia. This threat would be most immediate for Malaysia, with its heavy reliance on semiconductor assembly. However, important threats may also emerge for semiconductor manufac-

24. Applied Materials, the world's leading vendor of semiconductor production equipment, for instance, expects to increase its sales in China from $100 million in 2000 to $1 billion by 2005 (*Far Eastern Economic Review*, November 1, 2001).

Table 3.3 Major Investment Projects in China's Semiconductor Industry, since January 2000

Firm	Location	Activities	Investment (billions of U.S. dollars)	Capacity and technology
Global flagships (100 percent affiliates)				
Intel	Waigaoqiao Free Trade Zone, Shanghai	Expansion of chip assembly and testing	0.302	Intel's new 845 chipsets used with Pentium 4 processors
AMD	Suzhou, Jiangsu Province	Assembly and testing	0.108	
Motorola	Tianjin	Integrated wafer fabrication, assembly, and testing	1.9	
Fairchild	China-Singapore Suzhou Industrial Park, Jiangsu Province	Assembly and testing for wide range of logic, discrete, and analog devices; plans to outsource more than 50 percent to local suppliers	0.010 (2002); total: 0.200; plans to use latest in information technology to enhance cost-efficiency and time-to-market	120,000 square feet (2002); total: 800,000 square feet
Philips	Dongguan, Guangdong Province	Assembly and testing		

(*continued*)

Table 3.3 Major Investment Projects in China's Semiconductor Industry, since January 2000 (*continued*)

Firm	Location	Activities	Investment (billions of U.S. dollars)	Capacity and technology
Contract manufacturers, primarily from Taiwan, China				
SMIC (Semiconductor Manufacturing International Corporation)	Shanghai	Silicon foundry		480,000 8-inch wafers at 0.25 micron
Shanghai Grace Semiconductor Manufacturing Corporation	Shanghai (joint venture between Formosa Plastics and Chinese group	Silicon foundry	1.6	
TSMC (Taiwan Semiconductor Manufacturing Company)	Investing in silicon foundry in Shanghai			
UMC (United Microelectonics Corp.	Investment in silicon foundry in Suzhou			
ASE (Advanced Semiconductor Engineering)	Hangzhou, Zhejiang Province	Integrated circuit assembly and testing	0.0028	
USI Electronics (Universal Scientific Industrial), a subsidiary of ASE	Shenzhen	Personal computer motherboards (Pentium 4) for IBM with USI Taiwan (China)		Capacity expansion from 6 to 10 lines; monthly production capacity increasing to 400,000 units
Siliconware Precision Industry	Shanghai (possible joint venture with Shanghai Huahang Group)	Integrated circuit assembly and testing		

Chinese companies				
Shanghai Hongli Semiconductor Manufacturing	Shanghai	Production of 8-inch and 12-inch wafers		
Sast Group	Shenzhen	Two wafer fabrication lines	1.2	40,000 wafers per month
Beijing Xunchuang IC	Beijing (joint venture that involves Chinese companies and Kingston plus Taiwanese [Chinese] investment fund [Asia-Pacific Technology Development Corporation])	Assembly and testing	0.200	
China Great Wall Computer Shenzhen	Pudong "Silicon Harbor," Shanghai (joint venture with Kingston Technology, a U.S.-based memory module vendor that owns 80 percent)	Assembly and testing		50,000 square feet
Beijing Huaxia Semiconductor Manufacturing	Beijing	Wafer fabrication		8-inch, 0.25 micron

turing and wafer fabrication in Korea, Singapore, and Taiwan (China) as well as in Japan. But it is unlikely that these countries will simply sit still and let shifting comparative advantage run its course. In fact, all major competitors in the global semiconductor industry are pursuing aggressive policies to upgrade their product composition and capabilities, giving rise to a major transformation of this industry (for example, see Macher, Mowery, and Simcoe 2002).

In addition, it will not be easy for China to realize its ambitious upgrading objectives. One reason is the relatively low level of development of China's semiconductor industry. An important indicator of this is the fundamental mismatch between China's exports and imports of semiconductors. In value terms, China imports more than 70 percent of all semiconductor devices that it needs. This is slightly better than the ratio found in Korea's semiconductor industry during the early 1990s (Ernst 1994b, 1998). On the one hand, most Chinese exports are lower-end devices involving fairly mature and basic process and manufacturing techniques. Imports, on the other hand, are much more sophisticated. Between 1995 and 2000, China's semiconductor imports grew at a compound annual rate of almost 92 percent, while exports over the same period rose less than 60 percent.

A further fundamental constraint to a rapid upgrading of China's semiconductor industry are the massive investments required in production equipment, facilities, infrastructure, R&D, and education and training. The projected expansion of international market share requires an upgrading in the product mix as well as in process and design capabilities. Out of China's 25 wafer fabrication lines in early 2001, only one used 8-inch wafers, while 21 lines used outdated 5-inch wafers (six lines) and even 4-inch wafers (15 lines).[25] However, by 2003 a few more 8-inch lines had been added and several 12-inch wafer plants are planned or being built—and will begin operating in 2004–05.

Table 3.4 provides a widely used proxy, developed by the U.S. Semiconductor Industry Association, for the investment costs of wafer fabrication. These estimates assume that sophisticated infrastructure and support industries exist within the relevant clusters or at least in close proximity. This is by no means the case in China. For instance, 90 percent of the materials used to produce the existing 8-inch wafer fabrication lines must be imported. And domestic manufacturers of semiconductor production equipment meet less

25. *China Electronics News*, March 27, 2001. The 8-inch line is a joint venture with Japan's NEC and the Shanghai Hua Hong NEC Electronics.

Table 3.4 Investment Costs of Wafer Fabrication

Process technology	Minimum investment requirements[a]
6-inch wafers (0.5–1.2 microns)	$200 million
8-inch wafers (0.35–0.5 microns)	$1.2 billion
12-inch wafers (less than 0.35 microns)	$2.5 billion

a. Estimates, courtesy of U.S. Semiconductor Industry Association.

than 10 percent of domestic requirements. In other words, China's weak semiconductor infrastructure and base of support industries clearly imply that the effective investment costs for upgrading its semiconductor industry may be substantially higher than indicated by the figures of the Semiconductor Industry Association.

In short, there are clear indications in the semiconductor industry of a shift in network location away from the traditional export platform sites in Southeast Asia and toward China. Yet this process may not be as quick and smooth as many observers appear to believe. This may provide enough breathing space for developing appropriate upgrading strategies in those economies that are most heavily exposed to the threat from China.

IMPLICATIONS FOR MALAYSIA'S UPGRADING PROSPECTS IN THE ELECTRONICS INDUSTRY

In light of the fundamental transformations of global production networks in East Asia's electronics industry, what can be said about Malaysia's upgrading prospects? Do these transformations provide new opportunities for relieving domestic upgrading constraints? And what policies and support institutions could help Malaysia to successfully implement upgrading strategies?

This section first highlights achievements and structural weaknesses of the Malaysian electronics industry that define and constrain its upgrading prospects. I then assess current policies that try to link cluster development and global network integration, discuss adjustments in linkages with global brand leaders, and ask to what degree linkages with contract manufacturers can broaden these opportunities. I conclude by exploring new opportunities for diversifying Malaysia's international linkages that could enhance the upgrading prospects of its electronics firms, focusing on carriers of knowledge exchange that complement formal global production networks.

Achievements

The progressive integration into global production networks has been a primary driver of Malaysia's success in the electronics industry. This integration started in the early 1970s with offshore chip assembly, primarily by U.S. semiconductor firms. The next stage, beginning in the early 1980s, centered on Japanese electronics makers that moved their export platform production for consumer electronics to Malaysia and other Southeast Asian locations. Since the late 1980s, Malaysia has been integrated into the production networks of American producers of computer-related equipment as well as those established by their Taiwanese (Chinese) subcontractors. The most recent stage involves the production of communication and networking equipment and the acquisition of existing flagship affiliates by global contract manufacturers.

The Malaysian government, through its Industrial Master Plan (IMP) (1986–95), has tried to reap as many benefits as possible from this fortuitous tailwind of FDI (Ministry of International Trade and Industry 1986). The guiding principle has been "outward industrialization," subordinated to the needs of global network flagships. The results have been impressive, in terms of production, exports, employment, and investment (table 3.5).

Within a relatively short period, Malaysia experienced a substantial expansion in capacity and international market share for electronics products. The heavy reliance on electronics exports acted as a powerful engine of growth. Although there were periodic disruptions, like the downturn in 1985–86 and, in particular, the Asian financial crisis in 1997–98, the overall balance was remarkably positive. During the last decade, from 1990 to 2000, Malaysia's electronics industry registered a compound average annual growth rate of almost 24 percent. During the same period, exports

Table 3.5 Performance of Malaysia's Electronics Industry Compared with Objectives of the Industrial Master Plan (millions of ringgits)

Area	Goals	Performance
Production	11.0	30.5
Exports	10.9	32.6
Employment	5.5	21.17
Investment value	6,298	25,985

Note: Figures for production, exports, and employment are average rates of growth between 1986 and 1992. Those for investment are the cumulative total of investment approved between 1986 and 1995.
Source: Ministry of International Trade and Industry (1996).

grew at an annual average of 25 percent, while employment grew almost 11 percent annually (figures courtesy of Ministry of International Trade and Industry, Kuala Lumpur, June 2001).

Global electronics brand leaders—and, more recently, their contract manufacturers—played an important role. The electronics industry is the major recipient of foreign direct investment, absorbing more than one-third of total foreign direct investment in manufacturing between 1996 and 1998 (MIDA 1999). About 100 large foreign affiliates effectively dominate this industry. Their share in manufactured exports (most of it electronics) increased sharply, from 40 percent in 1985 to 68 percent in 1992 (Takeuchi 1997, p. 9). The 18 members of the Malaysian-American Electronic Industry Association accounted for more than 14 percent of Malaysia's electronics exports in 2001 (*Business Times*, Kuala Lumpur, October 3, 2001).

Weaknesses

Despite these achievements, a shift in strategy is now overdue. Since the summer of 2000, the downturn in the global electronics industry has brutally exposed six structural weaknesses of Malaysia's electronics industry that define and constrain its upgrading prospects: (1) an asymmetric industry structure; (2) a heavy dependence on imports, due to weak domestic support industries and limited Hirschman-type linkages; (3) a heavy reliance on exports, especially to the U.S. market; (4) a highly concentrated composition of products, centered on low-end assembly operations; (5) a declining capacity to generate employment; and (6) a serious mismatch between the demand for and supply of skills. Although "outward industrialization" policies have provided Malaysia with substantial initial advantages in terms of export and capacity growth, these policies have failed to develop sufficient sectoral breadth and depth.

Asymmetric industry structure. Malaysia's integration into global production networks gave rise to the development of an asymmetric industry structure: multiple layers of electronics firms are distinguished by asymmetric control over resources and decisionmaking. At the end of 2000, roughly 900 electronics companies were registered in Malaysia, employing more than 400,000 workers. While Malaysian firms dominate in numbers, Malaysia's electronics industry continues to be shaped by strategic decisions of global flagships (both OEMs and major American contract manufacturers). In hierarchical order, four types of firms can be distinguished: at the top of the industry pyramid are global OEMs and contract manufacturers, followed by suppliers and contract manufacturers from Japan,

Korea, Singapore, and Taiwan (China), higher-tier local suppliers, and, at the bottom, lower-tier local suppliers.

In contrast to economies like Korea, Singapore, and Taiwan (China), Malaysia has failed to develop a broad and multi-tier base of support industries. There are, of course, a few widely quoted successes, almost all of them in Penang, such as BCM, Eng Teknologi Holdings, Globetronics, LKT, and Unico, that have positioned themselves as higher-tier local suppliers for leading OEMs (Best 2001; Rasiah 1995, 2003). These companies are moving to upgrade their capabilities to cope with the new opportunities and challenges that result from the transformations of global production networks.

However, the majority of the local suppliers possess few proprietary advantages and clearly qualify as lower-tier suppliers. They lack sufficient financial resources to invest in training (and retraining) and to invest in digital information systems and leading-edge equipment. This is so despite various promotional policies, focused on the smaller suppliers, introduced by the government. Possible explanations may include the proximity to Singapore and its base of sophisticated local suppliers, which may discourage flagships from using Malaysian suppliers.

This asymmetric industry structure has given rise to a lack of efficient domestic linkages and an inverted production pyramid—a huge and rapidly growing final product sector that rests on a weak and much smaller domestic base of support industries.

Import dependence. The result is a persistently high dependence on imports: rapid growth in the final products sector necessitates considerable imports of intermediates and production equipment. Between 1986 and 1992, imports of Malaysia's electronics industry increased at a rate of more than 24 percent, far exceeding the goal of 7.6 percent envisioned under the IMP. By the late 1980s, the Malaysian electronics industry had to import almost 43 percent of the intermediate goods that were required to produce one unit of final output, far more than Korea (37 percent) and Japan (8 percent; Takeuchi 1997, p. 7). This reflected the initial strategy to position Malaysia as a low-labor-cost assembly site. By the late 1980s, however, the government acknowledged that this strategy was no longer sustainable, as new lower-labor-cost locations emerged within Southeast Asia as well as in China and Mexico.

Yet Malaysia's dependence on imports of electronics components, especially semiconductors, kept increasing during the 1990s, both as a share of electronics imports and as a share of total merchandise imports (table 3.6).

This suggests a fundamental mismatch between the country's electronics exports and its imports, with negative terms-of-trade implications: while imports involve high-value-added core components, especially microprocessors and other imported semiconductors, Malaysia's component exports overwhelmingly consist of low-value-added final assemblies.

Export dependence. Malaysia's electronics industry remains heavily dependent on exports: in 2000 electronics manufacturing made up about 60 percent of Malaysia's total export value, of which 35–40 percent were exports from Penang. Malaysia is one of the eight economies that are most dependent on exports to the United States, six of these being from East Asia: Malaysia's exports to the United States (most of them electronics products) accounted for 24 percent of the country's gross domestic product in 1999. Malaysia's electronics industry is extremely vulnerable to a recession in major export markets, especially the United States. During 2001 when the U.S. electronics market was in free-fall, Malaysia's electronics production and exports dramatically declined, the former more than 25 percent and exports almost 19 percent.

Concentrated product composition. A highly concentrated product composition adds further to the country's vulnerability. By 1998 the share of electronics in merchandise exports had increased to almost 58 percent, up from 48 percent in 1993. And components accounted for 46 percent of all electronics exports, of which semiconductors alone accounted for more than one-third (table 3.7). Industrial electronics (including computer-related products and telecommunications equipment) accounted for slightly more than 30 percent, and consumer electronics accounted for about 15 percent. Recent figures from other sources show a further increase in the concentration of product composition. By 2000 the share of electronics exports in Malaysia's total exports had reached 70 percent (Zainal Abidin 2002, slide 6).

In historic terms, this constitutes an important improvement. Back in 1986, 84 percent of Malaysia's electronics exports were components (most of them assembled chips), with 14 percent consisting of consumer electronics and only 2 percent consisting of industrial electronics. Unfortunately, this impressive change was not sufficient to reduce the country's vulnerability to abrupt changes in the world market. An important reason for this vulnerability is that semiconductor exports generate very little local value added, as Malaysia only performs assembly and testing. A heavy dependence on assembly-type operations for a handful of products can be

Table 3.6 Dependence on Imported Components, 1993–98
(percent)

Economy and component	Electronics imports						Merchandise imports					
	1993	1994	1995	1996	1997	1998	1993	1994	1995	1996	1997	1998
Korea, Rep. of												
Components	57.0	53.2	54.1	55.3	60.2	71.6	9.5	9.5	9.7	10.0	12.0	16.6
Semiconductors	34.8	34.0	35.7	36.8	43.0	54.7	5.8	6.0	6.4	6.7	8.6	12.7
Taiwan, China												
Components	66.3	67.5	70.0	65.3	—	—	13.7	14.8	16.9	16.7	—	—
Semiconductors	41.2	43.9	46.5	43.1	—	—	8.5	9.7	11.2	11.0	—	—
Singapore												
Components	48.0	52.9	56.9	55.4	53.4	56.4	18.6	23.2	26.0	24.3	23.8	26.5
Semiconductors	25.0	30.5	35.6	33.9	33.6	37.3	9.7	13.4	16.2	14.9	15.0	17.5
Malaysia												
Components	76.5	78.7	78.1	76.7	72.8	78.5	25.3	28.3	28.8	29.6	28.3	36.7
Semiconductors	42.9	45.9	49.7	49.6	49.2	56.5	14.2	16.5	18.3	19.1	19.1	26.4

Thailand												
Components	55.1	55.9	59.1	60.0	58.7	—	9.6	11.6	12.4	12.6	14.1	—
Semiconductors	26.4	26.8	28.7	30.2	29.9	—	4.6	5.6	6.0	6.4	7.2	—
Philippines												
Components	37.8	40.3	43.7	54.7	34.9	—	5.8	6.6	7.7	20.7	9.4	—
Semiconductors	28.0	28.8	30.3	46.6	27.0	—	4.3	4.7	5.4	17.7	7.3	—
Indonesia												
Components	46.4	48.6	46.3	37.5	33.8	33.9	4.4	3.5	3.3	3.2	3.0	2.0
Semiconductors	4.5	4.0	3.8	2.9	2.1	2.4	0.4	0.3	0.3	0.2	0.2	0.1
Hong Kong, China												
Components	38.0	38.1	39.7	39.6	40.5	40.1	9.8	10.4	11.6	11.6	12.5	12.8
Semiconductors	18.8	18.7	20.7	20.2	20.3	19.6	4.8	5.1	6.0	5.9	6.3	6.2
China												
Components	40.9	43.1	44.3	49.6	55.0	53.1	5.3	6.3	6.9	7.4	9.2	11.4
Semiconductors	11.0	12.5	14.7	17.7	21.4	22.3	1.4	1.8	2.3	2.6	3.6	4.8

—Not available.

Source: UN Trade Database Comtrade.

Table 3.7 Revealed Comparative Advantage and Leading Export Products, 1993–98

	Revealed comparative advantage						Share in electronics exports					
Economy and product	**1993**	**1994**	**1995**	**1996**	**1997**	**1998**	**1993**	**1994**	**1995**	**1996**	**1997**	**1998**
Korea, Rep. of												
Electronic data processing	0.9	0.8	0.8	0.9	0.9	0.7	14.4	11.9	12.2	14.5	15.5	13.9
Storage	0.2	0.3	0.4	0.7	1.3	1.1	0.5	0.8	1.0	1.8	4.1	4.1
Components	2.4	2.7	2.8	2.7	2.8	2.7	50.1	56.2	62.4	60.8	62.3	63.4
Semiconductors	3.3	3.8	4.1	3.6	4.0	3.8	30.4	37.2	45.7	40.3	42.9	45.3
Consumer electronics	2.3	2.4	2.0	2.0	1.7	1.5	22.5	20.5	16.1	15.6	12.8	12.7
Telecommunications	0.9	0.9	0.8	0.8	0.6	0.5	3.0	2.7	2.4	2.4	2.1	1.9
Share of electronics in merchandise exports (percent)	28.0	29.7	30.9	28.8	29.2	28.3						
Taiwan, China												
Electronic data processing	2.5	2.7	3.0	3.4			39.4	39.0	41.6	45.0	44.62	45.29
Storage	0.3	0.3	0.5	0.6			0.8	0.6	1.0	1.3	2.10	1.97
Components	1.9	2.0	2.1	2.2			37.2	39.3	41.6	40.2	41.93	40.86
Semiconductors	1.6	1.8	2.0	2.2			13.7	16.8	20.4	19.6	22.05	21.77
Consumer electronics	1.4	1.5	1.2	1.1			12.9	12.0	8.5	6.8	6.42	5.98
Telecommunications	1.7	1.7	1.6	1.6			4.9	4.8	4.1	4.0	3.59	4.20
Share of electronics in merchandise exports (percent)	29.5	31.0	34.3	35.8								
Singapore												
Electronic data processing	4.6	5.1	5.1	5.4	5.5	5.2	40.7	38.6	39.4	42.8	44.1	44.6

Storage	12.9	13.4	12.8	15.3	12.4	11.4	17.6	15.7	16.0	18.8	19.3	19.8
Components	2.7	3.4	3.4	3.6	3.7	3.7	29.4	35.4	38.5	38.0	38.9	40.7
Semiconductors	3.2	4.0	4.2	4.5	4.8	5.0	15.7	19.4	23.8	23.9	25.0	27.4
Consumer electronics	3.1	3.2	2.9	2.9	2.3	1.7	15.9	13.7	11.6	10.4	8.7	6.8
Telecommunications	1.1	1.2	1.1	0.8	0.7	0.6	1.9	1.9	1.6	1.3	1.1	1.1
Share of electronics in merchandise exports (percent)	53.0	58.8	60.7	60.7	60.6	61.4						
Malaysia												
Electronic data processing	1.4	1.8	2.0	2.3	2.9	3.0	13.8	15.6	17.1	20.5	25.3	27.4
Storage	0.0	0.1	0.8	0.2	3.7	4.1	0.0	0.2	1.1	0.3	6.3	7.6
Components	3.7	3.7	3.6	3.8	3.9	3.9	44.6	42.6	43.9	44.6	45.1	45.7
Semiconductors	5.6	5.2	4.7	4.9	5.3	5.2	30.5	28.3	29.8	29.3	30.1	30.7
Consumer electronics	4.2	5.0	5.1	5.0	4.1	3.5	24.2	24.2	22.6	20.1	16.5	15.1
Telecommunications	2.1	2.4	2.0	2.0	1.7	1.5	3.8	4.0	3.2	3.3	3.1	2.8
Share of electronics in merchandise exports (percent)	47.6	52.5	54.9	54.7	55.8	57.5						
Thailand												
Electronic data processing	1.4	1.8	1.9	2.4	2.5		32.2	34.0	36.8	41.4	40.9	
Components	1.4	1.5	1.5	1.6	1.6		38.4	38.8	39.9	36.5	35.8	
Semiconductors	1.7	1.6	1.4	1.6	1.7		20.8	19.3	19.5	18.8	18.3	
Consumer electronics	1.3	1.6	1.4	1.5	1.7		16.7	16.4	13.3	12.0	13.1	
Telecommunications	1.2	1.1	1.0	1.2	1.2		5.2	3.9	3.7	4.1	3.9	
Share of electronics in merchandise exports (percent)	20.8	24.0	24.9	28.4	29.6							

Source: UN trade database Comtrade.

crippling, as those operations can be easily replicated in countries with low levels of education.

The government has tried to address this issue by investing in two new silicon foundries: Silterra (in Kulim Hi-Tech Park) and 1st Silicon (in Sarawak). To justify the heavy investment, an attempt has been made to use these fabrication plants as catalysts for the development of circuit design houses. It is too early to assess the success of these investments. Bad timing has been an important constraint—these facilities became operational during the recent downturn. However, there are positive signs. As the provision of silicon foundry services is becoming a commodity (see United Microelectronics 2001, on risk factors), there are new entry possibilities for low-cost foundries in China and Malaysia, while the industry leaders (IBM Microelectronics and Taiwan [China]'s TSMC and UMC) move up the ladder to combine design capabilities with advanced fabrication technology (12-inch wafers; Depeyrot 2002).

Declining capacity for employment generation. A capacity for job creation in Asia's thriving electronics sector has been a hallmark of the region's successful export-oriented industrialization. Since the 1997 financial crisis, however, the sector's capacity to generate employment has declined. Take Seagate, the leading U.S. disk drive manufacturer. Since the mid-1980s, the company was among the largest employers in Southeast Asia, topping the list in Penang and Singapore. That golden age of employment generation has long gone. Table 3.8 documents the company's massive destruction of manufacturing jobs in Asia.

In Malaysia's electronics industry, an estimated 150,000 to 165,000 jobs have been lost since the financial crisis (table 3.9). During 2001, the most recent year for which data are available, almost 19,000 workers were laid off. Malaysian labor market experts talk of a declining employment-generating capacity of the electronics industry: after earlier downturns, a substantial share of laid-off workers were rehired, but this no longer seems to be the case.

The latest unemployment report, prepared for the Penang State government, conveys some distressing findings (Too and Leng 2002). With job

Table 3.8 Manufacturing Employment of Seagate, Peak Year and 2002

Location	Peak year	2002	Job losses
Malaysia	24,000 (1987)	5,500	–18,500
Thailand	40,000 (1998)	18,000	–22,000
Singapore	20,000 (1998)	9,000	–11,000
Total	84,000	32,500	–51,500

Source: Company reports and 10K forms.

Table 3.9 Job Losses in Malaysia's Electronics Industry, 1998–2001

Year	Number of jobs lost
1998	83,900
1999	37,400
2000	25,600
2001[a]	18,900

a. January–August.

Source: Government figures, quoted in Hamid (2001).

losses of more than 16,000 during 2001, most of them in sectors related to the electronics industry, retrenchment has been dramatic, and it has hit primarily low-skilled, female production workers.[26] Particularly disturbing is the unusually high proportion of retrenched workers (62 percent) who could not be located, indicating a massive return of Malay females (in the 25–29 age range) to their villages. Export-led electronics manufacturing is unlikely to act again as an engine of employment growth.

Mismatch between the demand for and supply of skills. Finally, an increasingly important weakness in Malaysia's electronics industry is the serious mismatch between the demand for and supply of skills. Despite the recession, job vacancies reached nearly 90,000 nationwide in September 2001, with the biggest job openings in the "managerial and professional" categories in the electronics industry. Data collected in Penang show that a growing deficit of specialized information technology skills is an important qualitative constraint on Malaysia's upgrading possibilities in the electronics industry (DCT Consultancy Services 2002). This is especially true for engineers with degrees in electronics, mechanical engineering, quality control, testing, and chemistry. There are also important bottlenecks for mechanics, tool and die makers, and information technology professionals, especially system analysts. All of this indicates weak incentives for firms to invest in long-term assets, such as specialized skills.

This human resource bottleneck also has an important qualitative dimension. As Too and Leng (2002) document, having the right degrees in electrical and electronics engineering, information technology, and management does not guarantee entry into the labor market. They also find that

26. Production workers with limited skills account for three-quarters of the total rentrenched workers, while female workers account for almost two-thirds of total job losses.

the majority of unemployed graduates have not held a job of any kind since graduation. This reflects the perception of electronics firms that local university graduates have book knowledge but are ill equipped to deal with real-world problems on the shop floor and lack basic skills in communication, negotiation, and presentation. This has led to the emergence of a bifurcated labor market, where the winners pick all the stakes: there is intense competition for those engineers and managers who either graduated from overseas universities or worked for a foreign firm. Obviously, upgrading efforts will remain truncated as long as this skills mismatch is not addressed.

A Shift in Strategy (I): Clusters and Global Network Integration

Two policy initiatives are important for assessing Malaysia's prospects for upgrading in the electronics industry: the Second Industrial Master Plan (IMP2) and the Multimedia Super Corridor (Ministry of International Trade and Industry 1996; Multimedia Development Corporation 2002). Both represent attempts to overcome some of the structural weaknesses of this sector. Both point in the right direction but have had only limited success. The rest of this section explores whether the transformations of global production networks provide new opportunities for relieving the constraints on domestic upgrading.

The IMP2. The IMP2 document signals a fundamental change in Malaysia's industrialization strategy, away from assembly-based "outward industrialization" to value chain–based manufacturing, from sector-based to cluster-based development, and from performance targets to productivity-driven growth. The strategy is defined by two key concepts: manufacturing ++ and cluster-based development. In line with Porter (1990), manufacturing ++ highlights activities at both ends of the value chain, that is, R&D and engineering and inbound logistics, on the one hand, and outbound logistics and sales and marketing, on the other hand. It is argued that a move into knowledge-intensive support services like product development, process engineering, supply chain management, and select areas of R&D will enhance local value added and productivity. Cluster-based development implies that, based on existing strengths, especially in components and semiconductors, developing a dense web of domestic linkages will enhance value added and deepen domestic capabilities.

On paper, these two concepts represent the cutting edge in current policy debates on regional and technology policy (for example, Best 2001; OECD 1999). However, within Malaysia, there are four electronics clusters that differ quite substantially in their upgrading objectives and capabilities:[27] (1) Penang Island in the north and the Kulim Hi-Tech Park in the neighboring state of Kedah; (2) Selangor and Negeri Sembilan in the center; (3) the southern part around Johor, with close linkages to Singapore; and (4) the more recent Multimedia Super Corridor around Kuala Lumpur. Of these, the combined Penang-Kedah cluster has arguably been the most successful, with good chances for further upgrading.

IMP2 highlights four specific objectives: (1) foster the growth of leading local companies (Malaysian brands); (2) reduce dependence on imported inputs; (3) strengthen agglomeration economies by developing integrated manufacturing centers for global network flagships; and (4) develop cross-border clusters. Of these, the first two objectives are problematic, while the last two indicate a move in the right direction.

The first objective represents an outdated concept of industrial upgrading that assumes a fixed pattern of sequencing from low-end, assembly-type subcontracting to original brand name (OBM) manufacturing (for a typical example, see Hobday 1995). We now know that the transition to OBM is extremely difficult—even Taiwan (China)'s Acer has had only limited success (Ernst 2000b, 2000c). The limited achievements of the Proton City cluster in automobiles also indicate that this objective may be unrealistic. As for the second objective, much depends on whether the country succeeds in finding the right balance between reaping the benefits of imported foreign inputs (as described in Rodrik 1999) and developing local backward and forward linkages.[28]

Integrated manufacturing centers. The third objective contains some promising elements. Take recent developments in the Penang cluster. Rather than just giving in to requests of foreign companies for improved incentives, the state government has been pursuing a more selective approach: incentives are explicitly linked with the promotion of integrated manufacturing centers. The goal is to induce global flagships to move to Penang an "entire chain of operations for a particular product" (*Asia Computer Weekly Online*, October 22, 2001, p. 4). It is expected that this should enable the Penang cluster to upgrade from mere assembly and testing to knowledge support services, like sales and marketing, adaptive process engineering and tooling,

27. The designation of a particular location as a "cluster" raises tricky questions that are beyond the scope of this chapter. For instance, are any of these "clusters" just a collection of firms drawn to this location purely based on incentives given to local firms? And are there rationales other than policy for the observed level of agglomeration in the electronics industry in Malaysia?

28. Hirschman (1958) emphasizes the need to combine both effects.

financial planning, and, eventually, parts of R&D like design and development.[29] Table 3.10 provides examples.

Equally important is the effort by the Penang State government to develop a photonics industry cluster. Photonics is the technology of harnessing light for digital applications, covering CD-ROMs, fiber-optic communications, lasers, sensing and measuring devices, and liquid crystal displays. Penang has developed volume-manufacturing capabilities in all of these areas. Major global players in this sector, such as Agilent, Finisar, Osram, and Solectron, are already active in Penang. In August 2001 the Penang Photonics Consortium was established. Its main objective is to bundle existing activities into a dynamic cluster that could provide a broad range of contract manufacturing services in optical components. One of the main local players is Chahaya Optronics, a company that has received funding from the U.S.-based hard disk manufacturer, Komag, and leading venture capital firms.

The recent downturn in the global telecommunications industry drastically slowed the pace of these efforts. Firms are struggling to survive. But, in principle, the idea is sound, and the challenge is to be prepared, once demand

Table 3.10 The Integrated Manufacturing Center in Penang, 1999–2001

Year	Company and product	Remarks
1999	Komag: world largest supplier of thin-film disks	Relocates entire U.S. operation to Penang, except R&D and sales and marketing
2000	Dell Computer: personal computers, servers, and storage products	Establishes Penang as built-to-order hub for the Asia-Pacific region Reverses the decision one year later, when much of these activities are moved to Xiamen, China
2001	Quantum: hard disk drives	Plans to move to Penang entire manufacturing line for digital linear tape
2001	Intel: computer chips	For embedded 8-bit processors, Penang covers all value stages, including design of chip and motherboard Plans to locate two new design centers in Penang
2001	Motorola: software	Motorola's software centers in Penang and MSC receive ACI level-five certification (highest level of software certification)

29. Beginning in the late 1980s, Japanese flagships, like Matsushita, had relocated design and development activities to places like Malaysia and Taiwan, China (for example, Ernst 2000a; Ernst and Ravenhill 1999).

for optical components begins to grow. Penang's photonics cluster development is characterized by attempts to leverage multiple and diverse sources of knowledge and capital, both foreign and domestic, to create a critical mass for local clustering. For instance, the Penang Development Corporation has organized two working meetings with engineers and managers from Taiwan's Photonics Industry and Technology Development Association, with a twofold objective: to learn from Taiwan's experience and to develop joint projects. Equally important are linkages, developed by the Penang Photonics Consortium, with U.S. and Taiwanese (Chinese) venture capital firms. Finally, a concerted effort is under way, in cooperation with the Penang Skills Development Centre and several local universities, emphasizing photonics in their core curriculum.

Upgrading Linkages with OEMs

The fourth objective of the IMP2 is of greatest interest. Extending existing clusters beyond national borders originally was driven by two concerns: to ease the severe shortage of information technology skills by establishing joint growth triangles with neighboring countries that would attract low-cost engineers from throughout Asia. Yet competition among Asian economies for scarce information technology skills has drastically intensified, frustrating progress along these lines. It is time now to redefine the meaning of cross-border clusters and to ask how Malaysia's existing electronics clusters could reap greater benefits from participation in global production networks.

As part of the concept of integrated manufacturing centers, Malaysia was able to induce some OEMs to contribute to the development of specialized clusters. Such policies, which build on earlier successful policies in Singapore (Wong 2000) and Taiwan, China (Ernst 2000b), can play an important catalytic role and need to continue.

Yet, to a large degree, the outcome of these policies depends on sector-specific developments that are beyond the control of a mid-size country like Malaysia. Accumulated agglomeration economies matter, of course, in terms of human resources, infrastructure, and support industries (Best 2001; Ernst and others 2001). They also need to be continuously improved. Beyond that, the availability of incentives may tip the balance in favor of a particular location, but only if all the other conditions are in place. Otherwise, incentives will not have the desirable effect.

A brief comparison of the cases mentioned in table 3.10 illustrates this point. The decisions by Komag and Quantum reflect the relentless pressure within the hard disk drive industry to move volume manufacturing and support services to locations in close proximity to Singapore, the dominant global cluster for these activities (Ernst 1997a).

Footloose nature of foreign direct investment. An especially serious concern is that much of Malaysia's inward FDI remains highly "footloose" and prone to sudden decisions to relocate to lower-cost locations. Equally important is that global flagships forced to downsize to retain shareholder value in a recession are inclined to cut employment in export platform locations first, reflecting their flexible labor market regulations.[30] Table 3.11 provides examples of both cases.

Of particular interest is the recent decision by Dell to relocate its desktop production for the Japanese market from Penang to Xiamen, China

Table 3.11 Footloose Foreign Direct Investment in Malaysia, 1998–2001

Year	Company and product	Remarks
1998	Read-Rite: hard disk drives	Penang facility closed and relocated to the Philippines and Thailand; 4,000 jobs lost
2000	Seagate: hard disk drives	Shareholder-driven downsizing; facility in Ipoh closed; 2,000 jobs lost
2001	Seagate: hard disk drives	One plant in Prai closed; 4,000 jobs lost
2001 (April)	Motorola: software	10 percent of 4,000 employees laid off in plant in Sungei Way, Selangor; 400 jobs lost
2001	Intel: computer chips	Worldwide workforce reduced 5 percent; expected job loss in Malaysia: 500; massive expansion in China
2001	AMD: computer chips	1,300 jobs lost in Penang plant (52 percent of worldwide job cuts); massive expansion in China
2001 (August)	Dell: personal computers, servers, and storage products	Desktop production for Japan market relocated from Penang to Xiamen, China; Xiamen becomes exclusive supply base in China for Dell's complete product line; with exception of desktops, Penang remains hub for the rest of Asia-Pacific; main reason: limited flight connections between Malaysia and Japan
2001	Lucent: telecommunications equipment	Regional technology center closed in Malaysia; 150 jobs lost; workforce at manufacturing plant cut 50 percent

30. Thus far, contract manufacturers have been reporting only limited retrenchments. It remains to be seen to what degree the new limitations of contract manufacturer–based outsourcing may force contract manufacturers to lay off workers and close factories in Asia.

and to assign Xiamen to be the exclusive supply base in China for Dell's complete product line. Although Dell's two plants in Penang remain the build-to-order shipment hub for the rest of Asia-Pacific (with the exception of desktops), this constitutes a major blow for Malaysia. While immediate job losses are only 60 (out of a total of 2,000), more such redeployments may be in the offing. Dell gives three reasons for its redeployment to China: good and low-cost Chinese engineers, cheap land, and a limited number of flight connections between Japan and Malaysia. This indicates how unpredictable and fragile Malaysia's upgrading prospects are.

Upgrading opportunities: embedded software and RosettaNet. Linkages to OEMs also provide important opportunities for upgrading, including embedded software and RosettaNet.

For Malaysia, important upgrading opportunities reside in embedded software, a no-frills program used in a broad array of electronics systems that does only the specific task it is meant to perform. The program takes very specific inputs from its usage environment, processes these, and produces very specific outputs. Typical applications are car electronics, avionics, intelligent consumer products, communication and tracking devices, industrial automation, and medical equipment. In Malaysia, examples include the joint software development projects of Intel and Motorola.

These projects are in line with a general industry trend. There is a worldwide shortage of the specific skills required to develop embedded software. Embedded software requires very distinctive skill sets that are closer to hardware design than to mainstream software development. Essential prerequisites are experience in manufacturing and hardware design and state-of-the-art equipment and quality control. Places, like Penang, with accumulated experience in manufacturing and product design (even if it involves only product customization) apparently have some advantages over traditional locations for software outsourcing, such as in India, with less exposure to hardware design.

Another interesting example is the 5 million ringgit (RM) grant allocated in the 2002 budget to promote the adoption of RosettaNet e-business standards (interview with participants of Penang Skills Development Centre's seminar Jump Start Your e-Business with RosettaNet/XML, July 11, 2002). RosettaNet is a global consortium of more than 400 of the world's leading OEMs and contract manufacturers for electronic components, semiconductors, computers, and telecommunications equipment, working to create, implement, and promote open process standards for e-business. Malaysia is the fifth economy in Asia to join RosettaNet, after Japan, Korea, Singapore, and Taiwan (China).

Two tools are available to implement the RosettaNet initiative: incentives and participation in standard definition. Out of the RM 5 million grant, roughly 10 percent has been spent to set up the local operations of Rosetta-Net, with the Penang Skills Development Centre responsible for providing the backbone infrastructure. The remaining RM 4.5 million will be given out to eligible companies with no more than RM 100,000 per company. The grant will be administered by the Small and Medium Industry Development Corporation.

The idea is to involve major global network flagships that are already on the RosettaNet, such as Agilent, AMD, Cisco, Dell, Hitachi, Intel, Motorola, Quantum, Siemens, and Solectron. These flagships could then be used to pressure and cajole their local suppliers to upgrade their information technology infrastructure so that the local suppliers become eligible for the grants. Another criterion for receipt of a grant is financial strength, that is, the eligible company must finance out of its own funds another 50 percent of the project cost. It is an open question, however, whether smaller lower-tier suppliers can overcome the substantial constraints on their adoption of the RosettaNet standards.

Participation in defining the RosettaNet standards is probably the more immediately relevant tool. Six Malaysian electronics engineers, on loan to RosettaNet for two years, will work for six months at the California-based RosettaNet headquarters alongside American engineers to define XML-based specifications for the global electronics industry. The companies that provide these Malaysian engineers include global flagships (Intel, Microsoft), leading local suppliers (BCM Electronics, Globetronics Multimedia Technology), and two employees of the Malaysian Institute of Microelectronics Systems, a Web developer and a key public infrastructure developer. Obviously, these six Malaysian engineers will play an important role as multipliers and upgrading catalysts, once they return from their U.S. mission. They will also act as gatekeepers for these more knowledge-intensive linkages with global flagships.

Developing Multiple Linkages with Contract Manufacturers

To what degree can linkages with contract manufacturers broaden Malaysia's opportunities in information technology? Within Asia, two regions have experienced the greatest concentration of clusters of contract manufacturers: first Malaysia and Singapore (with a few additional sites in Thailand) and then, during the latter part of the 1990s, China (table 3.5). As for Malaysia and its neighboring countries, four important developments affect the opportunities for information technology: the arrival of major U.S. contract manufacturers, the acquisition of second-tier Asian contract manufacturers by major U.S. contract manufacturers, the mutation of component suppliers from Japan and Taiwan (China) into contract manufacturers, and the upgrading efforts of Malaysian higher-tier suppliers.

Arrival of major U.S. contract manufacturers. All the main U.S. contract manufacturers are now present in the Northern Penang–Kulim Hi-Tech cluster or in the Southern Johor–Singapore cluster. Solectron is present in Johor, Penang, and Singapore; Flextronics, in Johor and Singapore; Sanmina/SCI, in Penang and Singapore; Celestica, in Kedah's Kulim Hi-Tech Park; and Jabil Circuit, in Penang. There are also a few important investments elsewhere in the region, such as in Indonesia (Celestica), Malaysia's Kuching–Sarawak (Sanmina-SCI), and Thailand (Celestica, Flextronics, and Sanmina-SCI).

The arrival of leading global contract manufacturers thus far has created only limited upgrading opportunities for countries like Malaysia. Against initial expectations, a website search for the five leading global contract manufacturers, which was conducted in December 2001, did not provide evidence that operations in Asia had moved significantly beyond manufacturing. Compared to a few years ago, the main progress was an increasing sophistication in assembly technologies, especially multi-tier surface-mount technology, used for printed circuit board assembly. Most of these sites routinely provide support services related to manufacturing, with the exception of assets and logistics management. Typically, this also includes electrical and mechanical design services, global test services, printed circuit board layout services, and detailed process engineering (known as advanced manufacturing technology research). These services provide manufacturing solutions that enable a quick ramping up of volume manufacturing. But more recent interviews in late 2002, primarily with leading Taiwanese ODM companies, showed an increase in more knowledge-intensive activities, including electronic design, indicating a growing similarity in the distribution of capabilities across the Americas, Europe, and Asia (table 3.12).

A few locations, primarily in Penang and Singapore, are also involved in the introduction of new products. These two locations are witnessing the development of ODM capabilities, but on a very limited scale. Overwhelmingly, leading global contract manufacturers concentrate design (and especially circuit, advanced optical, and systems design) in Europe and the United States. One would, of course, expect such a disparity in design and product development, due to their high knowledge intensity. This, however, is changing, as Taiwanese (Chinese) contract manufacturers are providing such ODM services (Wu 2002). Industry observers expect that leading Taiwanese (Chinese) design firms will soon provide ODM services from their overseas network sites in China as well as in Penang and Singapore.

Table 3.12 Geographic Dispersion of Capabilities in Contract Manufacturing, 2001

Capability	Americas	Europe	Asia
Manufacturing and distribution			
Supply base and logistics management	•	•	•
Printed circuit board assembly	•	•	•
Complex systems assembly	•	•	•
Build-to-stock systems assembly	•	•	•
Build-to-order systems assembly	•	•	•
Configure-to-order systems assembly	•	•	•
Channel assembly	•	•	Penang (2002)
Systems integration and reconfiguration testing	•	•	a
Environment stress screening	•	•	a
Custom packaging	•	•	a
Logistics and distribution management	•	•	
Support services			
Repair systems and printed circuit boards	•	•	•
Product refurbishment and remanufacturing	•	•	•
Assets and logistics management	•	•	•
Product upgrades	•	•	•
Sustaining engineering	•	•	•
End-of-life manufacturing	•	•	•
Warranty processing	•	•	•
Technology			
Interconnection and packaging consulting	•	•	•
Process development	•	•	•
Reliability and failure analysis	•	•	•
Manufacturing technology roadmap	•	•	•
Test technology roadmap	•	•	•
Design			
Design of application-specific integrated circuits	•	•	b
Circuit design	•	•	b
Radio-frequency and wireless design	•	•	b
Mechanical design	•	•	b
Systems design	•	•	b
Test process design	•	•	b
Design validation	•	•	b
Product development			
New product introduction management	•	•	•
Component engineering	•	•	•
Design-for-manufacturability	•	•	•
Design-for-testability	•	•	•
Printed circuit board layout	•	•	•
Test development	•	•	•
Quick-turn prototyping	•	•	•
Quick-turn testing	•	•	•

a. Projected to change: focus on China.

b. Projected to change: Taiwanese ODM companies as trend-setters.

A comparison with locations in China, Hong Kong (China), and Taiwan (China) also demonstrates an important weakness of contract manufacturer locations in Southeast Asia. Very few final or system assembly activities are located in the Malaysia-Singapore clusters. Overwhelmingly, they are located in Europe and the United States, in close proximity to the traditionally dominant markets. But over the last few years, and especially in response to the recession in Europe and the United States, leading contract manufacturers have started to establish final assembly locations and build-to-order shipment hubs in China, Hong Kong (China), and Taiwan (China). The obvious motivation is to be as close as possible to the potentially huge Chinese market.

In short, the inflow of substantial investments from contract manufacturers thus far has produced some opportunities for Malaysia to move beyond its traditional focus on volume manufacturing. But these opportunities have not yet reached a level sufficient for a major push into more knowledge-intensive activities.

The acquisition of second-tier Asian contract manufacturers. A second important development in Malaysia's cluster is that leading U.S. contract manufacturers have recently rushed to acquire second-tier Asian contract manufacturers, primarily in Singapore, but also in Malaysia and elsewhere in the region. These acquisitions reflect the rapid concentration in the global contract manufacturer industry, which has been driven by M&A (see figure 3.2). The recession has further accelerated these trends toward concentration.

Important recent examples include Solectron, which acquired the Singaporean contract manufacturers Natsteel Electronics and Singapore Shinei Sangyo (the latter an affiliate of a Japanese component supplier); Flextronics, which acquired second-tier Singaporean contract manufacturer JIT Holdings and Li Xin Industries; and Celestica, which, through its acquisition of Singapore's contract manufacturer Omni (October 2001), acquired facilities in Indonesia, Malaysia, Singapore, and Thailand, with almost 9,000 employees.

To the degree that these acquisitions will result in plant closures and layoffs, they may constrain opportunities for industrial upgrading. However, they could have positive effects as well, if they insert new capital, customers, and management approaches.

Component suppliers from Japan and Taiwan (China). A third important development predates the arrival of American contract manufacturers, which has absorbed most public attention. Suppliers of parts and components from Japan and Taiwan (China), whose arrival in Malaysia goes back to the second part of the 1980s, catalyzed the development of Malaysia's local support industries (for example, Ernst 1997a; Takeuchi 1993). Japanese component manufacturers are concentrated primarily in the consumer electronics sector. Some of them, however, have also branched out into the computer sector. An interesting example is Kobe Precision (Malaysia), a company that, in October 2000, was acquired by one of the leading Malaysian contract manufacturers, Eng Technologi Holdings (*Business Times*, Kuala Lumpur, October 27, 2000).

Taiwanese (Chinese) firms have played an important role in Malaysia's computer industry. Prominent examples are Acer Peripherals and Iventech. Both companies became involved in Malaysia during the late 1980s. Over time, the Malaysian affiliates of these and other Taiwanese (Chinese) firms upgraded from simple volume manufacturing, according to designs owned by the global flagships, to sophisticated contract manufacturing for leading computer network flagships. In addition to manufacturing, the Malaysian sites of these Taiwanese (Chinese) affiliates now provide product and component design, supply chain management services, and other knowledge-intensive support services. A handful of large Taiwanese (Chinese) contract manufacturers, led by firms like Acer, Delta Networks, Iventech, Kinpo Electronics, and USI, have pioneered the use of ODM capabilities in Asia (report by Technology Forecasters, Alameda, Calif., quoted in *EMS Insight*, supplement on Circuits Assembly, September 2000). This apparently has forced major U.S.-based contract manufacturers to follow suit. For Malaysian firms that interact with affiliates of these Taiwanese (Chinese) contract manufacturers, the move to ODM capabilities may provide opportunities for industrial upgrading. This example illustrates that FDI policies and incentives should not target only the world industry leaders. Wherever possible, they should try to attract second-tier actors that are willing to bring along more knowledge-intensive activities.

Upgrading efforts of Malaysian higher-tier suppliers. There might be significant opportunities for higher-tier Asian suppliers to compete as low-cost niche contract manufacturers. Recent interviews in Penang (July 2002) indicate that leading higher-tier local suppliers all understand that they need to move up *within* the hierarchy of contract manufacturing arrangements, from low-end box-build and consignment arrangements to the provision of ODM and total solutions. However, they face major problems in sustaining and expanding their upgrading efforts. They all face the demanding challenge of pursuing *simultaneously* the following upgrading strategies, each of which requires major investments: establish a credible position as a low-cost niche contract manufacturer; develop a global presence, through overseas investment; achieve diversification and market segmentation; develop knowledge-intensive support services; and invest in design and R&D.

LC1, one of the most successful local companies, is building on existing strengths in contract manufacturing and the provision of ODM services in an effort to become a lower-cost "total solution provider" for carefully chosen niche markets. To do this with low overhead requires strong capabilities in six highly inter-dependent functions: manufacturing, quality, materials, procurement, engineering, and human resources. It is important to emphasize the systemic nature of the required capabilities.

The implementation of these upgrading options requires the development of a broad and diverse set of capabilities. Take manufacturing services. The move from printed circuit board assembly and box-build to testing necessitates the development not only of testing capabilities (which are scarce) but also of system engineering and maintenance capabilities. Furthermore, developing design and engineering capabilities requires substantial funds for R&D. Or take after-ship services. A seemingly mundane activity like repair requires technicians trained in failure analysis, while end-of-life program management requires capable supply chain managers. As for the upgrading of procurement and outbound logistics, substantial funds are required to gradually upgrade the necessary information systems.

In short, the ability to upgrade and become an Asian niche market contract manufacturer requires substantial investments in training, equipment, facilities, and, most important, R&D. LC1 identifies the following seven challenges that result from this strategy. Substantial improvements are required for supply chain management and efficient use of the company's assets. These are the most fundamental requirements for staying in this business. Yet their implementation requires substantial resources and attention from management. Challenges three to five constitute the medium-term challenge: the company needs to develop a strong portfolio of designs (so-called intellectual properties), it needs to capture new global niche market opportunities, and it needs to develop a global presence. Finally, the last two challenges highlight critical changes in industry organization, that is, the move toward flexible domestic supplier networks that can complement LC1's own capabilities and the overriding importance of human resource development as a constant process of acquiring new skills and knowledge.

A Shift in Strategy (II): International Knowledge Sourcing

Despite impressive achievements, Malaysia's knowledge base in the electronics industry remains too weak to sustain industrial upgrading into more knowledge-intensive activities. There is a heavy reliance on technological capabilities developed within affiliates of global flagships and their eventual spillover into local firms. This traditional pattern of transfer does not seem to work any longer.

Searching for new sources of productivity growth. In Penang, for instance, a disturbing slowdown in total factor productivity growth has been observed since 1995 (State Government of Penang 2001). Between 1995 and 1997, total factor productivity declined 0.5 percent for all manufacturing compared to an increase of 8.9 percent between 1990 and 1995. In the electronics industry, total factor productivity growth fell to 2 percent (from 14 percent during the earlier period), hardly sufficient for an industry that is supposed to be the engine of upgrading.

For all of Malaysia, most estimates put total factor productivity growth at about 1–2 percent annually (until 2000). This is way below the 3.2 percent minimum total factor productivity growth (for the period 2001–10) that is necessary if Malaysia wants to achieve the projected growth rate of 7.5 percent. Compared with historical patterns of productivity growth in industrial countries, this slowdown in productivity growth comes much too early. For Malaysia, such a massive slowdown in total factor productivity growth is certainly premature, in light of the limited progress made in its specialization by product and production stage.

In short, while FDI by major global electronics OEMs used to play a catalytic role in boosting Malaysia's productivity growth before 1995, it may no longer play that role. Except for China, and possibly India, OEMs are unlikely to increase their inward FDI in Asia. These changes are structural rather than cyclical, and they are here to stay. Hence, Malaysia needs to develop a set of alternative international linkages that could play a complementary role as external sources of productivity growth.

The multimedia super corridor. A widely known attempt to address this issue is the government's initiative to establish a $40 billion Multimedia Super Corridor that was supposed to leapfrog the country into the status of a fully developed nation by 2020 (Multimedia Development Corporation 2002). In 1996 the government hired McKinsey, the global consulting firm, to draft a blueprint for a 15-kilometer-by-50-kilometer strip intended to be Malaysia's answer to Silicon Valley. An unprecedented set of incentives, enshrined in the Bill of Guarantees, was offered to companies involved in the creation, distribution, integration, or application of multimedia products and services within the Multimedia Super Corridor.

According to the Multimedia Development Corporation (2002), these incentives include the following commitments:

- Provide a world-class physical and information infrastructure
- Allow unrestricted employment of local and foreign knowledge workers
- Ensure freedom of ownership by exempting companies with Multimedia Super Corridor status from local ownership requirements
- Give the freedom to source capital globally for Multimedia Super Corridor infrastructure and the right to borrow funds globally
- Provide competitive financial incentives, including pioneer status (100 percent tax exemption) for up to 10 years or an investment tax allowance for up to five years, and remove duties on the importation of multimedia equipment
- Become a regional leader in intellectual property protection and cyber laws
- Ensure no censorship on the Internet
- Provide globally competitive telecommunications tariffs
- Tender key infrastructure contracts to leading companies willing to use the Multimedia Super Corridor as their regional hub
- Provide a high-powered implementation agency to act as an effective one-stop super shop.

It would be hard to find a more aggressive list of incentives. By 2000, $3.7 billion had been spent, but the results were disappointing (however, there was some increase in investor interest in 2003). A leaked confidential report by the very same company that designed the project (McKinsey) concluded in February 2001 that the Multimedia Super Corridor "had not attracted much interest from global investors nor made an impact on the domestic economy" (Prystay 2001).[31] In the meantime, the debate has moved on to explore what one can learn from the experience to date (author's interviews with members of the National Information Technology Council, July 4, 2002, in Kuala Lumpur). Three conclusions emerge. First, initiating the Multimedia Super Corridor was a step in the right direction. However, it is now time to expand its geographic coverage and to extend Multimedia Super Corridor status to other electronics clusters in Malaysia. The Penang State government, in particular, has been lobbying for such an extension, arguing that this is necessary to attract specialized skills from overseas on a contract basis to overcome critical shortages.

31. Some 500 companies, including 45 international ones, are located in the corridor. But the McKinsey report describes the level of investment of those international electronics firms as "not very significant" (Prystay 2001).

Second, lavish tax incentives and massive investment in infrastructure are insufficient to bring about the development of dynamic clusters. Especially for the information technology sector, infrastructure is a highly perishable "public good": the infrastructure for the Multimedia Super Corridor was perhaps state-of-the-art when it was established a few years ago, but it became obsolete very rapidly. Third, perhaps the most important ingredient of successful cluster formation is missing: specialized skills and innovation capabilities. As emphasized throughout this chapter, the keys to success are incessant efforts on a massive scale to continuously upgrade existing skills and capabilities. The lack of depth and horizontal mobility in Malaysia's labor market increases the risk of individual investment in specialized skills. Therefore, the importing of scarce skills should be given greater emphasis. While this may seem obvious, this simple fact is frequently forgotten. In the case of Malaysia, the gap between the supply of and the demand for specialized information technology personnel has continuously increased, especially for engineers.[32] In short, as long as this critical human resources bottleneck is not overcome, there is little hope that the Multimedia Super Corridor will act as a "breeding ground for technological innovations, new businesses, and companies through the cooperation between the industrial and academic circles" (Takeuchi 1997, quoting from official background documents).

The following major priority areas for reducing the skills mismatch in the Malaysian electronics industry were identified during interviews with government agencies and leading companies (June-July 2002):[33] (1) a massive reskilling and retraining requirements of production workers, (2) graduates, especially for electrical and electronics engineering, information technology, and circuit design, who are able to combine hardware, software, and application knowledge, (3) experienced managers, especially for strategic marketing, upgrading management, and management of international linkages, (4) entrepreneurs who combine streetwise commercial and financial instincts with analytic capacity for strategic decisionmaking, (5) experienced and industry-savvy administrators who are willing to stick out their necks

32. On this indicator, Malaysia is continuously ranked at the bottom in the annual *World Competitiveness Yearbook* (Institute for Management Development various years), lagging substantially behind India, Korea, Singapore, and Taiwan (China).

33. In Kuala Lumpur, I am especially indebted to discussions with Dr. Zawai Ismail, director, Commerce Asset Ventures, who set up brainstorming sessions with relevant government agencies and venture capital firms. In Penang, I am especially indebted to discussions with Dr. Koh Tsu Koon, chief minister of Penang, Dr. Toh Kin Woon, Penang State executive councilor, Mr. Boonler Somchit, executive director of the Penang Skills Development Centre, Dr. Ganesh Rasagam, chief executive officer, DCT Consultancy Services, and Dr. Anna Ong, senior analyst, Socio-Economic and Environmental Research Institute.

and to do more than just follow the rules (this, of course, requires some incentive alignment), (6) incentive alignments for university professors and academics that encourage close interaction with the private sector (company internships and sabbaticals), (7) dense interactions with expatriate nationals who are based in Australia, Europe, the United States, or elsewhere in Asia, and (8) a capacity to bring in at short notice specialized experts from overseas who can bridge existing knowledge gaps and catalyze necessary changes in organization and procedures required to develop these capabilities locally.

Diversifying international linkages. In light of the transformations of global production networks documented in this chapter, Malaysia should exploit new opportunities for diversifying international linkages. Three complementary international linkages deserve particular attention: (1) links with foreign universities and research institutes; (2) links with information service and consulting firms, especially smaller, second-tier firms; and (3) links with informal global peer group networks that are playing an increasingly important role as carriers of knowledge.

Foreign universities and research institutes. Current transformations in the organization of global production networks, accelerated by the use of digital information systems, have substantially increased the mobility of knowledge. Digitalization implies that knowledge can be delivered as a service and built around open standards. This has fostered the specialization of knowledge creation, giving rise to a process of modularization, very much like earlier modularization processes in hardware manufacturing.

These developments may well create new opportunities for more aggressive forms of industrial upgrading that no longer need to avoid the creation of original knowledge. At this stage, this is largely uncharted territory, as these developments are very new and there is practically no research. Nevertheless, it is time for a country like Malaysia to strengthen linkages with overseas universities that can help to upgrade research, development, and design capabilities in Malaysian universities and public labs. The starting point is to correct the current policies. The focus thus far has been on a handful of global elite institutions that bring in their standard, routine information technology and business courses at very high cost. Instead, collaboration should focus on specific niche areas, in line with Malaysia's needs. Possible examples include certain areas of chip design, packaging technology, and photonics. Realistically, the search should move beyond the exclusive ranks of the Ivy League universities: there is ample choice of smaller, less well-known universities and research institutes that are more willing to develop innovative courses customized to the specific needs and capabilities of Malaysia's electronics clusters. These may include institutions in economies like India (the well-regarded Indian Institutes of Technology and the Indian Institutes of Information Technology), Korea (the Electronics and Telecommunications Research Institute, the Korea Advanced Institute of Science and Technology, the Information and Communications University of Korea, and other specialized research institutes), and Taiwan, China (the Industrial Technology Research Institute, the Electronics Research and Service Organization, and the Institute for Information Industry).

International consulting firms. Equally important is a reconsideration of linkages with consulting firms. For information technology, these firms now play a critical role in diffusing both codified and tacit knowledge. The problem is that this market is overwhelmingly dominated by a handful of giant corporations like IBM and consultancies like Accenture that grew out of global accounting firms. These firms thrive on the economies of scale in knowledge sharing (which information economists call network economies). As flagships of global information service networks, these firms provide a standard product wherever they go. Customization is possible only within the limits of a package of standard solutions. This approach to customization is extremely costly: customers are charged for the time required to adjust the package of standard information technology and to achieve effective implementation, and these costs are inflated by massive delays. The result is that new systems often come in late, over budget, and unable to solve problems they were meant to address. This has created a demand for smaller, specialized niche players that do not start from a package of standard solutions and that offer clients fixed-price projects. A wide choice of smaller, less well-known but proven information service and consulting firms is now available. This may include firms from some of the Asian economies that already have some experience with knowledge-intensive information services.

Informal peer group networks. Malaysia also needs to tap into an increasingly important carrier of international knowledge diffusion: transnational technical communities (Saxenian 2002) of technically skilled immigrants with business experience and connections in Europe, Japan, and the United States that play an important and complementary role to network flagships in global production networks. Such informal peer group global networks have created new opportunities for industrial upgrading in formerly peripheral economies around the world. By linking their home country with the world's centers of information and communication technology (encompassing Silicon Valley and other centers of excellence in less well-known places like Grenoble, Helsinki, Kista [Stockholm], Munich, Tel Aviv, and Tsukuba), these informal social networks transform what used to be a one-way brain drain into a two-way process of "brain circulation." These networks generate invaluable knowledge on trends in global markets and technology in a way that addresses the needs of domestic firms much better than linkages with global flagships or, for that matter, with global consulting firms. They also provide entrepreneurs and venture capitalists who can function in both worlds. This has created alternative and robust mechanisms for exchanging knowledge across geographic borders and firms. Examples include China, India, Israel, Korea, Taiwan (China), as well as Brazil and Mexico.

In Malaysia, the Penang cluster has obviously benefited from students who have studied engineering and management overseas, whether in Australia, Japan, the United Kingdom, or the United States, and who have returned with business experience and connections. Predominantly, these connections have been with global flagships like Intel and Motorola in semiconductors or with Matsushita and other Japanese flagships in consumer electronics. Overwhelmingly, the technology, skills, and knowledge generated by these immigrant engineers have focused on manufacturing-related activities. It is time for Malaysia to adjust this brain circulation to encompass new areas like knowledge-intensive support services, circuit design, and chip packaging.

CONCLUSIONS AND POLICY SUGGESTIONS

Based on operational definitions of the key characteristics of global production networks and industrial upgrading, this chapter has explored how East Asia, and especially Malaysia, could build on recent transformations in the structure, coordination, content, and location of global production networks to promote a continuous upgrading of the electronics industry. We have shown that the reintegration of geographically dispersed, specialized production and innovation sites into multilayered global production networks and the increasing use of information technology–based information systems to manage these networks are *gradually* reducing the constraints on the international diffusion of knowledge. Global production networks expand inter-firm linkages across national boundaries, increasing the need for knowledge diffusion, while information systems enhance not only the exchange of information but also the sharing and joint creation of knowledge.

This new mobility of knowledge has created greater opportunities for using network participation as a catalyst for upgrading Asia's electronics industries further. But realizing this potential has become more difficult for mid-size countries like Malaysia. While Malaysia has developed some of the most ambitious sets of policies and incentives, the country has a long way to go to exploit these new opportunities. The best choice arguably is to move forward in incremental steps and to build on existing strengths in assembly and volume manufacturing by adding knowledge-intensive support services.

While public policy documents frequently talk about "catching up" and "forging ahead" through technology leadership strategies (ISIS Malaysia 2002), the de facto strategies appear to focus on technology diversification, that is, the recombination of (mostly) known technologies. Defined as "the expansion of a company's or a product's technology base into a broader range of technology areas" (Granstrand 1998, p. 472), such strategies focus on products and services that draw "on several . . . crucial technologies which do not have to be new to the world or difficult to acquire" (Granstrand and Sjölander 1990, p. 37). The transformations in global production networks that this chapter has explored in detail may have created opportunities for Asia's leading electronics-exporting countries to engage in technology diversification that did not exist before. Asian electronics firms may also have important latecomer advantages, building on their accumulated capabilities to implement, assimilate, and improve foreign technologies, as technology diversification often involves the exchange of knowledge with foreign parties (Ernst 2003c).

Of critical importance is the absorptive capacity of local suppliers, that is, their resources, capabilities, and motivations. The absorptive capacity is shaped by pressures exerted by network flagships and by existing incentives. This chapter documents that, to stay on the global production networks, local suppliers must constantly upgrade their absorptive capacity by investing in their skills and knowledge base.

Equally important are attempts to strengthen the country's innovative capabilities through selective sourcing of international knowledge. Transformations in global production networks are gradually reducing the barriers to effective knowledge diffusion. As an immediate policy instrument, it is advisable to import critical skills from overseas. This could help to catalyze necessary reforms in the domestic innovation system. The timing may be good, as massive retrenchments in the U.S. and European electronics industries and a more hostile attitude toward foreign researchers, especially from the Middle East but also from Asia, may induce foreign researchers to work in Malaysia.

Adequate incentives are required to generate sufficient investments in the development of skills and capabilities, as illustrated, for instance, by the Nordic countries in Europe and by Singapore and Taiwan (China). Policies toward both OEMs and contract manufacturers need to move beyond "incentive tournaments." Infrastructure development is critical but needs to move beyond a widespread bias toward hardware. As illustrated in this chapter, successful upgrading within global production networks requires policies that support local firms through the development of local suppliers, (cofunded) development of skills, setting of standards, and provision of investment and innovation finance through a variety of sources, including venture capital and initial public offerings. Of particular importance for East Asia are new opportunities to tap into international flows of human capital and knowledge through informal peer group networks of technically skilled immigrants with business experience and connections in Europe, Japan, and the United States. These international social networks can play an important and complementary role as carriers of knowledge and capital to OEMs, contract manufacturers, and global consulting firms.

To reap the benefits of integration into global production networks requires a very active involvement of the state (that is, local, regional, and central government agencies as well as a variety of intermediate institutions). But this involvement is taking on a very different form from earlier top-down, "command economy" industrial policies. It also differs from the new-economy liberalization doctrine.

Traditional Asian developmental policies are no longer feasible. With their top-down approach, controlled investment finance, and reliance on state-owned enterprises or *chaebol*-type conglomerates, these policies are too rigid to cope with the complex challenges and opportunities of the global network economy that have been explored in this chapter. These policies cannot cope with the conflicting needs of multiple and increasingly vocal domestic actors. In addition, traditional developmental policies are unable to cope with the high uncertainty and rapid changes in technology and markets that are typical of the electronics industry. Finally, in light of their protectionist focus, these policies are unlikely to generate and benefit from international knowledge linkages that are of critical importance for upgrading Asian electronics industries.

Neither can the new-economy liberalization doctrine cope with the new opportunities and challenges for Asian electronics industries that are examined in this chapter. This doctrine claims that, except for education, infrastructure, and a few general incentives (for training and R&D), the state should get out of the way and let transnational technical and venture capital communities make the necessary investments in innovative capabilities (for example, Bresnahan, Gambardella, and Saxenian 2001). While such a selective approach is an important condition for breaking up the stalemate of the traditional Asian developmental policies, it is insufficient to cope with the new opportunities and challenges raised by the transformation of global production networks. It is ironic that the U.S. government does not do what the proponents of the new-economy doctrine claim it does. Driven by national security concerns and by fears of losing its technological leadership, the federal government spends $2 billion a year on information technology research, coordinated through a variety of agencies including the Defense Advanced Research Projects Agency, which has played a prominent role in the development of the U.S. computer and Internet industry (for example, Flamm 1988).

In short, for Malaysia's upgrading efforts in the electronics industry to succeed, new policy approaches are required (Ernst forthcoming c):

- Strengthen the state's steering and coordination capacity
- Provide public goods and create assets (infrastructure, bottleneck skills, training, and education)
- Facilitate access to and diffusion of knowledge; balance this with the need to protect intellectual property rights
- Encourage innovations in the financial sector
- Generate dialogue at various levels among multiple participants (local and foreign) in production and innovation networks
- Foster interactive learning and innovation
- Provide social protection and retraining options for the losers from innovation
- Facilitate international knowledge sourcing through corporate networks, institutional collaboration, and diverse social networks (global knowledge communities and expatriates).

REFERENCES

The word *processed* describes informally reproduced works that may not be commonly available through libraries.

Aberdeen Group. 2001. "Offshore Software Development: Localization, Globalization, and Best Practices in an Evolving Industry." Boston, Mass. Processed.

Almeida, Paul, and Bruce Kogut. 1997. "The Exploration of Technological Diversity and the Geographic Localization of Innovation." *Small Business Economics* 9(1):21–31.

Ariffin, Norlela. 2000. "The Internationalisation of Innovative Capabilities: The Malaysian Electronics Industry." Ph.D. diss., University of Sussex, Science and Technology Policy Research Unit.

Best, Michael H. 2001. *The New Competitive Advantage: The Renewal of American Industry.* Oxford: Oxford University Press.

Borrus, Michael, Dieter Ernst, and Stephan Haggard, eds. 2000. *International Production Networks in Asia: Rivalry or Riches?* London and New York: Routledge.

Boswell, David. 1993. *Subcontracting Electronics: A Management and Technical Guide for Purchasers and Suppliers.* London: McGraw-Hill.

Bresnahan, Timothy, Alfonso Gambardella, and Anna-Lee Saxenian. 2001. " 'Old Economy' Inputs for 'New Economy' Outcomes: Cluster Formation in the New Silicon Valleys." *Industrial and Corporate Change* 10(4):835–60.

Chen, Shin-Horng. 2002. "Global Production Networks and Information Technology: The Case of Taiwan (China)." *Industry and Innovation* 9(2). Special issue on global production networks.

DCT Consultancy Services. 2002. "Annual Survey of the Manufacturing Industries in Penang Development Corporation Industrial Areas." Report submitted to the Penang Development Corporation, Penang, Malaysia. Processed.

Depeyrot, Michel. 2002. "Taiwan Foundries Acquiring Design Capabilities." *Electronic Engineering Times*, September 25.

Ernst, Dieter. 1994a. "Network Transactions, Market Structure, and Technological Diffusion: Implications for South-South Cooperation." In Lynn K. Mytelka, ed., *South-South Cooperation in a Global Perspective.* Development Centre Documents. Paris: Organisation for Economic Co-operation and Development.

———. 1994b. *What Are the Limits to the Korean Model? The Korean Electronics Industry under Pressure.* BRIE Research Monograph. University of California, Berkeley Roundtable on the International Economy.

———. 1997a. "From Partial to Systemic Globalization: International Production Networks in the Electronics Industry." Data Storage Industry Globalization Project Report 97-02. University of California at San Diego, Graduate School of International Relations and Pacific Studies for the Sloan Foundation. Processed.

———. 1997b. "Partners in the China Circle? The Asian Production Networks of Japanese Electronics Firms." In Barry Naughton, ed., *The China Circle.* Washington, D.C.: Brookings Institution.

———. 1998. "Catching Up: Crisis and Industrial Upgrading, Evolutionary Aspects of Technological Learning in Korea's Electronics Industry." *Asia Pacific Journal of Management* 15(2):247–83.

———. 2000a. "Evolutionary Aspects: The Asian Production Networks of Japanese Electronics Firms." In Michael Borrus, Dieter Ernst, and Stephan Haggard, eds., *International Production Networks in Asia: Rivalry or Riches?* London: Routledge.

———. 2000b. "Inter-Organizational Knowledge Outsourcing. What Permits Small Taiwan Firms to Compete in the Computer Industry?" *Asia Pacific Journal of Management* 17(2, August):223–55. Special issue on knowledge management in Asia.

———. 2000c. "Placing the Networks on the Internet: Challenges and Opportunities for Managing in Developing Asia." Paper presented at the Second Asia Academy of Management Conference, Singapore, December 15–18. Processed. Also forthcoming in Bengt-Âke Lundvall and Keith Smith, eds., *Knowledge Creation in the Learning Economy.* Cheltenham: Edward Elgar.

———. 2001a. "Catching up and Post-Crisis Industrial Upgrading: Searching for New Sources of Growth in Korea's Electronics Industry." In Frederick C. Deyo, Richard Doner, and Eric Hershberg, eds., *Economic Governance and the Challenge of Flexibility in East Asia.* Lanham, Md.: Rowman and Littlefield.

———. 2001b. "Moving beyond the Commodity Trap? Trade Adjustment and Industrial Upgrading in East Asia's Electronics Industry." In Richard S. Newfarmer and Christina

Wood, eds., *East Asia: From Recovery to Sustainable Development.* Washington, D.C.: World Bank.

———. 2002a. "The Economics of Electronics Industry: Competitive Dynamics and Industrial Organization." In William Lazonick, ed., *The International Encyclopedia of Business and Management (IEBM), Handbook of Economics.* London: International Thomson Business Press.

———. 2002b. "Global Production Networks and the Changing Geography of Innovation Systems: Implications for Developing Countries." *Journal of the Economics of Innovation and New Technologies* 11(6):497–523.

———. 2003a. "Digital Information Systems and Global Flagship Networks: How Mobile Is Knowledge in the Global Network Economy?" In Jens F. Christensen and Peter Maskell, eds., *The Industrial Dynamics of the New Digital Economy.* Cheltenham, U.K.: Edward Elgar.

———. 2003b. "The New Mobility of Knowledge: Digital Information Systems and Global Flagship Networks." In Robert Latham and Saskia Sassen, eds., *Digital Formations in a Connected World.* Princeton, N.J.: Princeton University Press for the U.S. Social Science Research Council.

———. 2003c. "Pathways to Innovation in Asia's Leading Electronics Exporting Countries: Drivers and Policy Implications." Paper prepared for the conference the Common Future of the Twenty-first Century Pacific, Democratic Pacific Assembly, Taipei, Taiwan, September 18–20. Available as East-West Center Economics Working Paper 62. East-West Center, Honolulu, October. Processed. Forthcoming in *International Journal of Innovation and Technology Management (IJITM).*

———. Forthcoming a. "Global Production Networks and Industrial Upgrading: A Knowledge-Centered Approach." In Gary Gereffi and Eric Hershberg, eds., *Who Gets Ahead in the Global Economy? Industrial Upgrading, Theory and Practice.* Baltimore, Md.: Johns Hopkins University Press.

———. Forthcoming b. "Internationalisation of Innovation: Why Is Chip Design Moving to Asia?" *International Journal of Innovation Management,* special issue in honour of Keith Pavitt (Peter Augsdoerfer, Jonathan Sapsed and James Utterback, guest editors). Also published as East-West Center Economics Working Paper 62.

———. Forthcoming c. "Pathways to Innovation in the Global Network Economy: Asian Perspectives." *Oxford Development Studies.* Special issue in honor of Linsu Kim.

Ernst, Dieter, Tom Ganiatsos, and Lynn Mytelka, eds. 1998. *Technological Capabilities and Export Success: Lessons from East Asia.* London: Routledge.

Ernst, Dieter, and Paolo Guerrieri. 1998. "International Production Networks and Changing Trade Patterns in East Asia. The Case of the Electronics Industry." *Oxford Development Studies* 26(2): 191–212.

Ernst, Dieter, Paolo Guerrieri, Simona Iammarino, and Carlo Pietrobelli. 2001. "New Challenges for Industrial Districts: Global Production Networks and Knowledge Diffusion." In Paolo Guerrieri, Simona Iammarino, and Carlo Pietrobelli, eds., *The Global Challenge to Industrial Districts: Small and Medium-Sized Enterprises in Italy and Taiwan.* Aldershot, U.K.: Edward Elgar.

Ernst, Dieter, and Linsu Kim. 2002. "Global Production Networks, Knowledge Diffusion, and Local Capability Formation." In *Research Policy* 31(8–9):1417–29.

Ernst, Dieter, and John Ravenhill. 1999. "Globalization, Convergence, and the Transformation of International Production Networks in Electronics in East Asia." *Business and Politics* 1(1, April): 53–62.

Feenstra, Robert C. 1998. "Integration of Trade and Disintegration of Production in the Global Economy." *Journal of Economic Perspectives* 12(4):31–50.

Flamm, Kenneth. 1988. *Creating the Computer: Government, Industry, and High Technology.* Washington, D.C.: Brookings Institution.

Globerman, Stephen. 1997. "Transnational Corporations and International Technological Specialization." *Transnational Corporations* 6(2, August): 95–114.

Granstrand, Ove. 1998. "Toward a Theory of the Technology-Based Firm." *Research Policy* 27(5, September):465–89.

Granstrand, Ove, and Sören Sjölander. 1990. "Managing Innovation in Multi-Technology Corporations." *Research Policy* 19(1, February):35–60.

Grossman, Gene M., and Elhanan Helpman. 1991. *Innovation and Growth in the Global Economy.* Cambridge, Mass.: MIT Press.

———. 2002. "Integration versus Outsourcing in Industry Equilibrium." *Quarterly Journal of Economics* 117(1, February):85–120.

Hamid, Hamisah. 2001. "Lay-Offs Unlikely to Be as Severe as Three Years Ago." *Business Times*, October 3.

Hirschman, Albert O. 1958. *Strategy of Economic Development.* New Haven, Conn.: Yale University Press.

Hobday, Michael. 1995. *Innovation in East Asia: The Challenge to Japan.* Aldershot, U.K.: Edward Elgar.

Institute for Management Development. Various years. *World Competitiveness Yearbook.* Lausanne, Switzerland.

ISIS (Institute of Strategic and International Studies) Malaysia. 2002. *Knowledge-Based Economy Master Plan.* Kuala Lumpur, September.

Jabil Circuit. 2001. *Jabil 10K Report to the United States Securities and Exchange Commission.* New York, September 30.

Jones, Ronald, and Henryk Kierzkowski. 2000. "A Framework for Fragmentation." In Sven Arndt and Henryk Kierzkowski, eds., *Fragmentation and International Trade.* Oxford: Oxford University Press.

Lakenan, Bill, Daren Boyd, and Ed Frey. 2001. "Why Cisco Fell: Outsourcing and Its Perils." *Strategy + Business* 24(3rd quarter):54–65.

Lall, Sanjaya. 1997. "Technological Change and Industrialization in the Asian NIEs: Achievements and Challenges." Paper presented at the international symposium on "Innovation and Competitiveness in Newly Industrializing Economies," Science and Technology Policy Institute, Seoul, Korea, May 26–27. Processed.

Lazonick, William. 2000. "Understanding Innovative Enterprise: Toward the Integration of Economic Theory and Business History." Unpublished mss., University of Massachusetts, Lowell and European Institute of Business Administration (INSEAD), Fontainebleau, May. Processed.

Lipsey, Richard G. 2001. "Understanding Technological Change." East-West Center Working Paper, Economics Series 13. East-West Center, Honolulu, February. Processed.

Macher, Jeffrey T., David C. Mowery, and Timothy S. Simcoe. 2002. "eBusiness and the Semiconductor Industry Value Chain: Implications for Vertical Specialization and Integrated Semiconductor Manufacturers." In *Industry and Innovation* 9(3, December):155–81. Special issue on global production networks.

MIDA (Malaysian Industrial Development Authority). 1999. *Malaysia's Manufacturing Sector: Into an Era of High Technology.* Kuala Lumpur, Malaysia.

Ministry of Information and Communication. 2002. *IT Korea 2002*. Seoul, Korea.

Ministry of International Trade and Industry. 1986. *Industrial Master Plan 1986–95*. Kuala Lumpur, Malaysia.

———. 1996. *Second Industrial Master Plan*. Kuala Lumpur, Malaysia.

Multimedia Development Corporation. 2002. "Unlocking the Full Potential of the Information Age." Available at www.mdc.com.my/package/. Processed.

Navaretti, Giorgio B., Jan I. Haaland, and Anthony Venables. 2002. *Multinational Corporations and Global Production Networks: The Implications for Trade Policy*. Brussels: European Commission, Directorate General for Trade, March.

Nelson, Richard R. 1990. "Acquiring Technological Capabilities." In Hadi Soesastro and Mari Pangestu, eds., *Technological Change in the Asia-Pacific Economy*. Sydney: Allen and Unwin.

Nelson, Richard R., and Sidney G. Winter. 1982. *An Evolutionary Theory of Economic Change*. Cambridge, Mass.: Bellknap Press.

OECD (Organisation for Economic Co-operation and Development). 1999. *Boosting Innovation: The Cluster Approach*. Paris.

Ozawa, Terutomo. 2000. "The 'Flying-Geese' Paradigm: Toward a Co-evolutionary Theory of MNC-Assisted Growth." In Khosrow Fatemi, ed., *The New World Order: Internationalism, Regionalism, and the Multinational Corporations*. New York: Pergamon.

Pavitt, Keith. 2003. "What Are Advances in Knowledge Doing to the Large Industrial Firm in the 'New Economy'?" In Jens F. Christensen and Peter Maskell, eds., *The Industrial Dynamics of the New Digital Economy*. Cheltenham: Edward Elgar.

Penrose, Edith T. 1995. *The Theory of the Growth of the Firm*. Oxford: Oxford University Press.

Porter, Michael. 1990. *The Competitive Advantage of Nations*. London: Macmillan.

Prystay, Chris. 2001. "Malaysia Seeks to Lift Super Corridor." *Asian Wall Street Journal*, August 24.

Rasiah, Rajah. 1995. *Foreign Capital and Industrialization in Malaysia*. London: St. Martin's Press.

———. 2003. "Export Experience and Technological Capabilities: Evidence from East and Southeast Asian Firms." Unpublished ms. United Nations University, Institute for New Technologies, Maastricht, The Netherlands. Processed.

Richardson, G. B. 1990. *Information and Investment: A Study on the Working of the Competitive Economy*. Oxford: Oxford University Press.

Rodrik, Dani. 1999. *The New Global Economy and Developing Countries: Making Openness Work*. ODC Policy Essay 24. Baltimore, Md.: Johns Hopkins University Press for the Overseas Development Council.

Romer, Paul M. 1990. "Endogenous Technological Change." *Journal of Political Economy* 98(5, pt. 2):S71–S102.

Rugman, Alan M. 1997. "Canada." In John H. Dunning, ed., *Governments, Globalization, and International Business*, ch. 6. London: Oxford University Press.

Rugman, Alan M., and Joseph R. D'Cruz. 2000. *Multinationals as Flagship Firms: Regional Business Networks*. New York: Oxford University Press.

Saxenian, Anna-Lee. 2002. "The Silicon Valley Connection: Transnational Networks and Regional Development in Taiwan, China, and India." In *Industry and Innovation* 9(2, August). Special issue on global production networks, information technology, and local capabilities.

Simon, Denis. 2001. "The Microelectronics Industry Crosses a Critical Threshold." *China Business Review* (November–December):42.

State Government of Penang. 2001. *The Second Penang Strategic Development Plan, 2001–2010.* Penang, Malaysia.

Sturgeon, Timothy. 2002. "Modular Production Networks: A New American Model of Industrial Organization." *Industrial and Corporate Change* 11(3):451–96.

Takeuchi, Junko. 1993. "Foreign Direct Investment in ASEAN by Small- and Medium-Sized Japanese Companies and Its Effects on Local Supporting Industries." *RIM Pacific Business and Industries* 4(22).

———. 1997. "The New Industrialization Strategy of Malaysia as Envisioned in the Second Industrial Master Plan." *RIM Pacific Business and Industries* 3(37).

———. 2001. "Comparison of Asian Business by Japanese and American Companies in the Electronics Sector." *RIM Pacific Business and Industries* 1(3).

Toh, Kin Woon. 2002. "The Political Economy of Industrialization in Penang." Unpublished ms. Processed.

Too, T., and T. P. Leng. 2002. "Unemployment Situation in Penang." *Economic Briefing to the Penang State Government* 4(4).

United Microelectronics. 2001. "Form 20-F Report." Submitted to the U.S. Securities and Exchange Commission, December 31. Processed.

Wong, Poh Kam. 2000. "Riding the Waves: Technological Change, Competing U.S.-Japanese Production Networks, and the Growth of Singapore's Electronics Industry." In Michael Borrus, Dieter Ernst, and Stephan Haggard, eds., *International Production Networks in Asia: Rivalry or Riches?* London: Routledge.

Wu, Nancy. 2002. "Design Trends and the EDA Market in Taiwan: A White Paper." Report prepared for the Electronic Engineering Times, Design and Engineering Group, Gartner Dataquest, February. Processed.

Zainal Abidin, Mahani. 2002. "Issues on Malaysia's Competitiveness." Slide presentation at the seminar on Asian competitiveness, Penang State Government, September 30. Processed.

CHAPTER 4

PRODUCTION NETWORKS IN EAST ASIA'S AUTOMOBILE PARTS INDUSTRY

Richard F. Doner, Gregory W. Noble, and John Ravenhill

East Asia, with the exception of Japan, is a far less significant player in the auto parts industry than in other manufacturing sectors, most notably electronics. Nonetheless, the rapid evolution of Asia's auto industry has created opportunities for parts producers to expand production both for local assembly and for export. Recent investments in the region by foreign assemblers aim to produce components and vehicles for export throughout regional and global networks as well as for domestic sale. Indeed, assemblers face an imperative to export because of significant local overcapacity created by excessive investment driven by protectionist barriers and, more recently, by excessively optimistic expectations of rapid growth in the region's auto markets. Intense cost pressures and opportunities for profits in nonmanufacturing segments of the auto industry are causing assemblers worldwide to outsource an increasing proportion of parts and components. They are seeking to increase local parts production to avoid foreign exchange fluctuations, meet short delivery times, adapt to local demand, and make use of cheaper local inputs. In sum, domestic overcapacity, combined with the efforts of assemblers to balance the benefits of localization with the need for centralization and economies of scale, has created significant opportunities for the export of locally produced automotive components.

Translating these opportunities into efficient businesses is not always smooth, however, because barriers to entry for locally owned parts producers are rising significantly. Although trade protection in East Asia economies in attenuated and uneven form, reliance on profitable, protected domestic markets is less of an option for domestic producers than in the past. Governments are reducing tariff levels and abandoning trade-related investment measures (TRIMs) under pressure from requirements of the World Trade Organization (WTO) and, for Southeast Asian states, from obligations of the Association of South East Asian Nations (ASEAN) Free Trade Area. Faced with new competitive challenges, assemblers are demanding that parts suppliers match the levels of quality control, productivity, and cost-effectiveness prevailing in industrial economies. At the same time, they are demanding that first-tier suppliers take responsibility for the design as well as the manufacture of modules and not just individual components. Locally owned firms that lack financial, technological, and managerial resources will not survive these demands; indeed, many have already begun to disappear or to merge with newly arrived foreign component producers. The role of foreign capital in the region's production of auto parts has grown and will probably continue to do so. Although the new environment offers significant opportunities for local production, policymakers will have to meet the challenge of dealing with an industry that is increasingly denationalized.

The central message of this chapter is that the location-specific assets that the economies examined here—China, the Republic of Korea (hereafter Korea), Malaysia, Taiwan (China), and Thailand—can offer to foreign automakers and to first-tier global suppliers vary significantly, and thus so do their strategic options. Small domestic markets, weak indigenous technical capacities, and modest automotive-related human resources mean that Malaysia, Taiwan (China), and Thailand can only optimize within the regional and global frameworks created by the strategies of global assemblers. However, even within such "constrained optimization," these three countries have potentially attractive options in product niches in both original equipment manufacture (OEM) and replacement equipment manufacture (REM), particularly in the development of regional markets. Larger domestic markets in China and Korea—and, in the Korean case, established local assemblers and significant experience with auto parts exports—provide these two countries with greater opportunities to mold the strategies of global assemblers to national priorities. This might translate into more assembler acceptance of de facto protection and more willingness to transfer technology and, therefore, not only more extensive local production of parts but also greater opportunities for indigenous producers, possibly in joint ventures with the dominant multinational first-tier suppliers.

But potential resources do not translate automatically into leverage. The capacity to convert strategic options into reality depends on policies, institutions, and politics. Effective policies in areas such as technical training, infrastructure, marketing, trade policy, finance, and ownership can greatly magnify the attractiveness of location-specific assets and thus a country's leverage vis-à-vis multinational assemblers and parts producers. Public sector expertise and cohesion, public-private sector linkages, and coordination within the private sector are all important influences on a country's capacity to formulate and implement effective policy. And politics, especially the distributional consequences of automotive policies, influences both the strength of relevant institutions and the effectiveness of implementation.

This chapter develops these themes as follows. First, it provides a brief overview of the region's auto industry, highlights its weakness through a brief comparison with the region's booming electronics industry, and identifies the auto industry "drivers" that account for the differences. Then it identifies the new drivers that are creating opportunities and challenges for regional auto parts production and presents concise overviews of the industry in China, Korea, Malaysia, Taiwan (China), and Thailand. A final section identifies the location-specific assets of these five economies, their strategic options, and the policies, institutions, and politics influencing their ability to pursue these options.

Finally, we offer a caution: East Asian automotive industries are in an active process of transformation. Whether it is Japanese and U.S. firms shifting all Asian-based production of pickup trucks to Thailand, substantial new investments by foreign companies in China, Malaysia's transfer of management control of Perodua to Daihatsu, General Motors' purchase of Daewoo, or the resurgence of Samsung Motor under Renault's control, the geographic and corporate contexts of East Asian parts production are becoming both more extensive and more interdependent than anyone would have predicted a decade ago. The objective of this chapter is not to make predictions but rather to identify potential areas of growth and the factors determining whether such potential will be realized.

AUTOMOTIVE PARTS PRODUCTION IN EAST ASIA

In contrast to their stellar performance in the electronics sectors, East Asian economies remain relatively small players in the global auto parts industry. In this section of the chapter, we first review the record of East Asian economies in auto parts production and export and then contrast it with the integration of East Asian economies into global production networks in electronics.

(Uneven) Weaknesses

With the exception of Japan, East Asia remains a relatively small player in the global production of automotive parts. Exports in 1999 from the five economies in this study accounted for only 4.3 percent of total world exports of auto parts (calculated from the U.N. Commodity Trade Statistics Database, known as U.N. Comtrade).[1] This modest global position reflects the relatively small role of parts production within each national economy and, with some exceptions, the lack of international competitiveness of this industrial sector in East Asia. In none of the economies do auto parts account for more than 3 percent of total export earnings, a figure lower than the share of this commodity in overall global exports (figure 4.1), even though the value of parts exports has risen over the past decade (figure 4.2). Only Korea exports significant volumes of automotive products to Canada and the United States, the most open and competitive of the large markets, but Korea's market share in the two countries (1 and 3 percent, respectively) is dwarfed by that of Mexico (6 and 17 percent). Korean exports to Japan are inconsequential, and while Taiwan (China) and Thailand both export to Japan, each accounting for 2 percent of Japanese automotive imports, their individual shares of the Japanese market are less than that of Mexico (3 percent; WTO 2002, p. 63).

Further evidence of the region's weaknesses in auto parts production is found in generally negative automotive trade balances (table 4.1), the continued high levels of protection that governments perceive are required by the industry (table 4.2), and the low-value-added character of exported auto parts. Exports are dominated by natural resource–based products, such as

Figure 4.1 Share of Auto Parts in Total Exports, by Economy, 1999

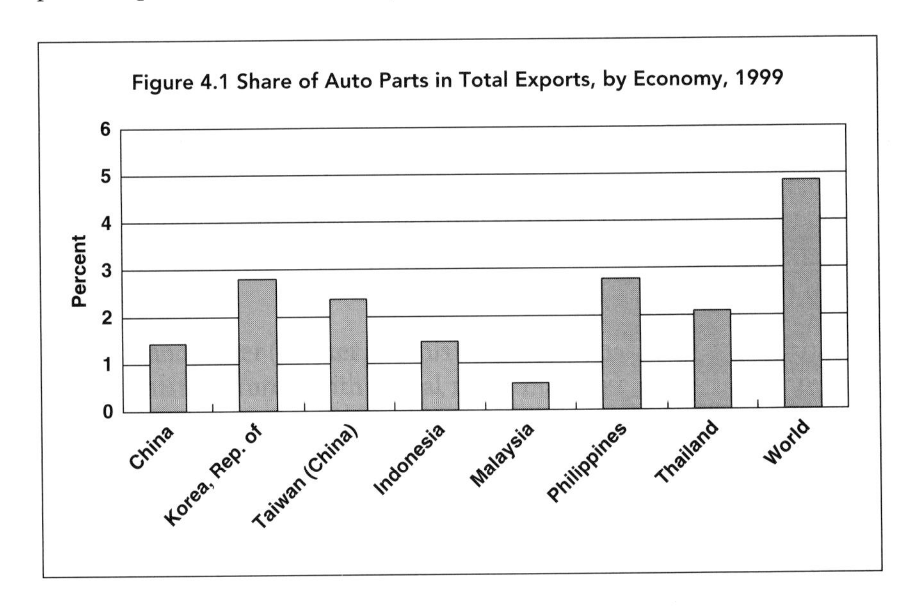

1. Available from the United Nations Statistics Division at http://unstats.un.org/unsd/comtrade/.

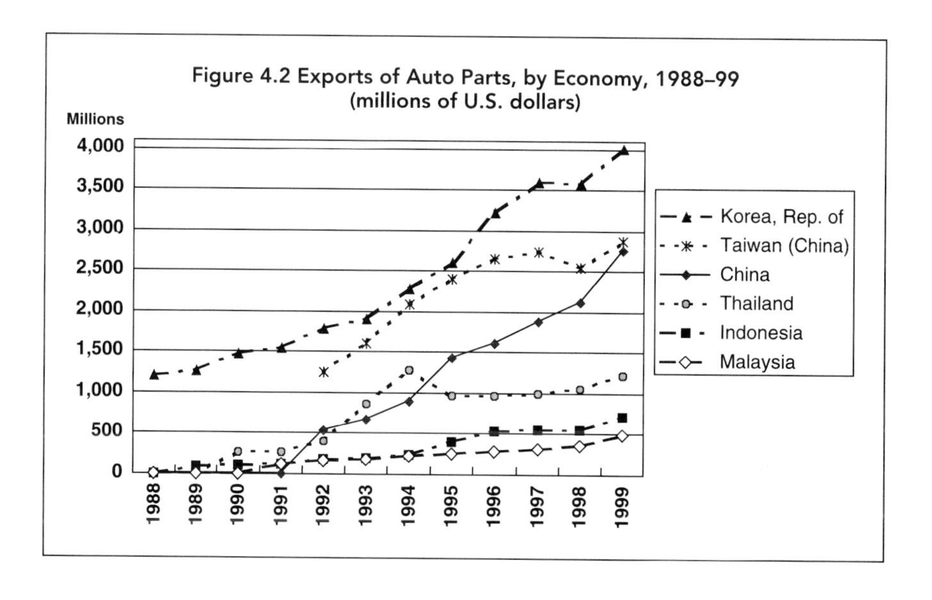

Figure 4.2 Exports of Auto Parts, by Economy, 1988–99 (millions of U.S. dollars)

tires, and labor-intensive goods often produced by subsidiaries of Japanese or Western component firms, such as batteries and wire harnesses (table 4.3). There are, to be sure, some notable bright spots in this picture. Korea and Taiwan (China) both run trade surpluses in auto parts. Korea benefits both from a sizable domestic market as well as from a significant replacement market based on Korean-made vehicles sold abroad and has developed substantial expertise in some components, such as climate control systems. Taiwan (China) has begun to exploit the replacement market, with its products now accounting for 80 percent of global production of aftermarket sheet-metal body parts (Liang 2002). Overall, however, the industry continues to be characterized by low levels of technology and low-skill production.

Table 4.1 Balance of Trade in Auto Parts, by Economy, 1999 (thousands of U.S. dollars)

Economy	Exports	Imports	Trade balance
China	2,757,721	2,885,638	–127,917
Indonesia	710,734	721,305	–10,571
Korea, Rep. of	4,008,995	2,160,772	1,848,223
Malaysia	485,116	635,464	–150,348
Philippines[a]	818,023	417,344	400,679
Singapore	1,037,477	1,468,775	–431,298
Taiwan, China	2,874,750	1,716,435	1,158,315
Thailand	1,220,796	1,420,653	–199,857

a. Data for 1998.

Source: UN trade database Comtrade.

Table 4.2 Tariff Protection of Auto Assembly and Parts Industries, by Economy, 2002 (percent)

Economy	Assembled cars	Auto parts
China	70–80	9–50
Indonesia	75–200	25
Korea, Rep. of	8	8
Malaysia	50–140	5–30
Philippines	30	3–10
Taiwan, China	60	7.5–25
Thailand	30–80	20–33

Source: APEC tariff base (http://www.apectariff.org); for Thailand, see "Car Tax" (2002).

Contrast with Electronics

The weakness in auto parts exports stands in marked contrast to the region's experience in electronics. From 1995 to 1999, East Asia accounted for more than 37 percent of global electronics exports, compared with less than 3 percent of automotive vehicles and parts. The region's electronics industry, especially industrial electronics and components, has taken the form of networks composed of cost-efficient, specialized, interdependent production nodes linked to, and in many cases created or nurtured by, U.S. firms interested in exporting products back to developed-country markets (Borrus, Ernst, and Haggard 2000; McKendrick, Doner, and Haggard 2000; Scott 1987). Until the mid- to late 1990s, in contrast, East Asian auto production took the form of "hub and spoke" arrangements composed of inefficient, largely self-contained, and thus duplicative facilities producing for protected, national markets and communicating only with the headquarters of dominant Japanese assemblers, not with other production sites in the region (Hatch and Yamamura 1996). In electronics, local production facilities were sources of globally competitive components for use in final products exported to industrial markets. In the auto industry, local assembly facilities relied on a combination of major functional components imported principally from Japan and locally produced or assembled components whose quality and price typically failed to meet global standards.[2] These differences reflected different drivers in the auto industry. High transport costs, extensive protection, divergence in the preferences of local assemblers, limited interest of Western firms in small Asian markets, and relatively slow technological change combined to produce a regionally fragmented structure of production.

2. See the discussion of parts quality in chapter 5 of this volume.

Table 4.3 Principal Auto Parts Exports, by Economy, 1999 (percent)

Economy and product	Share of auto parts exports
China	
Tires	27
Nickel-cadmium accumulators	8
Brake parts	8
Wiring sets	7
Filament lamps	6
Wheels	5
Korea, Rep. of	
Tires	37
Lead acid accumulators	5
Body parts	4
Engines	2
Malaysia	
Nickel-cadmium accumulators	24
Tires	16
Lamp parts	6
Wheels	5
Lead acid accumulators	3
Engine parts	3
Oil filters	3
Taiwan, China	
Tires	14
Engine parts	10
Lamp parts	9
Wheels	9
Bumpers	3
Accumulators	3
Thailand	
Tires	24
Wire harnesses	22
Diesel engines	5
Engine parts	5
Wheels	5
Radiators	4

Source: Calculated from UN trade database Comtrade.

This structure provided local auto parts producers in most countries with neither extensive economies of scale nor incentives for technological improvement. Only Korea, whose protection of the auto industry was coupled with an emphasis on exports and whose parts exports followed its vehicle exports, became a partial exception to this pattern of domestically oriented and inefficient parts sectors. In electronics, low transport costs, low rates of protection, the irrelevance of local consumer preferences, and Western firms' active interest in exporting to developed-country markets all combined with rapid technological change to encourage the growth of efficient, locally based component production.

CHANGING AUTO INDUSTRY DRIVERS AND MULTINATIONAL STRATEGIC RESPONSES IN EAST ASIA

New auto industry drivers—the entrance of Western assemblers and intensified competition for the region's growing markets, rapid technological changes, and shorter product cycles—have prompted new assembler strategies and the evolution of more complex production networks. The resulting picture involves good and bad news for East Asian parts production: increased opportunities exist for local parts manufacture, but they are associated with steeper entry barriers for local producers.

Industry Drivers

Global assemblers must now contend with an increasingly challenging set of competitive pressures. The first involves sources of new market growth. The market for new cars in the Triad (Japan, the United States, and Western Europe) has been relatively flat for the past 10 years, with average demand growing less than 1 percent a year. Countries outside the industrial world have become the source of significant growth: East Asian automobile markets averaged around 15 percent annual growth rates before 1997, and several have recovered to pre-crisis levels. The Asia-Pacific region (excluding Australia and Japan) is projected to account for some 45 percent of incremental global industry volume from 1998 to 2006 "and will remain a catalyst for vehicle manufacture and supplier growth for years to come" (Burwell and Ferris 2000). The largest markets, China and Korea, are predicted to double in the next decade and together to reach Japanese volumes (table 4.4). And these projections may underestimate the potential for growth in China. Although ASEAN markets are smaller, their projected annual growth rates are impressive—between 10 and 20 percent (Veloso 2000, p. 24).

Global automakers, much like the producers of silicon wafers and electronic components and products, also have to contend with market fragmentation and overcapacity (current excess capacity is roughly 24 million units, the equivalent of 96 assembly plants). Capacity utilization rates worldwide have fallen since the early 1990s from 80 to 69 percent (PWC 2000, p. 2).

Table 4.4 Current and Prospective Vehicle Sales of Asian Economies, 1999, 2005, and 2010

Economy	1999	2005	2010
Japan	5,861	6,816	7,000
Korea, Rep. of	1,275	1,948	2,650
China	1,925	3,238	4,673
Thailand	218	687	1,253
Taiwan, China	423	560	638
Malaysia	289	504	747
Indonesia	94	439	696
Philippines	74	244	533
India	830	1,432	2,209
Pakistan	67	99	114
Australia	787	898	1,051
Others	102	148	186
Total	11,944	17,014	21,750

Source: Standard & Poor's DRI, cited in Veloso (2000, p. 24).

Within the region, relative scale economies are achieved mainly in Korea, where in the past the state limited the number of assemblers and the financial crisis brought consolidation of the parts industry (although significant overcapacity remains, necessitating a heavy dependence on export markets). China has more than 120 vehicle assemblers, many producing fewer than 15,000 units annually (Veloso 2000, p. 29). In ASEAN, protective measures favoring assembly over parts manufacture have resulted in a proliferation of assembly plants, most of which produce well under the 200,000 units traditionally assumed necessary to achieve efficient scale economies in a single plant (Abrenica 2000, p. 4; Legewie 2000, p. 220). Some scale economies are obtained in Thailand, where tax policies and local demand have generated a large market for one-ton pickups, and in Malaysia, where heavy protection has allowed the "national car"—Proton—to hold on to about half of the local market with sales of 155,000 in 2003. But overinvestment has produced excess capacity in all countries: utilization rates in East Asia were estimated to lie between 65 and 75 percent as of 2000 (Veloso 2000).

Assembler Strategies

Whereas auto assemblers tended to concentrate on dominating their home region through the 1980s, in the 1990s they made significant investments throughout the world, especially in emerging markets, to tap into the most rapidly growing sources of demand (Fine, LaFrance, and Hillebrand 1996, p. 24; Veloso 2000, p. 7). This trend is clear in East Asia, where the region's potential has intensified the competition among a larger number of auto manufacturers (Noble 2001). Japanese assemblers, long dominant in the ASEAN countries, expanded investments there in the 1990s, especially in Thailand. Most significantly, Western firms—assemblers, first-, second-, and third-tier suppliers, and service providers—have recently entered or reentered East Asian markets. Western firms have invested both on their own and through tie-ups with Asian OEMs. These alliances are designed to help Western firms to broaden their brand and product portfolios and to draw on Asian firms' expertise in the manufacture of attractive and affordable small vehicles. The principal focus of Western initiatives has been China and Thailand, although Korea has also attracted significant investments in the post-crisis environment.

Consolidation is an important part of assembler strategies as these firms search for scale economies in design, development, and manufacture. Auto manufacturers are merging and are developing strategic alliances to gain access to volume-generating markets, to expanded skill sets and competencies, and to innovation. PricewaterhouseCoopers anticipates that only six assemblers will account for some 80 percent of total vehicle output in the next decade, and one Japanese source predicts that only those assemblers producing 4 million units a year will survive (Matsushima 1999 and PWC 2000). A similar consolidation is occurring in East Asian operations.[3]

An emphasis on exports and support for trade liberalization are two additional components of emerging assembler strategies in East Asia. Exports were already a priority of the newly entered U.S. OEMs, but the 1997 crisis led them to increase exports to compensate for the drop in local vehicle sales (de Jonckheere 1998, p. 6). Recently, Japanese auto assemblers have begun to follow the trend of electronics producers in using offshore facilities for production for export back to the Japanese market. These export efforts have been undertaken through regional and global sourcing networks that allow OEMs to contend with, and even take advantage of, currency shifts (de Jonckheere 1998; Takayasu and Mori, chapter 5 of this volume).

Complementing these export efforts is intensified support by assemblers for regional trade liberalization, which is perceived as essential for the creation of efficient production networks. While East Asia's traditionally high

3. Of all the Japanese producers, only Honda and Toyota remain independent, although they have signed agreements with General Motors to share parts and technology. Two Korean auto companies—Daewoo and Samsung Auto—were acquired by General Motors and Renault, respectively. Hyundai has absorbed Kia. A process of consolidation and the disappearance of independent Asian companies is thus accompanying the entry of new competitors.

levels of automotive protection probably will not disappear completely,[4] all countries are reducing the level of tariff support given to the auto industry. ASEAN countries are now engaged in two sets of liberalization efforts. One is the ASEAN Industrial Cooperation (AICO) program—a bilateral complementation scheme launched in 1996 and aimed primarily at the auto industry. Under this scheme, firms pay only 0–5 percent tariffs if 40 percent of the product's value originates in another, participating ASEAN country. The second is the ASEAN Free Trade Area (AFTA), have cut automotive tariffs to 0–5 percent by 2003 (but Malaysia has successfully sought an exemption until 2005). Global automakers operating in East Asia have expressed strong support for both of these initiatives.[5] Indeed, foreign assemblers and suppliers, led by the Japanese, have established some 75 bilateral exchange programs under AICO. Honda, Mitsubishi, Nissan, Toyota, and the giant parts supplier Denso have been leaders in developing multicountry complementation efforts in Southeast Asia (Legewie 2000, p. 230).

Even as they attempt to increase scale economies, automakers must satisfy changing region- and country-specific tastes and conditions. "Platform" strategies are a key response to this challenge, and their adoption has implications for parts suppliers. These strategies, in which automakers use a small number of underbody platforms as the basis for a greater number of vehicle models, are detailed elsewhere in this volume (chapters 2 and 5). Here we emphasize three aspects of most platform strategies: consolidation, specialization, and localization. To cut the costs of platform development and to encourage component sharing among models, automakers, especially U.S. firms, have begun to reduce their number of platforms even as they strive to maintain flexibility by broadening the definition of a platform (for example, PWC 2000).[6] Corporate consolidation is facilitating the reduction of platforms. For example, platform sharing between Chrysler and Mitsubishi will allow Mitsubishi to slice its number of light-vehicle platforms from 12 to six or seven (Treece and Sherefkin 2001, p. 53). This

4. As a recent analysis of the Chinese auto market has noted, "Few people think Beijing will allow a flood of imports to swamp its domestic carmakers, any more than have South Korea, Malaysia, Indonesia—or for that matter, Japan" (Brooke 2002, sec. W, p. 1). Governments are likely to turn increasingly to less opaque nontariff barriers as a means of providing support to local industries.

5. There are thus some differences among the assemblers with regard to trade liberalization. U.S. firms tend to place more emphasis on the success of AFTA, whereas the Japanese have moved quickly to make use of AICO (for example, Yap 2001).

6. R&D expenditures in the OECD auto industries rose from 2.4 to 3.2 percent of production costs between 1972 and 1992 (Humphrey 1998, p. 3).

process of platform development and consolidation is clearly in evidence in East Asia, where, for example, Honda will produce the Odyssey based on the same platform as the Accord.[7] Similarly, Renault (including Renault Samsung) and Nissan will share platforms in the future.

National specialization is an important component of platform strategies. The Japanese have begun to promote specialization through the complementation efforts noted above. Equally notable is the emergence of Thailand as a hub for pickup truck production by almost all assemblers. In fact, Isuzu, Nissan, and Toyota have stated their intention to make Thailand the production site for all one-ton pickups produced outside of North America (Chatrudee 2002).

Finally, wherever possible, assemblers are attempting to localize the manufacture of components. Localization can help assemblers to facilitate just-in-time supply of local assembly facilities, to reduce foreign exchange risks, to meet local tastes and requirements, and, when local conditions permit, to reduce costs through local procurement (chapter 5 of this volume). Moreover, this localization is occurring through assembler "deverticalization." As they focus on capturing more of the *downstream* value chain (between assembler and final customers), assemblers are outsourcing the manufacture and, in some cases, the design of platform components.

Implications for Parts Suppliers: The Promises and Pitfalls of Globalization

The increased transaction costs generated by such highly demanding outsourcing have had important consequences for assembler-supplier relations. U.S. assemblers have moved away from relations with a large number of suppliers toward closer involvement with a smaller number of larger, richer, more technologically competent producers. Whereas Japanese OEMs historically maintained very close ties with a small number of suppliers, faced by new competitive pressures, they have begun to impose

7. In China General Motors is building a small car—the Buick Sail—off an Opel Corsa platform that was first modified for production in Brazil and then modified again for China. Volkswagen is producing the Audi A6 and the Passat in two different locations in China, appealing to two different price segments and customers. The two vehicles are built on the same platform. In Thailand Ford and Mazda produce two different brands of pickup—virtually identical except for cosmetics—off the same platform. Also in Thailand, General Motors and Isuzu are planning to produce two pickup trucks based on the same S-10 platform. The platform was jointly developed (Dunne 2001). Ford is assembling a new common sedan in four ASEAN plants and developing a regional component supply network based on the combined volume of the four facilities (de Jonckheere 1998, p. 4).

tougher requirements on existing suppliers and to source from new, sometimes Western, producers of parts. In both instances, heightened responsibility for design development by suppliers has meant that consolidation in the parts sector is proceeding in parallel with consolidation of the assemblers. A small number of independent global "mega suppliers" is emerging.[8]

These new strategies have raised the barriers facing developing-country firms attempting to supply parts and components to OEMs. In the past, assemblers designed and manufactured core components while outsourcing easy-to-produce, detail-controlled parts. In some cases assembler subsidiaries even had some autonomy to introduce changes in design. This provided local producers with opportunities to enter, to acquire competencies, and gradually to ascend automotive supply chains (Humphrey 1998, p. 10). Now, however, assemblers are relying on suppliers to produce and even design standardized, core components based on common platforms. Given the technological weaknesses of most developing-country suppliers, which preclude them from meeting the assemblers' requirements, assemblers have begun turning to what Humphrey terms the "follow source"—the large firm already producing the part elsewhere in the world. In developing-country operations, first-tier firms such as Bosch, Delphi, Denso, TRW, Visteon, or Yazaki are being asked to supply the local subsidiary of the assembler and will "be responsible for ensuring that the rest of the supply chain meets the assembler's standards" (Humphrey 1998, p. 5).[9]

Intensified competition among assemblers for East Asia's growing markets, combined with technological and managerial innovations and with increased currency instability, is encouraging greater localization of parts production within expanding regional and global production networks. The extent and nature of local parts production will vary according to two OEM firm-specific strategies and such industry- and product-specific factors as parts' value-to-weight ratios, position in assembly sequence, and degree of standardization.

8. In some systems, such as brakes and seats, the global industry is dominated by three or four global producers. PricewaterhouseCoopers predicts, "The 1,500 Tier 1 suppliers in 1998 will whittle down to about 150 large system integrators and 450 direct suppliers who will compete largely on price" (PWC 1999). By 2000 the five largest suppliers of automotive safety parts accounted for roughly 80 percent of global market share, up from around 55 percent in 1995 (McMaster 2001, p. 11).

9. First-tier suppliers sell directly to OEMs. They are typically capable of producing components, modules, or entire systems (for example, Lear produces complete interior systems; Delco produces electronics systems; Dana produces drive trains; Bosch produces ABS systems). For lists of U.S. suppliers, see Brunnermeier and Martin (1999, ch. 2).

Other things being equal, standardized and easily shipped parts such as printed circuits and switches will be sourced globally; parts such as alternators, heat exchangers, wiring harnesses, or fuel tanks that are large or more difficult to transport will be sourced regionally; and in-line parts supplied to the vehicle manufacturer in the sequence that vehicles are assembled will be supplied locally, often by firms in close proximity to the assembly site (Brooker Group 1997, p. II-8).[10]

How host countries fit into and influence these calculations depends on location-specific assets, such as market size and demand characteristics, OEM network presence, preexisting supplier base, human resources, and government policies.

The possibility for developing countries is that the localization component of assembler strategies will translate into denationalization and into intensified competition for becoming a regional hub in the new production networks. For instance, as of late 1997 Brazil—a large domestic market with a sizable base of local suppliers—had only one locally owned firm among the 13 largest component producers (Humphrey 1998, p. 10). In Thailand, almost all of the 250 first-tier suppliers are foreign subsidiaries or affiliates (Brooker Group 2002, p. 208). Even in Korea, the post-crisis opening to FDI has facilitated significant acquisitions of larger Korean parts companies by Western first-tier suppliers. With assemblers attempting to squeeze first-tier suppliers, and they, in turn, pressuring their component manufacturers to reduce costs, profit margins are almost certain to be smaller at the lower rungs of the value chain. And with markets for first-tier suppliers being dominated by foreign firms, the possibilities for technological learning and upgrading for local companies may be reduced unless appropriate public policies and incentive structures are put in place.

CASE STUDIES

This section of the chapter provides a more detailed examination of the policy framework for auto parts in our five economies and reviews their recent performance in domestic production and in exports.

10. Another hybrid scenario, this one proposed by a major producer of safety components, anticipates local and national sourcing of unique parts: local assembly of some parts in the firm's component assembly plant (10 percent of total parts), regional sourcing of region-specific parts (30 percent), and global sourcing of standard parts (60 percent). See McMaster (2001).

China

The lure of China's potentially huge market, large foreign investments induced by the government's protective policies, and the accumulation of experience and infrastructure in export-oriented manufacturing have created a formidable base for the growth of the Chinese auto industry. Nonetheless, the time required and the political and social turmoil involved in reaching global competitiveness should not be underestimated: China's market for passenger cars is still modest, and even amidst overall growth, pervasive excess capacity will ensure that many small parts producers fail and that large numbers of workers lose their jobs.

Demand for and production of autos. Volume production, mostly of trucks, began in the 1950s at the First Auto Works (FAW) in Changchun; FAW and a host of vertically integrated suppliers located in the same complex provided most components. In the 1960s fears of conflict with the Soviet Union and Maoist doctrines of local self-reliance spurred the creation of a host of new assemblers and parts suppliers in inland provinces. Economic opening in the late 1970s led to joint ventures with foreign auto companies, notably Volkswagen. In the 1990s General Motors and Toyota garnered assembly licenses by pledging to establish local design capacities and cultivate a network of local suppliers. As the government pressed to achieve economies of scale, the production of passenger cars rose steeply over the 1990s and reached 2.07 million units in 2003. High tariffs and other protective measures largely discouraged imports. Nevertheless, excess capacity and padded work forces remain pervasive, and the capabilities of firms vary drastically by location, size, and ownership.

Parts production. Production of parts grew quickly in the 1990s, doubling in value between 1994 and 1999. Parts production is geographically dispersed. The leading production network is centered in Shanghai (including Jiangsu and Zhejiang); other significant production areas are the northeast, Hubei, and Sichuan-Chongqing. Beijing-Tianjin and Guangzhou host major assembly sites and boast strong industrial bases, so they could emerge as centers for parts production in the future. Many producers are linked to state, provincial, or municipal conglomerates. Design capacities of local firms vary but are generally low; quality and delivery times, though greatly improved, are still weak, particularly at lower tiers of the supply pyramid. Even joint ventures are often plagued with excess labor.

Of approximately 1,540 firms supplying auto and motorcycle parts, about one-fourth have attracted foreign capital (Ministry of Machinery

Industry 2000, pp. 10, 314). Whereas foreign firms can own at most 50 percent of assembly operations, foreign participation in parts manufacture ranges from technology transfer through joint ventures to wholly owned subsidiaries. Firms with foreign participation are especially important in the production of more demanding parts, such as piston rings (two of 21 firms account for 65 percent of total production of this component), fuel injectors (one of six firms, but with 70 percent of production), and air conditioning (six of 10 firms, with 73 percent of production; Fourin 1999). They are also a major force in the production of standardized items such as tires and batteries.

Trade. From 1994 to 1999, imports of parts doubled, while exports nearly tripled. Before long, China is likely to generate a trade surplus in parts. Major exports include tires, wheels, wire harnesses, glass, and other heavy or labor-intensive items, but complex parts such as transmissions and axles are also exported in small numbers. Japan is the major source of imported autos and auto parts, followed by Germany, the United States, and Taiwan (China); the United States is the leading export market, followed by Japan, Germany, Hong Kong (China), and a mixture of European and developing countries (Ministry of Machinery Industry 2000, pp. 179–88).

With China's entry into the WTO, local content requirements were eliminated and tariffs began to drop, but the industry will be cushioned by the slow pace of liberalization and remaining promotional and protectionist policies, including taxes, controls on FDI, and local protectionism. Local production of mainline vehicles and their parts will continue to grow as manufacturers adapt their vehicles to the Chinese market and take advantage of inexpensive labor to produce for export markets.

Production networks. China contains a number of imperfectly integrated component production networks. Joint ventures and wholly owned operations from Taiwan, China (officially 166 as of 2002) tend to be efficient and produce with acceptable quality, often for export markets, but they rarely extend beyond low-medium levels of sophistication (tires, batteries, radios, steering wheels, lights, piston rings, brakes, crank shafts). Until recently, Japanese parts producers entered China only reluctantly, to support local assembly, producing such parts as cylinder heads and blocks, cam shafts, universal joints, steering systems, alternators, starters, and brake master cylinders. Industry leader Denso, for example, has only five plants in China (three in Tianjin). With the exception of Denso and other suppliers linked to Toyota, Japanese firms have been criticized for supplying only dated technology, but they do pass along quality control and on-time production techniques to their subcontractors. Western supply firms have concentrated in the Shanghai area. Nine of Delphi's 15 Chinese plants, for example, are located in or around Shanghai, producing wire harnesses, car audio equipment, shock absorbers, drum brakes, generators, fuel pumps, and other low- to mid-technical products as well as the more sophisticated fuel injection and engine management systems (EMS) at which the company excels. All plants export, with a group average of one-third. Other major global component firms with significant investments include Borg-Warner (transmissions), Bosch (spark plugs, electronic fuel injection systems, oil pumps), and T&N (cylinder gaskets, piston rings).

Summary. China is likely to continue experiencing rapid, but uneven and painful, growth. Older, state-owned firms located in inland areas and producing only for the domestic market face a dark future. Prospects are brightest for Shanghai-area firms supplying parts for small cars and light commercial vehicles, with capital and technical assistance from foreign owners or partners.

Korea

Korea has been more successful than any other industrializing country in developing a national automobile industry. The Automotive Industry Promotion Law, promulgated in 1962, part of the First Five-Year Development Plan, marked the birth of the modern automobile industry. It was subsequently nurtured by the full panoply of trade and industrial policy instruments that the Korean state had at its disposal. Imports of cars were prohibited—a ban that was not lifted for a quarter of a century. The government provided subsidized loans and tax incentives for investments as well as export subsidies, including export promotion loans that enabled Korean cars to be sold in foreign markets at less than half the domestic market price. Imported components were exempted from tariffs, but the state encouraged assemblers to increase local content by working with local suppliers. Immediately before the financial crisis, car production in Korea (all from domestically owned companies) was the fourth largest in the world, and half of this output was exported. But repeated state attempts to consolidate the industry failed, and the consequent overcapacity left the industry vulnerable to the collapse in domestic demand that followed the onset of financial crisis, with dire consequences for the parts industry.[11]

11. For further consideration of the accomplishments and failures of the Korean automobile strategy, see Ravenhill (2002).

Parts production. Of the economies included in this study, Korea has by far the largest parts industry in terms of overall turnover and exports. By 1999 the turnover of component companies was estimated to have grown to close to $10 billion (Economist Intelligence Unit 2001, p. 78). The domestic aftermarket accounted for about one-fifth of this total, a market then dominated by the assemblers.[12] Subsequently, the component manufacturing arms of Korea's largest assemblers were spun off into independent companies: Hyundai-Mobis and Daewoo Precision (reconstituted in mid-2002 after General Motors' acquisition of Daewoo).

Parts producers in Korea benefit from one of the region's largest domestic markets for assembled cars, with annual sales approaching 1.5 million vehicles, of which slightly more than 1 million are passenger cars. Even at the end of the 1990s, more than a decade after the government lifted its ban on imports, Korean companies enjoyed the luxury of a de facto sanctuary in the domestic market.[13] Korea has one of the most liberal trade regimes for auto imports in Asia, with assembled vehicles and most auto parts subject to an 8 percent tariff (table 4.2). The most significant barriers to auto imports today are the strong nationalist sentiment against purchasing foreign cars (with the limited exception of the luxury end of the market), the price competitiveness of small cars produced by domestic assemblers, and these companies' well-established distribution networks.

Besides the large domestic market, the auto parts industry also benefits from Korea's strong national innovation system, engineering and metalworking skills, and backward linkages with the steel industry. At the onset of the financial crisis in 1997, small- and medium-size enterprises dominated the industry, accounting for 95 percent of a total of approximately 1,400 parts manufacturers. These firms typically had limited capital and rudimentary technology and produced solely for the domestic market. Indigenous technological development was confined primarily to the 50 largest parts suppliers. In many instances, local development rested on improvement on technologies previously produced under license. Research and development (R&D) expenditures remain low and heavily concentrated in a few companies: only 1.4 percent of the industry's overall turnover was invested in R&D in 1999. Korean exports of car parts were derived almost entirely from these larger firms and from the assemblers

12. Only at the beginning of 2000 were component makers permitted to sell their products directly to maintenance shops or consumers rather than indirectly through the assemblers.

13. In 2000 imports of all motor vehicles totaled 11,168, and motor vehicle registrations totaled 1,441,628 (data on the Korean Automobile Manufacturers Association's web page: http://www.kama.or.kr).

themselves. Daewoo, Hyundai, and Kia provided about one-third of total exports, while another 10 companies contributed a further quarter of export earnings.

Trade. As Korean auto exports grew, so did exports of auto parts: the value of auto parts exports tripled between 1990 and 2000 (when they constituted 2.8 percent of total exports, up from 2 percent in 1988). Exported parts were divided almost equally into three categories: those used in assembly overseas, those used for the aftermarket for Korean cars, and those used for the aftermarket for cars produced by other (foreign) makers. By 1999 Korea enjoyed a large surplus in its trade balance in auto parts. More than one-third of all exports went to the North American market, the largest single export market for Korean cars. Asia accounted for 25 percent and Europe for 21 percent of the balance of parts exports.

Most auto parts exports remain at the low to medium end of the spectrum of technological complexity. Tires are by far the single most valuable export, constituting close to one-third of the total value. Other significant exports include air compressors, batteries, air conditioners and parts, bearings, wheels, clutch covers, window regulators, and steering parts. Korean assemblers still rely heavily on imports for the more technologically demanding components, including transmissions and engines (with the exception of Hyundai, which has developed its own engines for some models), safety systems, and electronic equipment.

Production networks: post-crisis consolidation. The economic crisis in the second half of the 1990s exacerbated the pressures that global consolidation was exerting on the domestic auto parts industry. The bankruptcies of four of the five domestic assemblers—Kia and Ssangyong (these two before the onset of the financial crisis), Samsung, and Daewoo—had flow-on effects for many of their suppliers, which were unable to withstand the nonpayment of invoices. Between 300 and 500 small- and medium-size enterprises exited the auto parts industry following the bankruptcies of the assemblers. The financial crisis dramatically changed the ownership structure of many of the larger Korean parts companies. Bankruptcies among some of the largest conglomerates forced divestment of some auto component subsidiaries. Even more significant, it opened the way for foreign investors to acquire significant shareholdings in Korean components producers—often at fire-sale prices (the advantage of low stock market prices being compounded by a favorable exchange rate). More than 100 parts suppliers have transferred 50 percent or more of their equity to foreign investors since the crisis began. This has facilitated the entry or the expansion of Korean operations of the leading global first-tier suppliers, including Autoliv, Britax, Delphi, FAG, Valeo, and Visteon.

The financial crisis hastened the consolidation of the Korean components industry and led to further differentiation between larger firms, on the one hand, increasingly with foreign equity as well as technology licensing agreements, and small- and medium-size enterprises, on the other. Restructuring and the opening to foreign investment, in turn, have facilitated the negotiation of new technological partnerships.

Prospects. The size of Korea's domestic auto market (even though domestic growth is expected to be modest), coupled with its continued dominance by local assemblers, provides Korean auto parts makers with a springboard for penetrating global markets, as does the increasing aftermarket for parts for Korean cars on foreign roads. The future of the Korean components industry inevitably rests in large part on the success of the assemblers. And here the prospects appear much brighter than many believed at the time of the crisis: Hyundai and Kia have dramatically increased their exports, especially to North American markets, going a long way toward overcoming the negative brand image that previously afflicted Korean-built cars. General Motors intends to use its new Daewoo subsidiary to supply small vehicles not only to the region but also to Europe and the United States. Renault has committed itself to investments that will substantially raise the output of Renault Samsung.

Nevertheless, investment in foreign assembly plants by Hyundai and Kia (with plants planned for China, Europe, and the United States) may reduce both the total exports of assembled autos from Korea and the opportunities for auto parts exports. Hyundai and Kia are encouraging a number of their larger suppliers to establish subsidiaries close to their overseas operations: more than 20 are planning to establish plants adjacent to Hyundai's facility in Alabama. China has also attracted more than 40 Korean auto parts companies, seeking to utilize local lower-cost labor to maintain their presence in the production of low-end components.

As domestic assemblers seek to increase their presence in global markets, their insistence on modularization and on Korean suppliers reaching international standards will force the domestic auto parts industry to become more competitive if it is to survive. Korean-owned companies face a potential pincer movement. At the low-tech end, they are not price competitive with products from China and Taiwan (China). At the high-tech end, the emphasis on modularization provides advantages to the transnational first-tier suppliers.

A possible strategy for the larger Korean parts producers, which takes advantage of the significance of the domestic market, is outlined by Hyundai-Mobis president and chief executive, Park Jeong-in: "Our weakness is we lack the technology, and we don't have time to learn it. We aim to select top-ranked makers of automotive components and offer joint venture production bases in Korea in exchange for the chance to supply parts to Hyundai and Kia" (quoted in "Hyundai-Mobis" 2001). Hyundai-Mobis is the largest supplier to Hyundai and Kia and is focused entirely on the production of modules, referring to itself as a "tier-0.5" company.

Further consolidation will be required if the Korean parts industry is to become competitive internationally. Too many Korean companies produce on too small a scale to be competitive—and their comparative inefficiencies are handicapping Korea's domestic assemblers. The pressure for modularization will inevitably hasten industry consolidation: Hyundai-Kia, which now has a joint procurement operation, plans to reduce its suppliers from the current figure of 750 to fewer than 200. Many more of the small- and medium-size enterprises are likely to disappear. Some current first-tier suppliers will drop to the second tier. Hyundai Motors' research unit estimates that a further 200,000 jobs will be lost through restructuring in the auto industry.

With a large portion of Korean auto parts exports now coming from foreign-invested companies, a key question that remains unanswered is what role the new joint ventures will play within the overall production networks of the largest players in the global industry. Will they be used primarily for local production to supply the Korean market, or will they be integrated into sourcing for global markets? And what technology will be transferred to the Korean subsidiaries? The history of joint ventures in the Korean automotive industry has not been a particularly happy one for local firms, which have been denied access to leading-edge technologies and prevented by their joint venture partners from exporting. Whether this will change in Korea's more positive climate for foreign investment (with a significant shift to wholly or majority-owned subsidiaries) and for protection of intellectual property rights remains to be seen.

Malaysia

Production networks in auto parts in Malaysia have been shaped by government policies aimed at promoting national champions in the assembly industry and at increasing the participation of ethnic Malays in commerce, objectives that have sometimes proved less than completely compatible.

Government promotion and ongoing high levels of protection of auto assemblers have enabled the two national car companies, Proton and Perodua, to capture more than 90 percent of the domestic car market. Although Malaysia has the largest domestic car market in ASEAN, with annual sales of close to 350,000 vehicles, and their domination of the domestic market gives Proton and Perodua top position in all car sales in ASEAN (together they shared 28 percent of the regional market in 2000 compared with Toyota's figure of 21 percent), overall production volumes remain too small for most local parts makers to be internationally competitive.[14] Although some parts firms have entered into joint ventures with foreign partners and into technology licensing agreements, government restrictions have provided a rather inhospitable climate for foreign investors—in marked contrast to Malaysia's generous treatment of investments in the electronics sector.[15]

In the early 1980s government disappointment with the slow progress of local auto firms, coupled with its desire to promote heavy industry to reduce dependence on commodity exports, led to a decision to develop a nationally owned auto industry. The first national car project, Proton, a joint venture with Mitsubishi, was launched in 1983. With heavy protection, sales of Proton vehicles soared: within three years of its first production Proton had secured 73 percent of the domestic market. Exports began in 1986. Volumes remain tiny, and sales have been heavily concentrated on the U.K. market. Proton cars have a negative brand image outside Southeast Asia, often being described as depending heavily on previous-generation Mitsubishi technologies. In 1993 the government launched a second national car project, Perodua, to produce small vehicles. Perodua was established as a joint venture between Daihatsu and its trading company partner—Mitsui of Japan—and three local companies.

Besides the objectives of increasing local content and achieving technological spillovers, the government also hoped to use the national car projects to increase the participation of ethnic Malays in an industry that had previously been dominated by Chinese entrepreneurs. Proton was instructed to purchase components from ethnic Malay–owned companies; it was not permitted to employ experienced labor laid off by other (Chinese-

14. Proton had 18 percent of the ASEAN market; Perodua had 10 percent. In Malaysia in 2000, Proton had 63 percent of the domestic car market, and Perodua had 30 percent.

15. The Malaysian government has permitted few foreign companies in the auto industry to establish majority-owned local subsidiaries. Volvo, with 100 percent ownership of its assembly plant in Selangor, is an exception. Ford, Honda, and Toyota have all been confined to a 49 percent stake in joint ventures.

dominated) assemblers and component suppliers after the establishment of the national car producer (Rasiah 1996).[16]

Trade policies. The national car companies receive the highest levels of protection in the region. Tariffs on motor vehicles were increased in 1998 in response to the economic crisis. The current rates on assembled passenger car imports range from 140 to 300 percent, depending on engine size. National car firms also are protected by various nontariff barriers, including reductions in excise taxes and import licensing. Malaysia has also maintained extensive local content requirements, including a list of components that must be purchased locally by all car manufacturers or assemblers and a percentage local content requirement (abolished on January 1, 2002, as part of an agreement with the WTO).

Malaysia successfully sought an extension until 2005 of the time period under AFTA for lowering its tariffs on auto imports from other ASEAN members (40 percent local content required) to the 0–5 percent range. Some reports have suggested that the government will continue to protect the local assemblers after this date through various discriminatory tax measures. Some relief from Malaysian duties for parts producers in other ASEAN countries is provided through the AICO scheme (of the 63 joint ventures in the auto industry currently approved under the scheme, 29 involve companies or their subsidiaries in Malaysia).

The local car parts industry receives direct assistance through government measures aimed at promoting the participation of ethnic Malay entrepreneurs in the economy. The Vendor Development Programme guarantees local producers a market with the national car companies. By 1998 the Vendor Development Programme for Proton had 188 suppliers responsible for 4,319 parts, including metal-stamped and pressed parts, plastic injection molded parts, wire harnesses, wipers, lamps, radio–cassette players, and air conditioners. By guaranteeing a market, the Vendor Development Programme removes any pressure for suppliers to become internationally competitive.

Exports. One of the strongest indicators of the lack of international competitiveness of the auto parts industry in Malaysia is its small volume of exports. By 1999 the auto parts industry contributed only half of 1 percent of total Malaysian exports, constituting a smaller share in overall export earnings than in the other economies included in this study and, indeed, than

16. In 1984 ethnic Malay equity in the auto assembly industry was 30 percent compared to the 43 percent held by Chinese Malaysians. In 1988, 94 percent of Proton employees were ethnic Malays, the vast majority of whom had no previous experience in the auto industry (Jomo 1994, pp. 266, 285).

in other ASEAN countries such as Indonesia and the Philippines that have much smaller assembly industries than does Malaysia.[17] Tires traditionally constituted the single most valuable auto parts export for Malaysia. In 1999, however, their position was supplanted by nickel cadmium accumulators. Other major exports are similarly low tech. Although the export market for Malaysian auto parts is diversified, ASEAN is by far the single most important outlet, accounting for more than a quarter of total exports, a substantial portion of which can be assumed to be spare parts for Proton and Perodua cars sold in other ASEAN countries. Europe is the second most important market, again linked to Proton sales in the United Kingdom.

The Malaysian auto parts industry. Despite or, probably more accurately, because of the very high levels of *unconditional* protection afforded by the government, the auto parts industry in Malaysia remains small scale, inefficient, and lacking in technological competencies. The government estimates that 350 companies are involved in the manufacture of automotive components. Most parts manufacturers operate on a very small scale. Few have more than 200 employees. In total, they employ around 17,000 workers, only slightly more than 1 percent of all employment in manufacturing in Malaysia. The larger firms typically are joint ventures with foreign partners.[18]

Despite the government's objective of increasing the value added by the domestic auto industry through localizing the production of parts, the trend in value added per unit of output in the parts and accessories industry as well as the manufacturing and assembly industry has been downward, largely because of the high import content of locally produced parts (Tyndall n.d.).

The principal problems faced by the auto parts industry in Malaysia can be summarized as (a) a lack of scale economies, (b) relatively high labor costs, (c) shortages of skilled labor, (d) lack of research and development, and (e) ongoing dependence on imported components. Government policies have either contributed to many of these problems or failed to address them effectively. The decision to use the national car project to promote ethnic balancing exacerbated the industry's problems of small scale and lack of skills. Meanwhile, the government has done little to alleviate the problems within the national system of innovation. Locally trained engineers are in short supply and are likely to find a more attractive career path in the electronics industry than in auto parts.

17. The definition of auto parts used in this study does not include car radios or compact disk players, which are a substantial export for Malaysia.

18. Tyndall provides a list of parts manufacturers and information about their products and number of employees at http://www.asean-auto.org/mal/sheet0017_1.htm.

The government has failed to provide institutions to assist the domestic auto industry to develop technology. No specific research institution exists for the automotive industry; existing institutions merely provide facilities for testing components. The government thus has not compensated for the lack of research and design capabilities of the local industry, a weakness deriving in part from the small size and limited budgets of local firms. The industry itself is poorly organized in comparison with its counterparts in other economies in the region. The principal business association, the Malaysian Automotive Component Parts Manufacturers Association, has only two full-time employees and plays a negligible role in facilitating technological upgrading. Malaysian auto components companies generally appear poorly equipped to cope with the increasing use of electronic commerce in the industry.

Despite the government's requirements for high levels of local content and its claims that these requirements are being met, the reality is less encouraging. Leutert and Sudhoff (1999, p. 258) estimate that the actual ratio of local content may be as low as 35–40 percent. Moreover, Proton suppliers have not kept up with the trend in the industry toward providing complete systems: 60 percent of Proton suppliers merely provide single parts. The parts industry consequently does not reflect to any degree the trend elsewhere toward suppliers being organized into various tiers. And because only a minority of its suppliers are capable of operating on a just-in-time basis, Proton is compelled to carry higher levels of inventories than most of its international competitors. The after-sales market is dominated by components supplied by Mitsubishi.

Prospects. Current regulations in Malaysia, though intended to promote the auto parts industry, often appear to have undermined its competitiveness. Because it has been insulated from market forces, Malaysia's auto parts industry has not participated in the consolidation that has occurred elsewhere in the region following the financial crises. Protection of the industry through import licensing and the provision of guaranteed markets for parts makers with the national car companies have encouraged a climate in which rent seeking thrives. Although the Malaysian government can be expected to continue to drag its feet on liberalization of trade in both assembled vehicles and auto parts, there being no significant domestic political constituency currently arguing in favor of such liberalization, it will eventually have to comply with AFTA and the WTO commitments on TRIMs. To do so will place the parts industry in some jeopardy: it is caught in a pincer movement because it competes on the basis neither of low wages nor of advanced technology.

The future of the parts industry inevitably will be closely tied to that of the national car producers. Whether Proton can survive in a globalizing industry without a strong foreign partner is questionable. Former Prime Minister Mahathir indicated his willingness to sell a 30 percent stake in Proton to a foreign manufacturer. Proton is reported to be actively seeking a foreign partner, but the investment climate in the automobile sector in Malaysia remains uncertain. There has also been discussion in the press of the possibility that Proton's plant will be made available to assemble foreign cars. Some observers consider the decision at the end of 2001 to allow Daihatsu to take management control of Perodua's production facilities as an attempt by the government to set up a test case to see what will happen if a foreign producer is allowed to take over a national car company. If deemed successful, it may provide a model for the future of Proton ("New Era" 2001).[19] The future of the national car makers themselves, however, will be determined in part by the competitiveness of their parts suppliers. Current government attempts to use the auto parts industry to pursue social and political objectives handicap the national car makers. If forced to choose between maintaining an assembly industry (probably) in alliance with foreign car manufacturers or attempting to prop up an inefficient locally owned auto parts industry, the government is likely to opt for the former, as strongly suggested by its decision to allow Daihatsu to assume control of Perodua.[20]

Taiwan (China)

Taiwan (China) is a medium-size, medium-tech auto market with extensive ties to Japan and emerging links to China and (to a lesser extent) Southeast Asia. Local firms export a large volume of replacement parts, especially to the United States, and a small but growing amount of original equipment, particularly to Southeast Asia and Japan. WTO entry will shrink, but not eliminate, the original equipment market.

---

19. Perodua plans to reach a production capacity of 300,000 units a year by 2005. Under the new arrangement with Daihatsu, it will also manufacture Daihatsu brand cars (Daihatsu lacks a production plant in Southeast Asia). As the current producer of the lowest-priced car in the region and, in partnership with Daihatsu, the only producer of cars with engines smaller than 1 liter in Southeast Asia, Perodua may carve out a niche for itself.

20. Malaysian sources suggest that Daihatsu gave a commitment to Perodua that it will provide existing Malaysian suppliers with an opportunity to meet its cost and quality requirements and supply assistance to facilitate technological upgrading. It also made it clear, however, that if local suppliers are unable to match its requirements, it will source components elsewhere ("New Era" 2001).

Demand for and production of autos. In the late 1990s congestion, slowing economic growth, the belated maturing of urban mass transit systems, and the popularity of motorcycles depressed auto sales from their peak of 575,000 units to little more than 420,000. Passenger cars have established a dominant position, but small utility vehicles remain popular. Local assemblers have increased their share of the market to more than 80 percent and have developed modest independent design capabilities. Production is fragmented among about a dozen assemblers, led by affiliates of four Japanese companies (Honda, Mitsubishi, Nissan, and Toyota) and Ford.

Automobile policy. After decades of vacillation, in the mid-1980s the government of Taiwan (China) embarked on a slow process of liberalization. By 2000 tariffs on passenger cars had fallen to 30 percent (although strict quotas limited imports from Japan and Korea). Tariffs on parts averaged about 15 percent, while local content requirements were reduced to 40 percent. The government provided some promotional help with research and gave locally designed components a reduction of three percentage points off the stiff commodities tax. The government also strongly promoted "central-satellite relations" between assemblers and their suppliers.

Parts production. Taiwan (China) has about 2,200 parts firms, of which about 300—almost all linked to Japan by capital investment or technology licenses—produce original equipment. By some estimates, perhaps 100 of the OEMs would be competitive even without protection. In 2000 parts production reached about $4 billion. Areas of strength include tires, steel and aluminum wheels, batteries, bumpers and other body parts, lights, and some engine parts. One of the weakest areas is transmissions, most of which are imported. Quality levels and design skills are considerably more advanced than in China or Southeast Asia but lag behind those in Japan. Quality standard 9000, ISO 9002, and ISO 14000 certification is common among large firms, but not small ones. Research and development spending increased rapidly over the 1990s, reaching 1.3 percent of sales by 1999.

Despite the pervasive influence of Japan, production networks are quite different from those in Japan or Korea. Low volumes and fragmented assembly have precluded the development of distinctive Japanese-style *keiretsu:* parts firms supply all assemblers evenly. Nevertheless, the clustering of assemblers and parts suppliers in the industrial region just south of Taipei has facilitated the development of a sophisticated division of labor.

Trade and WTO accession. Taiwan's (China's) auto industry is still protected by significant tariffs, quotas, and nontariff barriers, although those have loosened in recent years, especially for parts. Parts makers made up for a modest loss of domestic original equipment orders with increased exports. As a result, both imports and exports grew relative to production, and local firms began to run a surplus in auto parts by 1996. Production is diverse, but replacement parts and accessories such as body parts, tires, batteries, lighting, and bumpers lead the way. The ratio of exports to imports increased even in technically demanding areas such as engines and transmissions. In 1999 and 2000 surpluses averaged about 15 percent of production value. North America is far and away the largest export market, but China and other Asian destinations are catching up. The government has actively promoted exports to Japan and (especially) its overseas production bases by allocating licenses to import some Japanese cars to firms that increase their "reverse" exports to Japan. After steady progress, the sister firms China Motors and Yulong (with large minority investments by Mitsubishi and Nissan, respectively) have nearly balanced their trade with the rest of the Japanese production networks.

The elimination of local content requirements following accession to the WTO should reduce local production of some items. However, the slow pace of liberalization for local assemblers and the increased skill and export prowess of the parts exporters should cushion the blow considerably.

Outward foreign direct investment. Taiwan's (China's) auto parts firms are unusual in combining a high degree of reliance on transnational companies, overwhelmingly from Japan, with a high degree of outward investment. Both on their own and in conjunction with investments in China and Southeast Asia by the leading assemblers Yulong and China Motors, local parts firms have aggressively invested abroad, both to cut the costs of land and labor and to increase economies of scale. As of early 1999, Taiwanese (Chinese) auto firms had 106 ventures in China, largely divided between Fujian (home to the CMC and Yulong assembly operations and just across the Taiwan Strait) and China's economic heart: the Shanghai-Jiangsu-Zhejiang area. For a time Japanese assemblers largely ignored Taiwan (China). More recently, as Japanese firms have been forced to rationalize operations, they seem to have concluded that local firms provide reasonable quality at a reasonable price and that they can effectively complement operations in Southeast Asia and help to manage investments in mainland China. The auto industry, like Taiwan (China) more generally, appears to be finding its niche between China and Japan.

Thailand

By the mid-1990s the Thai auto industry had become the strongest in Southeast Asia in volume of production (559,000 units in 1996), number of parts producers (roughly 1,200 firms), and range of automotive parts produced.[21] This growth reflected roughly two decades of investments by Japanese OEMs, producing mostly for the protected Thai domestic market. In the early 1990s this supplier base and accompanying infrastructure, along with partial trade and investment liberalization and a growing domestic economy, drew export-oriented Western assemblers and, eventually, their first-tier suppliers. Despite being hit hard by the 1997 crisis, the industry is characterized by growing clusters of assemblers and suppliers generating significant growth of autos and auto parts exports. The major downside of this picture is the continuing general weakness of Thai-owned producers.

Demand for and production of autos. Thailand's automotive market grew rapidly from 1981 until the Asian economic crisis. The volume of domestic sales rose from 78,000 in 1986 to 304,000 in 1990, to 572,000 in 1995, and to 589,000 in 1996. After falling sharply to 144,000 in 1998, sales rose to 255,000 units in 2000, while production hit 459,000 units.[22] During this same period, the volume of domestic production rose from 74,000 in 1985 to 304,000 in 1990, to 525,000 in 1995, and to 559,999 in 1996. The difference between domestic sales and production reflects the growth of Thai auto exports, which increased from 68,000 units in 1998, to 126,000 in 1999, to 153,000 in 2000, and to 175,000 in 2001.[23] Auto parts exports grew as well. The structure of vehicle demand and production also merits note. The majority (58 percent) of vehicles produced in the last two decades were one-ton pickup trucks, which also dominated export sales (Brooker Group 2002, p. 106), with pickups accounting for more than 70 percent of all vehicles produced in Thailand in 2000 (see chapter 5 of this volume). The country's large domestic market for pickup trucks (second-largest in the world after the United States) has also been important in enabling Thai suppliers to achieve scale economies.

21. Production volumes are from the Automotive Industry Club of the Federation of Thai Industries, cited in Brooker Group (2002, table 3.14). Numbers of firms are from Brimble and Sherman (1999, p. 21). Brooker Group (2002, p. 208) cites a figure of 1,600 parts firms. Takayasu and Mori (chapter 5 of this volume) also note that Thailand has the most developed raw materials industry in Southeast Asia.

22. Sales volume from the Auto Industry Club of the Federation of Thai Industries are cited in Brooker Group (2001, pp. 2–13). Production volumes from the Auto Industry Club of the Federation of Thai Industries are cited in Brooker Group (2002, table 3.14).

23. Vehicle figures from the Thai Automotive Industry Association and MMC Sittipol are cited in Brooker Group (2002, p. 103, diagram 3.17). Vehicle exports rose from less than 5 percent of total production in 1995 to almost 40 percent in 2001 (Brooker Group 2002, p. 103, diagram 3.18).

Japanese brands have dominated the Thai market, accounting for more than 90 percent of sales throughout the 1990s, declining only slightly to 88 percent in 2001 (Toyota Motor Thailand, cited in Brooker Group 2002, p. 108). And although some 15 assemblers operate in Thailand, the market is dominated by Toyota (28 percent in 2001), Isuzu (24 percent), Honda (13 percent), Nissan (12 percent), and Mitsubishi (8 percent). Western firms entering the market in the mid-1990s, especially BMW, Daimler-Chrysler, Ford, and General Motors, operate largely through Japanese partners and Japan-designed vehicles.[24]

Automotive policy. Thai auto policy has been a balancing act between protection and product-specific incentives, on the one hand, and gradual liberalization, on the other. The auto industry expanded initially as a result of import substitution industrialization efforts beginning in the late 1960s. Key policy components included tariff protection, local content requirements (54 percent for passenger cars, 60–70 percent for one-ton pickups), and limits on new assemblers, facilities, and models. High protection encouraged the growth of local parts producers, which numbered more than 1,000 firms. Also aiding local suppliers was a special, low tax rate on pickup trucks that, until 1991, resulted in pickup truck prices that were about half the price of mid-size passenger cars. More suitable to Thailand's rural areas, pickups offered parts firms higher scale economies in lower-technology parts (such as brake drums). Pickups were, in a sense, Thailand's national car, albeit one not limited to one brand, as was the case for Malaysia's national car programs.

Beginning in around 1990, as Thailand's economy experienced rapid growth, the government began to loosen constraints on the industry. Liberalization included tariff reductions in 1991 and 1992; permission for existing assemblers to increase capacity and models; permission for investments in new assembly plants and parts production in 1994; provision of tax incentives to encourage automotive exports in 1994; lifting of local content requirements in January 2000 to comply with WTO regulations; and ending of special decentralization incentives to encourage clustering.[25]

24. Both Ford models produced in Thailand were designed by (Ford-owned) Mazda in Japan. General Motors operates through Isuzu.

25. Other measures included liberalization of taxi registration, which raised demand for 1,600 cubic centimeter passenger cars and a requirement that all producers install catalytic converters, which compelled assemblers to reduce their stock of autos prior to the deadline (see Nipon 1999, pp. 11–12).

In addition, the government expanded infrastructure and investment incentives for industrial estates in the eastern seaboard area and port facilities for container vessels at Laem Chabang (Unger 1998, ch. 6). Building on the overall growth of the Thai economy, these measures encouraged overall market expansion and attracted Western assemblers as well as first-tier suppliers such as Arvins, Bosch, Dana Spicer, Delphi, Johnson Controls, Lear, TRW, and Valeo.

Among other consequences of these investments, two merit special note at this point. First, they provided further impetus for the development of automotive clusters, already initiated by Japanese assemblers and suppliers in the early 1990s.[26] Second, they expanded capacity. By 1999 the auto industry had a total capacity of roughly 920,000 units (Ministry of Industry, cited in Nipon 1999, table 5). Because the domestic market was relatively small, this expanded capacity added to the pressure for export expansion.

There are limits to this liberalization, however. To help local parts producers prepare for the end of local content requirements, in 1997 the government initiated a number of tariff adjustments to encourage local assembly and sourcing.[27] In addition, the Thai parts industry continued to be plagued by a biased tax structure for finished products and raw materials that raised the cost of locally made products relative to equivalent imported goods.[28]

Parts production. Local production of parts expanded with the industry's overall growth. By 2001 parts "made" in Thailand accounted for roughly

26. These include the Laem Chabang Cluster (Mitsubishi along with suppliers such as Thai Summit and Siam Michelin); Gateway Cluster (Toyota, with only a few suppliers); Eastern Seaboard Industrial Estate (Ford-Mazda and General Motors), and several mini clusters and stand-alone facilities (Thai Automotive Institute presentation and website). Takayasu and Mori (chapter 5 of this volume) characterize the Thai cluster as the largest automobile industry cluster in ASEAN.

27. Adjusted tariff rates increased the gap between tariff rates applying to assembled cars and to completely knocked down passenger car kits but cut the differentials for commercial vehicles. In 1999, in a further move to help parts producers, the government raised the tariff rate for passenger cars and pickup trucks from 20 to 33 percent, making imported parts some 5 percent more expensive than locally produced ones (Brooker Group 2001, p. 2.4; Brooker Group 2002, pp. 209–10; Nipon and Chayanit 2000, p. 6). More recently, Thailand is preparing to restructure its tariffs to protect local parts firms prior to tariff reductions required by AFTA. After tariffs on finished vehicles are cut to 0–5 percent from 30–80 percent to comply with AFTA, duties on completely knocked down kits will also be cut to 0–5 percent from 20–33 percent to encourage local assembly. Further, tariff cuts on completely knocked down vehicles will be selective (Watcharapong 2002).

28. The problem is that tariffs for many raw materials are higher than those levied on finished parts. See, for example, Nipon (1999, p. 21).

60 percent of the components used in vehicles produced in Thailand (Brooker Group 2002, p. 205). The variety of locally produced products is the broadest in Southeast Asia, ranging from engines, to wiring harnesses, to radiators, to body parts, to tires. Auto parts exports expanded significantly, with the value of parts exports rising more than tenfold during the 1990s, from $78 million in 1991 to $871 million in 1995 and to $1.4 billion in 1999 (Brooker Group 2002, p. 214). From 1998 to 2001, exports were dominated by wire harnesses, engines, ignition systems, and tires.[29] Much of this export growth seems to have come from non-Thai producers. According to one Japanese executive, half of all parts exports came from just two Japanese electronics firms: Denso and Thai Arrow (cited in Nipon and Chayanit 2000, p. 7). Thus, in 1999, only 30 percent of Thai exports came from locally owned firms; the other 70 percent came from foreign and Thai-foreign joint ventures.[30]

The minimal Thai role in exports reflects weaknesses apparent even before the 1997 crisis. A 1995 study found that Thai-owned firms' level of production technology was a C+, below the international OEM requirement of a B grade (Nipon 1999, table 9; see also chapter 5 of this volume). This rating reflects generally low productivity in assembly (11 cars per worker compared with the global benchmark of 45), weaknesses in precision tool making, as well as a lack of personnel trained in general design and engineering. These problems are, in turn, a function of broader gaps in vocational training and persistently low enrollments in technical areas such as computing, engineering, mathematics, and science, especially compared with Korea and Taiwan (China).[31] With regard to tertiary technical enrollment, Thailand's percentage of population enrolled in 1997 was 0.19 percent, almost the same as in 1985 (0.16 percent), whereas enrollment in Korea and Taiwan (China) doubled from levels that in 1985 were already three to four times higher than Thailand's 1995 figures (from 0.86 to 1.65 percent in Korea and from 0.59 to 1.06 percent in Taiwan, China; Bell 2003, p. 14). Indeed, a 2003 report concluded that "the scale of resources and capabilities for exploiting technology and generating innovation lags far behind what might be expected on the basis of international

29. Data on market destinations of these exports are available only for "general auto parts," which do not include the most important export products: for example, engines, tires, and electrical ignitions (Brooker Group 2002, p. 215).

30. According to Brooker Group (2002, p. 214), in 1996 exports accounted for less than 2 percent of sales of one of the largest Japanese component producers operating in Thailand. By 2000 the percentage was 24 percent.

31. For a recent overview of the mismatch between work force skills and company needs, see Ketels (2003). On vocational weaknesses, see Ritchie (2001).

comparisons" with regional competitors.[32] The industry's trade balance and real level of local content reflect the weaknesses of local firms. In 1996 Thailand's total trade deficit was $14.6 billion, of which $3.8 billion, or 28 percent, was due to the vehicle and parts section (Brooker Group 1997, p. IV-2). And actual local content—that is, real value added of vehicles assembled in Thailand—was much lower than the official 54–70 percent, being closer to 20 percent according to one estimate (Veloso, Soto Romero, and Amsden 1998, p. 17).

The Thai parts sector has experienced sharp consolidation since 1997. Because most local firms were severely undercapitalized before 1997, the crisis hit them hard. After the government lifted the 49 percent ceiling on foreign shareholding in 1997, many local firms were taken over by foreign partners that also provided access to new export markets. The result was a shakeout in which all of roughly 230 first-tier suppliers survived, albeit under foreign ownership, while some 600 smaller, largely Thai-owned, producers went out of business. By 2001–02 the roughly 1,200 Thai parts producers operating prior to the crisis had been reduced to some 500–600 firms.[33]

Prospects. Given the small size of its domestic market, Thailand must optimize within the regional and global frameworks created by Western and Japanese assemblers. But the country also has important assets. Owing to its import substitution policies, product-specific incentives, local parts producers, infrastructure development, and recent, gradual liberalization, Thailand has become the site of several foreign-dominated automotive clusters. Given East Asia's expected market growth, these production sites are well poised to play important roles within the region.

Relatively low-technology, high-volume vehicles offer significant opportunities for local parts production. Since several firms—Ford, General Motors, Isuzu, Mitsubishi, Toyota—are committed to using Thailand's domestic market as a springboard for regional and global exports of pickup trucks, parts for pickup trucks, including engine components, constitute a major opportunity for local parts production. Recently, for example, Ford, which exports pickups from Thailand to 130 countries around the world,

32. Not only do Thailand's current capacities lag, but its commitment to developing these capacities also "lags behind the efforts of [Korea, Taiwan (China), and Singapore] around 10–20 years ago when they had levels of economic development similar to those of Thailand now" (Bell 2003, p. 5).

33. The total of Thai firms is 600 according to Brimble and Sherman (1999). The Thai Automotive Industry Association lists roughly 525 producers as of 2001 (cited in Brooker Group 2002, p. 206). The number of first-tier survivors is from Brooker Group (2002, p. 208).

announced plans to expand its Thai facilities for regional sales, since the company expects Asia to account for a substantial portion of the growth in total global auto sales in the coming decade (Pichaya 2002). Another source of demand for local parts is the production, largely by Japanese firms, of entry-level, global platform–type vehicles for the Asian market. These include low-priced vehicles, such as Toyota's multipurpose vehicle, suitable for developing countries, as well as low-cost passenger vehicles built on global platforms (so-called Asia cars) produced by Honda and Toyota. The Japanese have already begun to integrate these projects into regional component exchanges through AICO. Successful liberalization under AFTA will strengthen Thailand's position as a base of regional production.

With regard to particular types of parts, expanding the range of electronics components may be possible given Thailand's extensive consumer and industrial electronics sectors and the fact that electrical parts are already significant export items for the country's auto industry. This is advisable since Thai wage rates are no longer highly competitive relative to those of regional competitors (for example, China), and producers of labor-intensive products such as wire harnesses may opt to go elsewhere. Building on its long experience in assembly, the country also has strengths in press parts and related supporting industries such as dies, molds, and jigs. Unlike more technologically sophisticated components, these products offer opportunities for entry and technological upgrading based on the particular needs of developing-country markets. It is worth note that a Thai company has emerged as one of the world's best suppliers of low-volume tooling.[34] Finally, Thai parts firms unable to break into OEM markets may pursue a "Taiwan (China) model" by exploiting aftermarket and REM opportunities.[35]

The preceding has several important consequences. First, as shown in chapter 5 of this volume, the growth opportunities for local parts manufacturers are shrinking. The best OEM options for local firms lie, at least for the medium term, in becoming second- or third-tier suppliers or providers of tooling (jigs, molds, dies) and of intermediates and raw materials, such as plastics, leather goods, glass, paint, rubber goods, and petrochemical products (Ki 2002). The replacement market offers real opportunities, but even here, Thailand is something of a "middle child": lacking broad technologi-

34. The firm, Aapico, recently won the contract to design, supply, and test jigs for the worldwide assembly of Mercedes's new luxury E-Class. See Crispin (2002); Deyo and Doner (2001).

35. The Thai Ministry of Industry recently announced a policy to reduce dependence on OEM production and follow "Taiwan's successful model" of REM exports to the U.S. market ("Auto Parts Exports" 2002).

cal competencies, but no longer able to build on inexpensive labor (Brooker Group 1997, p. V-13). The degree of indigenous participation in Thailand's growing role as a regional production base thus depends on broader public policies and institutions devoted to human resources, small- and medium-size enterprise financing, acquisition of technology, information technology, logistics, and general value chain management. Efforts in most of these areas have already begun.[36]

Even absent indigenous participation, Thailand's role as a base of regional production is likely to grow as a result of its preexisting assets, the weaknesses of other Southeast Asian industries, and the pressures on global producers to localize supplier and support industries. But, given China's market, labor costs, and growing technological assets, Thailand cannot afford to rely on existing assets and investments alone. Sustained growth will require considerable attention to the same areas required for the development of local firms: human resources, information technology, logistics, and other types of infrastructure development.

ASSETS, MARKET STRATEGIES, AND POLITICS

The strategic options open to Asian governments in adjusting to this new environment of great opportunities but increasing barriers to entry depend in good measure on the location-specific assets that they bring to negotiations. But the translation of assets into desired outcomes depends on policies, and the effectiveness of those policies is, in turn, a function of institutions and political factors.

Assets

First and most important are the size of the local market and its prospects for future growth. Autos are complex, heavy products whose appeal is deeply influenced by local consumer preferences and government policies, from gas taxes and road quality to safety and environmental standards. Even absent tariffs and other trade barriers, automakers usually prefer to produce locally when possible, but the profitability of local production is greatly affected by the size of the market. For assemblers or large component suppliers considering new investments, market growth relative to existing capacity is also crucial. Of the five economies discussed in this chapter, China offers the most attractive domestic market, while Korea, whose

36. An interesting example is the establishment of a joint effort linking Thai parts producers and Singaporean precision engineering firms looking for lower-cost production sites (Nareerat 2002b).

domestic market is largely saturated, has considerable potential as a regional hub for exports; these economies are best placed to attract new investors. They are followed by Thailand and then Malaysia, while wealthy but saturated Taiwan (China) is the least promising. To the degree that ASEAN free trade is realized, Malaysia and Thailand will become more attractive sites for foreign investors, since similarities in geography and climate make it possible to sell identical or similar cars from facilities in these countries to the rest of Southeast Asia.

A second major consideration is the size and strength of the pool of existing assemblers and suppliers and the degree to which they are located in economically efficient geographic clusters. The countries vary with regard to the presence of OEMs, with China's pool of assemblers growing, Korea's and Taiwan's (China's) fairly strong, and Thailand's and Malaysia's weak. With regard to clusters, the gap across our five economy cases is not great. China, with its continental reach, is somewhat more complicated than the others, although the location of most regional clusters near the coast ameliorates the situation somewhat. All of the others have significant clusters of suppliers within easy reach of sea transport, usually concentrated in one part of the economy (for example, for Taiwan, China, the area south of Taipei, and for Malaysia, the area near Kuala Lumpur) or in two (for example, Busan and Seoul). Thailand's four clusters, especially the agglomeration of assemblers and suppliers in the eastern seaboard, constitute an especially attractive asset.

Cross-economy differences with regard to the numbers and manufacturing capacities of local suppliers are somewhat greater. Parts firms in Korea and Taiwan (China) are at the upper end of the regional scale with regard to technological competence. Thai firms, although numerous, are generally weak in the kinds of manufacturing capacities required by assemblers. The situation is generally similar for local Chinese suppliers. Malaysian parts producers are probably the weakest in the region as a result of high levels of protection and distortions generated by ethnic preferences. Despite some strengths, in none of the economies covered here, with the partial exception of Korea, do local suppliers constitute a major asset, except at the lower tiers of the supply pyramid. The strength of local suppliers is also affected by the strength, stability, and flexibility of each economy's financial system. None of the East Asian financial systems lacks problems, as the financial crisis of 1997 and the more chronic problems in Taiwan (China) and Japan have demonstrated. China, the largest of the five, is most severely beset with nonperforming loans and has the fewest alternatives to the traditional banks, while Malaysia, the smallest of the five, probably has the soundest banking sector.

Finally, physical and human infrastructure clearly differentiates the richer from the poorer economies. Korea and Taiwan (China) have the best transportation systems, the highest penetration rates for the Internet and broadband telecommunications, and the most sophisticated engineers. Malaysia and Thailand are weaker (but improving) in areas such as transport and communications, but they present idiosyncratic problems: a very weak educational system in the case of Thailand and continuing ethnic tensions (and educational and economic gaps) in the case of Malaysia. The most interesting case—and the most difficult for foreign investors to read—is China. On the one hand, it is the poorest and the most unevenly developed of the five, unemployment and other potentially explosive social problems abound, and local protectionism still remains a barrier to the integration of the domestic auto industry. On the other hand, at least along the coastal provinces, the improvement in physical infrastructure, including the creation of an extensive highway system, is remarkable. And since the first stirrings of the industrial revolution over two centuries ago, the capacities of local manufacturers have probably never increased more rapidly than they have in China over the last two decades, a tribute not only to the determination of China's policymakers and the diligence and flexibility of Chinese managers and workers but also to the increasingly streamlined global system of technology transfer.

Market Strategies

In principle, these location-specific assets suggest somewhat different leverage and options. Large and growing markets in China and Korea, and significant experience with auto parts export in Korea, mean that these two are in the best position to help local firms jockey for a better position in the global division of labor. Conversely, smaller domestic markets mean that Malaysia, Taiwan (China), and Thailand will find it more difficult to shape comparative advantage, with Malaysia and Thailand burdened further by weak indigenous technical capacities.[37] But in spite of the constraints they face, even these economies have options.

We can identify four categories of options for national auto suppliers, three of which involve market orientation: domestic or export market, original or replacement equipment manufacturing, and product-specific

37. For example, owing to delays in implementing AFTA's multilateral tariff reductions, automakers that selected Thailand as their primary regional hub might opt to divert some investment to other countries in the region to take advantage of lower (bilateral) tariffs under the AICO scheme. Such a development would result in a network of smaller plants, a possible increase in investment to China, and a further stimulus to localization, since AICO exchanges require 40 percent ASEAN content (Nareerat 2002a; Srisamorn 2002).

niches. The fourth has to do with the degree of local versus foreign presence in locally based parts production. China and Korea have the opportunity to expand exports from companies benefiting from economies of scale based on large domestic markets. In China's case, the growth of the domestic market provides significant opportunities for production and export of parts for lower-end cars and light commercial vehicles, although, as noted in the case study, the value added of these exports might be stratified among local operations linked to Taiwanese (Chinese), Japanese, and Western parts producers (see, for example, Murphy and Lague 2002). In Korea there are opportunities both for domestic OEM and exports of REM parts for Korean cars sold overseas. New linkages with global assemblers may provide new opportunities for marketing Korean-built cars and parts in regional and extraregional markets (for example, Latin America). The challenge for local parts firms is to increase their level of technological competency to enable them to tap into these new openings. As Taiwanese (Chinese) firms shift production of many OEM parts to China, their best options for domestic production seem to be in replacement parts such as bumpers and metal stampings for export markets. This is an area in which locally owned firms have managed to hold their own despite gradual liberalization. Whereas continued growth in their domestic market is likely to provide Thai producers with some OEM sales, the greatest opportunities are for OEM exports to regional markets in components for pickup trucks. Except for lower-tier components, this process will be dominated by foreign producers. Malaysia is the weakest player among the five. At best, there are opportunities for lower-value-added parts for national cars sold in the domestic market and, potentially, for regional sales of Proton, a small-volume passenger car that might be marketable elsewhere in the region. Whether this occurs, of course, will depend on trade, ownership, and other policies to which we now turn.

Policies

In this section of the chapter we review how policies in specific areas—trade, ownership, finance, infrastructure, and industrial policy more generally—have exerted an impact on the auto parts industries in our five economies.

Trade. Regional and global trends toward freer trade are powerful. Everywhere from ASEAN to the Republic of Korea tariffs, quotas, and nontariff barriers are declining; with the accession of China and Taiwan (China) in late 2001, virtually all of the Asian economies have joined the WTO. Still, within the general trend of liberalization, some room for government maneuver remains. Taiwan (China), for example, continues its path of consistent but slow liberalization, while Thailand, feeling pressure from the rush of foreign investment to China and benefiting from a strong assembler-supplier complex, is moving more quickly. The hand of government weighs heavily in China and Malaysia, but (with the partial exception of some leftovers from the financial crisis) more lightly in Korea.

Ownership. A crucial issue concerns policies and informal attitudes toward ownership and autonomy. Local ownership can provide greater opportunities for skill acquisition and technology transfer as well as for more independent corporate strategies. The automotive history of East Asia strongly suggests that the best-performing local firms have limited foreign ownership and diversified sources of technology rather than depending on foreign parent companies for capital or technology (Toyota versus Nissan in Japan, Hyundai versus Daewoo in Korea, Yulong–China Motors versus Liu Ho in Taiwan; see, for example, Amsden 1989; Cusumano 1985). At the same time, independence is no guarantee of success, and with increasingly liberal trade and a powerful trend toward global consolidation, it is unrealistic to expect that Proton, for example, could repeat the success of Toyota. Even local-foreign joint ventures, which can be difficult to manage, are increasingly giving way to full foreign control. The overall trend is clearly toward liberalization of ownership requirements: to compete effectively in regional and global markets, assemblers and major parts suppliers are demanding and getting a freer hand. In assembly, this is illustrated by General Motors' takeover of Daewoo in Korea and the acquisition of effective control over Malaysia's Perodua by Toyota's subsidiary Daihatsu as well as the increasing control by Western firms of their Japanese affiliates where such linkages are proving to be commercially successful. Still, some room for maneuver remains for the countries with the best location-specific assets, especially China and Korea.[38]

38. The Korean government, for example, ignored antitrust considerations in permitting the acquisition of Kia by Hyundai, thus greatly strengthening the position of Hyundai as de facto national champion. Another telling case is the September 2002 joint venture between Nissan and China's second-largest assembler, Dongfeng. In return for an initial investment of more than $1 billion and detailed promises to continue investing in China, Nissan received the right to assemble a full range of vehicles, something China had not granted to foreign producers in the past. Nissan did not, however, gain majority control, and press reports suggest that the speed with which it will reduce Dongfeng's bloated work force will be substantially slower as a result (*AutoAsia*, September 19, 2002; *Business Week*, October 7, 2002). Although the motivation may be largely domestic—sustaining employment among the urban working class, which has traditionally been the bedrock of support for the Communist Party—the effect of such agreements may well be to send "costly signals" that Sino-foreign joint ventures will inevitably be committed to sustaining and expanding production. In a region troubled by domestic overcapacity, such signals could deter investments in Korea, Taiwan (China), and Southeast Asia.

Systematic information on policies regarding the ownership of parts producers is less readily available, but Taiwan (China) and Thailand seem to be moving toward complete liberalization of ownership requirements. Yet some diversity in ownership patterns evidently persists, as reflected in the overseas operations of Denso, Japan's largest producer of auto parts.[39] China and Korea appear to be the two partial exceptions to the trend toward complete foreign control, although the number of cases is too small to allow definitive conclusions.

Finance. A third important policy area is finance, particularly banking. The vast majority of auto parts producers remain dependent on banks for financing; only the largest suppliers have the skill and scale to finance their operations primarily through offerings of stocks or bonds. The quality, stability, and accessibility of local banks thus are crucial to the viability of the auto industry. Although weaknesses plague all Asian banking systems, cross-economy variations are emerging. China is especially troubled, and the speed at which new banks, including foreign banks, enter the market may become an important issue for local companies. Thailand's banking system also remains weak. Korea, in contrast, is turning around what had been one of the least effective banking systems in the region. An additional source of financing is support from Japan—both from the parent companies of local subsidiaries and from trading companies—and from Japanese foreign aid, a good chunk of which has found its way toward local subsidiaries of Japanese firms, especially in the aftermath of the financial crisis. Local parts producers not closely affiliated with the Japanese may thus find themselves at a disadvantage if financial problems create another credit crunch.

Infrastructure. With the slowdown of growth in Southeast Asia and massive spending on roads, railways, and ports in China, transportation congestion has generally eased, at least for a few years. But the generation and transmission of electricity have sometimes failed to keep pace, even in relatively advanced areas such as Taiwan (China). To some degree, the private sector can itself deal with energy issues. With advances in technology, it is increasingly feasible for large firms to generate electricity on site, perhaps selling excess capacity to the national power grid. But this may create dilem-

39. Denso has adopted full or almost full ownership in most markets, particularly for its largest plants, but 50-50 joint ventures or even minority shareholdings are not uncommon, particularly in Asia. As of 2001, Denso's shares in three Korean subsidiaries were 40, 51, and 100 percent, respectively; in Thailand they were 51.3 and 100 percent; and in China they were 30, 40, 51, 85.9, and 94.2 percent (Denso 2001, pp. 8–9).

mas for small producers not clustered near larger firms. Telecommunications facilities may also manifest some clustering tendencies, as suppliers find it economical to extend fiber optic cables to large and tightly packed industrial users near urban areas. Smaller and richer areas—Hong Kong (China), Korea, Singapore, Taiwan (China)—are pulling away from the developing countries, despite rapid installation of broadband access throughout the region. Again, China is in between—uneven and behind, but rapidly improving. And again, gaps are also opening up by firm size, as virtually all large suppliers can offer products on the Internet, but few smaller companies can do so. In electronics, Dell and other foreign assemblers are requiring that all procurement be handled through e-commerce; as similar trends spread to autos, economies where the entire supply base can work effectively through the Internet will find themselves with a major competitive advantage.

Full implementation of e-commerce is dependent not solely or even primarily on physical infrastructure but also on skills and organizational infrastructure. Indeed, skills may become the great differentiator between winners and (relative) losers in the auto business. Technical training and upgrading of local skills is largely a "win-win" situation for foreign assemblers and their local hosts and suppliers (see, for example, the agreement between General Motors and Thailand; Noble 2001) but implementation remains difficult, particularly in countries with weak or uneven educational systems. Takayasu and Mori, chapter 5 of this volume, also refer to the weaknesses of Thai auto component suppliers with respect to information technology. Japanese assemblers have proven more willing to provide technical assistance to their suppliers (even in North America, suppliers report that the Japanese are significantly more helpful), but they also face a greater language and cultural gap, since few locals know Japanese, while Japanese managers and technicians often have weak English skills. Less noticed, but important for the long run if firms and workers are to be convinced to invest heavily in specific skills that could become vulnerable to market vicissitudes at the firm or industry level, is the provision of retraining opportunities and employment protection for auto parts workers (for example, see Estevez-Abe, Iversen, and Soskice 2001).

Implementation of training programs and many other collective policies, such as format standardization, electronic commerce, and interactions with other industries, often are greatly facilitated by effective industry associations. Here again, the Japanese role has been particularly important. Japanese advisers have taken the lead in establishing many industry associations, Japanese agencies have often insisted on working through local associations, and Japanese-affiliated firms are usually the leading participants in both national and regional auto associations (Doner 1997). To improve industry associations requires financial support, capable staff, and the willingness of governments to communicate with and delegate many tasks to the associations. Korea and Taiwan (China) are well ahead of Southeast Asia, while China is hampered by its size and diversity.

Industrial policy. Improving physical, human, and organizational infrastructure is a difficult but essential task for East Asian economies. The degree to which regional governments should focus on particular products, technologies, or firms—that is, articulate an explicit or implicit industrial policy—remains a matter of considerable controversy. The usual arguments in favor of industrial policy—promoting economies of scale, scope, learning, and agglomeration as well as overcoming coordination dilemmas, market externalities, and weaknesses in local financial markets—all apply to autos, and first-mover advantages, once less true of autos than of electronics, may be growing. Increasing oligopolization of global assembly and component production also makes host economies reluctant to trust to markets that are less than perfectly competitive.

Given these considerations, some version of industrial policy is likely to persist, despite the challenges and dangers involved, such as daunting information requirements, difficulty in keeping up with rapid market developments, infant industries that fail to grow up, politicization and rent seeking, and a world trading system that is much less willing to acquiesce in government policies of protection and promotion than it was when Japan and Korea developed their auto industries. Such persistence may be particularly the case for China and Korea, which export significant quantities of auto parts to world markets.

Such industrial policies will probably take a soft form that does not attempt to target highly specific products if such targeting is inconsistent with the strategies of subsidiaries of global firms. Instead, policies will be designed to promote and expand comparative advantage, even if such advantages are the result of highly targeted policies from the past. In Thailand, for example, tax policy has created an overwhelming preference for oneton pickups that has allowed assemblers to achieve economies of scale in pickup production. Along with assembly, production of pressed parts and jigs for pickups is also concentrating in Thailand. Building on this base, the Thai government has initiated policies to make auto parts production, including for the replacement market, one of the country's target growth industries ("Competitiveness" 2002).

To the extent that economies attempt to build on (somewhat fortuitous) outcomes of past policies as opposed to imposing sectorally neutral policies, they must address two further issues. The first is whether governments should encourage local parts producers to concentrate on OEM or the production of replacement parts. At first glance, production of aftermarket goods appears more appropriate for developing economies, since these tend to be price sensitive and to impose only modest technical requirements. For many years, Taiwan (China) exported more auto parts than Korea, despite the much larger size of the Korean auto industry, precisely by concentrating on replacement parts. The size of the no-brand replacement market is limited, however, and pressure by Western consumers to prevent insurance companies from paying for only the lowest-quality replacement parts may shrink it further. In recent years, the expansion of exports of Korean cars has propelled the export of both OEM and replacement parts, so that Korean auto parts exports now easily outpace those of Taiwan (China). Production of OEM parts, however, requires tighter coordination with global suppliers and first-tier component suppliers, and, as noted, requirements for quality and interoperability are higher. Local firms hoping to break into OEM markets, and even foreign firms looking to expand local production to cut costs, may well petition governments for help on a variety of fronts, from technology to marketing. At the same time, Taiwan's (China's) experience, in which public-private sector linkages were critical to certification and "national reputation," suggests that local firms succeeding in developed-country REM markets also require help in areas such as quality control and delivery.

A second issue is whether governments should orient policies toward smaller parts makers at the bottom of the production pyramid or toward medium and larger firms in the first or second tier. Opportunities for rapid productivity growth and high-paying jobs are greater near the top of the pyramid, but establishing a major position in the regional and global production networks of global assemblers, including increasing responsibility for design, will be difficult for most local suppliers, which, once again, are likely to petition governments for support. At the same time, promotion of small- and medium-size enterprises itself poses important challenges in areas such as information diffusion, quality certification, technical training, and financing.

Thus, while the general trend of trade and industrial policies in the region is clearly toward more liberalization and openness, the incentives for companies to continue to seek and for governments to supply some degree of protection and, most critically, promotion are not likely to disappear. The policy challenge will be to do so intelligently and with realistic goals well suited to factor endowments. Successfully meeting this challenge is a function not of some abstract "political will," but rather of the distributional and institutional factors influencing the formulation and implementation of policy.

Politics: Interests and Institutions

Local auto parts producers tend to press for protection against imports and market downturns: production is generally labor intensive and, at the higher end, involves large and often highly asset-specific investments. Nevertheless, corporate reorganization has proceeded at a rapid pace in recent years, and Asian governments have generally acquiesced. Korea and Thailand are the clearest cases. In Taiwan (China), General Motors, Honda, and Peugeot have all dumped their long-time joint venture partners and put more emphasis on imports, while many parts firms have moved much or all of their production to China. Even though urban workers have traditionally constituted the core support of the Chinese Communist Party, the Chinese government has allowed and even pushed state-owned enterprises to shed labor and explicitly favors the top three assemblers over all others.[40] Even Malaysia may be the exception that proves the rule: the government has relinquished majority ownership in Perodua, and former Prime Minister Mahathir, the godfather of the domestic industry, signaled his intention before his retirement to ease the ethnic component of industrial policy; in any event, a fundamental review of auto policy is likely now that Mahathir has retired. Resistance to liberalization has not evaporated, but it is relatively muted: organized labor is a significant force only in Korea, the reluctance of Taiwan (China) to allow unimpeded trade and investment with China has had only a minor restraining impact on autos, and so forth.

But distributional issues are far from resolved. For if particularistic pressures in the auto industry are proving weaker than expected in the face of regional and global trends toward liberalization and consolidation, it is less clear that East Asian governments have developed the coalitional bases to develop competent bureaucracies with effective and relatively transparent ties with the private sector. Thailand, for example, has resisted protectionism but has proved woefully weak at provision of crucial public goods and, despite considerable prodding from Japan, has been inconsistent in its attempts to build institutional links to a well-coordinated private sector. Korea has a stronger bureaucracy but is still struggling to

40. Toyota's joint venture to build engines in Tianjin, for example, has used a variety of methods to cull employment by 20 percent, and management is determined to cut another one-third in the next couple of years (Noble interview, Tianjin, April 2002).

build an open and effective relationship with the giant *chaebol* business groups, a task now complicated by the presence of major transnational partners. Democratization in a context of personalized and factionalized politics has yet to dampen the dramatic policy swings associated with a one-term presidential cycle.

While both Thailand and Malaysia have tended to promote one particular product niche, they have differed in a number of important ways. In assembly, Thailand's emphasis has been on small commercial vehicles—one-ton pickups—whereas Malaysia's has been on passenger vehicles. Also in assembly, Thailand has pursued a private sector–led strategy with few limits on foreign ownership, whereas Malaysian efforts have been state led and fairly restrictive with regard to foreign ownership. In part owing to its more open policies regarding FDI and in part due to its lack of discrimination against ethnic Chinese–owned parts producers, Thailand's indigenous parts sector is larger and more dynamic than Malaysia's. The result has been a greater Thai willingness to pursue liberalization, both regional and global. But if the two countries differ with regard to the size and strength of their parts sectors, their support for local parts firms, their trade policies, and their approaches to foreign investment, they resemble each other in their indigenous technological weaknesses, their relatively small domestic markets, and the fragmentation of their policies and institutions devoted to technological improvement.

Institutional weaknesses are likely to become all the more damaging as the region's economies attempt to implement the kinds of policies, especially in areas such as infrastructure and human resource development, necessary to meet the growing demands of assemblers and their first-tier suppliers. Several features make these issue areas especially challenging. First, effective policies in these areas require information that is technical and dynamic. Second, the required number of participants is relatively large, and the breadth of participation is relatively wide. For example, the increasing importance of first- and second-tier components firms from Japan and the West suggests that governments will need to enter complex negotiations with a wider variety of foreign firms and will no longer be able to depend on relations with a handful of global assemblers. Third, areas such as technical training tend to be plagued by free rider and associated collective action problems. Fourth, distributional tensions may worsen as liberalization, rationalization, and consolidation result in the disappearance of local firms and the rise of unemployment. Finally, the kinds of efficiency-related measures required to benefit from today's automotive production networks exhibit long time-to-payoff. They do not lend themselves to quick fixes. Investments in infrastructure, human resources, and technology diffusion often take time to implement and even longer to yield tangible benefits.

All of these considerations mean that effective policies must be backed up by institutions that can collect, exchange, and diffuse information, coordinate large numbers of interests, provide selective benefits and punishments, and operate on the basis of long time horizons. Absent such capacities, it is unrealistic to expect significant local participation in and, especially, benefits from regional and global auto value chains.

CONCLUDING OBSERVATIONS

Great opportunities beckon the auto parts industry in East Asia, but exploiting those opportunities is becoming more difficult. Strong growth in demand for autos and parts in the region combines with global trends toward liberalization, consolidation, and delegation of manufacturing to lower parts of the production chain to open up lucrative opportunities to local firms with the skills and savvy to assume vital roles in regional and production networks.

Liberalization and consolidation have created losers as well as winners, particularly among smaller local firms and those without foreign capital or technological links, but so far, protectionist impulses have remained surprisingly weak. Governments have generally accepted the verdict of the market and have made only limited efforts to moderate the pace of liberalization.

Less progress has been made in the positive task of creating capacities in the public and private sectors to boost local skills and technology to take advantage of the opportunities created by growth and liberalization. In much of the region, design and management skills remain weak, and e-commerce is far from universal. In recent years the Northeast Asian economies, while still modest actors in the global industry, have developed some impressive location-specific assets. Korea, in particular, combines a large market with solid engineering capacities. China's uneven development is more than offset by rapid growth in demand and skills, the lure of its vast population, and the commitment of its leadership to industrial upgrading. The picture in Southeast Asia, where many locally owned firms have failed and others are likely to succumb to regional and global liberalization, is less sanguine. Even Thailand, an emerging center of regional assembly of cars and pickups, lacks the organizational capacity and human skills necessary to meet global standards for quality, design, speed, and reliability of delivery.

Nevertheless, assemblers continue to prefer local assembly and procurement of parts whenever production volumes allow. With combined efforts from local governments, industry associations, aid donors, and international economic organizations, Southeast Asian producers still have ample opportunity to develop new capacities to supply local, regional, and global markets, both independently and in conjunction with the leading global assemblers and particularly the two or three dozen first-tier suppliers that are increasingly responsible for the manufacture of automobiles and their parts. Whether such efforts occur and succeed will depend in large measure on political dynamics: partisan competition, coalition formation, and their impact on bureaucratic development.

REFERENCES

The word *processed* describes informally reproduced works that may not be commonly available through libraries.

Abrenica, Joy V. 2000. "Liberalizing the ASEAN Automotive Market: Impact Assessment." Pacific Economic Cooperation Council project on the ASEAN auto industry. Processed.

Amsden, Alice H. 1989. *Asia's Next Giant: South Korea and Late Industrialization.* New York: Oxford University Press.

"Auto Parts Exports: Strategy Shift for Local Firms." 2002. *Nation*, June 25.

Bell, Martin. 2003. "Knowledge Resources, Innovation Capabilities, and Sustained Competitiveness in Thailand. Final Report." University of Sussex, Science and Technology Policy Research Unit, January. Geneva Processed.

Borrus, Michael, Dieter Ernst, and Stephan Haggard, eds. 2000. *International Production Networks in Asia: Rivalry or Riches?* London: Routledge.

Brimble, Peter, and James Sherman. 1999. "Mergers and Acquisitions in Thailand: The Changing Face of Foreign Direct Investment." Paper prepared for the UN Conference on Trade and Development. Geneva. Processed.

Brooke, James 2002. "Japan Carves out Major Role in China's Auto Future." *New York Times*, July 11, sec. W, p. 1.

Brooker Group. 1997. "Automotive Industry Export Promotion Project: Thailand Industry Overview. Final Report." Report prepared for the Thai Ministry of Industry, Bangkok. Processed.

———. 2001. "Thailand." Paper prepared for a project on the international competitiveness of Asian economies: a cross-country study. Bangkok, October. Processed.

———. 2002. "Thailand's Automotive Industry." Bangkok. Processed.

Brunnermeier, Smita, and Sheila A. Martin. 1999. "Interoperability Cost Analysis of the U.S. Automotive Supply Chain: Final Report." Research Triangle Institute, Research Triangle Park, N.C. Processed.

Burwell, Michael, and Lisa Ferris. 2000. "Road to Riches." *Daily Deal (New York)*, August 2. Available at LexisNexis: www.TheDeal.com. Processed.

"Car Tax: Major Overhaul to Precede AFTA." 2002. *Nation*, May 20.

Chatrudee, Theparat. 2002. "Toyota Targets 100 Percent Local Pickup Content in Four Years Awaiting Final Word on Shift to Thailand." *Bangkok Post*, August 28.

"Competitiveness: PM Unveils Global Drive." 2002. *Nation*, May 21.

Crispin, Shawn. 2002. "Fast Lane to Success." *Far Eastern Economic Review*, September 12.

Cusumano, Michael A. 1985. *The Japanese Automobile Industry: Technology and Management at Nissan and Toyota.* Cambridge, Mass.: Harvard University, Council on East Asian Studies.

de Jonckheere, Terry. 1998. "Asian Crisis: Challenges and Opportunities." Presentation to Asian Autos after the Crash: Where to from Here? Ford Motor Company, Dearborn, Mich.; University of Michigan, Ann Arbor, Office of Study of Automotive Transportation. Processed.

Denso. 2001. "Corporate Profile 2001." Kariya, Japan. Processed.

Deyo, Fredrick, and Richard F. Doner. 2001. "The Enclave Problem: Challenge of Flexible Production in a Weakly Coordinated Market Economy." In Fredrick Deyo, Richard Doner, and Eric Hershberg, eds., *Economic Governance and the Challenge of Flexibility in East Asia.* Boulder, Colo.: Rowman and Littlefield.

Doner, Richard F. 1997. "Japan in East Asia: Institutions and Regional Leadership." In Peter J. Katzenstein and Takashi Shiraishi, eds., *Between Two Worlds: Japan in Asia.* Ithaca, N.Y.: Cornell University Press.

Dunne, Michael. 2001. "When the Barriers Come Down." Presentation to Automotive News International Asia-Pacific Congress, March 28–29. Processed.

Economist Intelligence Unit. 2001. "Automotive Forecast Asia." London. Processed.

Estevez-Abe, Margarita, Torben Iversen, and David W. Soskice. 2001. "Social Protection and the Formation of Skills: A Reinterpretation of the Welfare State." In Peter A. Hall and David W. Soskice, eds., *Varieties of Capitalism: The Institutional Foundations of Comparative Advantage.* Oxford: Oxford University Press.

Fine, Charles H., John C. LaFrance, and Don Hillebrand. 1996. *The U.S. Automobile Manufacturing Industry.* Washington, D.C.: U.S. Department of Commerce, Office of Technology Policy.

Fourin. 1999. *Kaigai Jidōsha Chōsa Geppō.* (Monthly Survey of Global Car Manufacturing) 172. Nagoya, December.

Hatch, Walter, and Kozo Yamamura. 1996. *Asia in Japan's Embrace: Building a Regional Production Alliance.* Cambridge, U.K.: Cambridge University Press.

Humphrey, John. 1998. "Assembler-Supplier Relations in the Auto Industry: Globalization and National Development." Unpublished mss. Institute of Development Studies, University of Sussex, Brighton. Processed.

"Hyundai Mobis, Bosch Ink Brake Technology Deal." 2001. *AutoAsia Online*, November 16.

Jomo, K. S. 1994. "The Proton Saga: Malaysian Car, Mitsubishi Gain." In K. S. Jomo, ed., *Japan and Malaysian Development: In the Shadow of the Rising Sun*, pp. 263–90. London: Routledge.

Ketels, Christian H. M. 2003. "Thailand's Competitiveness: Key Issues in Five Clusters." Unpublished mss., Harvard Business School, Cambridge, Mass. Processed.

Ki, Woo. 2002. "Auto Sector Rises Again." *Nation*, October 11.

Legewie, Jochen. 2000. "Driving Regional Integration: Japanese Firms and the Development of the ASEAN Automobile Industry." In Verena Blechinger and Jochen Legewie, eds., *Facing Asia: Japan's Role in the Political and Economic Dynamism of Regional Cooperation*. Munich: Iudicium.

Leutert, Hans-Georg, and Ralf Sudhoff. 1999. "Technology Capacity Building in the Malaysian Automotive Industry." In K. S. Jomo, Greg Felker, and Rajah Rasiah, eds., *Industrial Technology Development in Malaysia: Industry and Firm Studies*, pp. 247–73. London: Routledge.

Liang, Quincy. 2002. "Taiwan Leads the World in Sheet-Metal Auto Body Parts." Available at http://www.taiwanfurtniture.net/linerpt/20000302063.html. Processed.

Matsushima, Noriyuki. 1999. "Japan's Auto Industry Plunged into an Era of Great Realignments." Interview with the director of Equity Research, Nikko Salomon Smith Barney, June 15. Available at http://www.fpcj.jp/e/gyouji/br/1999/990615.html. Processed.

McKendrick, David, Richard F. Doner, and Stephan Haggard. 2000. *From Silicon Valley to Singapore: Location and Competitive Advantage in the Hard Disk Drive Industry*. Palo Alto, Calif.: Stanford University Press.

McMaster, T. D. 2001. "Autoliv." Presentation to Automotive News International Asia-Pacific Congress, March 28–29. Processed.

Ministry of Machinery Industry. 2000. *Zhongguo Qiche Gongye Nianjian 2000* [*China Automotive Statistical Yearbook 2000*]. Beijing.

Murphy, David, and David Lague. 2002. "Time for a Tune-up." *Far Eastern Economic Review*, July 4, pp. 24–27.

Nareerat, Wiriyapong. 2002a. "ASEAN Free Trade: AFTA Delay Stifles Auto Plans." *Nation*, July 17.

———. 2002b. "Joint Vehicle-Parts Council: S'pore Agrees to Form Auto Club." *Nation*, October 31.

"New Era for Perodua." 2001. *Autoworld.com.my*, December 6. Available at http://www.autotrade.com.my/EMZine/Review/article_t2.asp?awReviewID=846&aw Parent ID =846. Processed.

Nipon, Paopongsakorn. 1999. "Thailand Automotive Industry (Part I)." Thailand Development Research Institute, Bangkok. Processed.

Nipon, Paopongsakorn, and Wangdee Chayanit. 2000. "The Impact of Technological Change and Corporate Reorganization in the ASEAN Automotive Industry." Pacific Economic Cooperation Council, project on the ASEAN auto industry. Processed.

Noble, Gregory W. 2001. "Congestion Ahead: Japanese Automakers in Southeast Asia." *Business and Politics* 3(2, August):157–84.

Pichaya, Changsorn. 2002. "Ford Mulls $100m Boost." *Nation*, November 8.

PWC (PricewaterhouseCoopers). 1999. "Major Auto Mergers Drive Sweeping Changes in the Parts Industry, According to PricewaterhouseCoopers Survey." Available at www.pwc.com. Processed.

———. 2000. "The Second Automotive Century Executive Summary." Processed.

Rasiah, Rajah. 1996. "Rent Management in Proton." Working Paper 2. University of Kebangsaan, Institut Kajian Malaysia dan Antarabangsa (IKMAS), Sarawak, Malaysia. Processed.

Ravenhill, John. 2002. "From National Champions to Global Partners: Crisis, Globalization, and the Korean Auto Industry." In William W. Keller and Richard Samuels, eds., *Crisis and Innovation: Asian Technology after the Millennium*, pp. 108–36. Cambridge, U.K.: Cambridge University Press.

Ritchie, Bryan. 2001. "Innovation Systems, Collective Dilemmas, and the Formation of Technical Intellectual Capital in Malaysia, Singapore, and Thailand." *International Journal of Business and Society* 2(2):21–48.

Scott, Alan J. 1987. "The Semiconductor Industry in Southeast Asia." *Regional Studies* 21(2):143–60.

Srisamorn, Phoosuphanusorn. 2002. "Brooker Confident over Local Industry's Outlook." *Bangkok Post*, July 17.

Treece, James B., and Robert Sherefkin. 2001. "Platform Sharing Key to Profits." *Automotive News* 75(5919):53.

Tyndall, Paramjit Singh. n.d. "The Malaysian Automotive Industry." Pacific Economic Cooperation Council project on the ASEAN auto industry. Available at http://www.asean-auto.org/Malaysia.pdf. Processed.

Unger, Daniel. 1998. *Building Social Capital in Thailand: Fibers, Finance, and Infrastructure.* New York: Cambridge University Press.

Veloso, Francisco. 2000. "The Automotive Supply Chain: Global Trends and Asian Perspectives." Background paper for a project on international competitiveness of Asian economies: a cross-country study. Massachusetts Institute of Technology, Cambridge, Mass. Processed.

Veloso, Francisco, Jorge Soto Romero, and Alice Amsden. 1998. "A Comparative Assessment of the Development of the Auto Parts Industry in Taiwan and Mexico: Policy Implications for Thailand." Massachusetts Institute of Technology, Cambridge, Mass. Processed.

Watcharapong, Thongrung. 2002. "Major Overhaul to Precede AFTA." *Nation*, May 20.

WTO (World Trade Organization). 2002. *International Trade Statistics 2001.* Available at http://www.wto.org/english/res_e/statis_e/its2001_e/chp_4_e.pdf.

Yap, Chips. 2001. "Will Malaysia Ever Open Up?" *Automotive Resources Asia*, October 15. Available at www.auto-resources-asia.com/experts. Processed.

CHAPTER 5

THE GLOBAL STRATEGIES OF JAPANESE VEHICLE ASSEMBLERS AND THE IMPLICATIONS FOR THE THAI AUTOMOBILE INDUSTRY

Ken'ichi Takayasu and Minako Mori

East Asian automobile industries have experienced dramatic changes since the late 1990s. These far-reaching changes are being driven by powerful forces that include trade and investment liberalization, strategic partnerships among vehicle assemblers in the advanced industrial economies, and the rising standard of Asian automobile industries.

In this chapter, we analyze the major changes that are occurring in East Asian automobile industries from four interrelated perspectives. First, the East Asian and global strategies of assemblers based in advanced economies are having an even greater influence on automobile industries in the East Asian economies than previously. The aim of those strategies is to enhance competitiveness by taking advantage of trade and investment liberalization to improve the efficiency of production networks in East Asia.

Second, vehicle assemblers recognize the location-specific assets of East Asian economies, including markets, technologies, human resources, and industry clusters (see chapter 4 of this volume). Faced with escalating international competition, they are giving priority in their investment allocations to the economies with the best assets. Their aim is to build competitive production networks. When the governments of target economies imposed various restrictions on investment, assemblers merely needed to manufacture whatever vehicles they could under these limitations at prices that allowed them to earn a profit in local markets. In the environment created by trade and investment liberalization, assemblers cannot survive against competition unless they build structures that allow them to provide markets with products that meet international standards of quality and price.

Third, an extremely important policy issue from the viewpoint of East Asian economies is the positioning of their economies in the international production networks of vehicle assemblers and primary parts manufacturers based in advanced industrial economies. This is true not only of members of the Association of South East Asian Nations (ASEAN) but also of the Republic of Korea, which has its own internationally active vehicle assemblers, and China, which has a vast market and extensive industrial infrastructure. Even in these economies, many domestic vehicle manufacturers have formed strategic partnerships with manufacturers from advanced industrial economies.

Fourth, the present situation offers both opportunities and challenges to automobile industries in East Asian economies. The expansion of foreign assemblers' activities in these economies will bring increased business opportunities for local parts manufacturers. To turn these potential business opportunities into reality, however, parts manufacturers will need to meet international requirements in terms of such factors as quality, price, and delivery. Domestic manufacturers that are unable to reach these standards will either fall by the wayside or be forced to reposition themselves as secondary or tertiary parts suppliers for foreign parts manufacturers. From a national economic perspective, increased investment by foreign vehicle assemblers and parts manufacturers will lead to the expansion of production and exports and to higher added value and export earnings. Yet if this growth is driven by foreign manufacturers, it will not necessarily raise the standards of domestic manufacturers. For these reasons, East Asian governments need to develop policies that make their assets more attractive, attract investment by foreign manufacturers, and enhance the competitiveness of domestic manufacturers.

In Thailand, for example, the focus of efforts to develop the automobile industry is shifting from the protection of domestic industry to the development of human resources and infrastructure needed to link the global strategies of foreign assemblers to the advancement of the domestic automobile industry.

This chapter seeks to clarify the ways in which the regional and global strategies of vehicle assemblers have influenced automobile industries in East Asian economies by focusing on the dramatic expansion of investment in Thailand by Japanese vehicle assemblers. Japanese assemblers have been involved in vehicle manufacturing in Thailand since the 1960s, and they have made a major contribution to the accumulation of production technology. In recent years the choice of which type of vehicle to manufacture has become a highly significant question in relation to the expansion of investment in Thailand. Ultimately, the regional and global strategies of assemblers are what determine decisions about which vehicles will be produced in which economies and at what prices and quality standards. An assembler will decide to start production of a new model in Thailand after making decisions on a range of other factors, including (a) the choice of vehicle type, based on analyses of consumer preferences and the structure of vehicle markets, (b) the choice of manufacturing method, including the selection of a platform and technology to be transferred from Japan, and (c) the role of Thailand in East Asian and global production and sales networks in determining where the vehicles should be produced and sold.

There are two reasons for our decision to focus on Thailand. First, Thailand has experienced the growth and development of its domestic automobile industry as a consequence of policies that actively welcomed investment and participation by foreign assemblers and parts manufacturers. Second, Thailand has become a major base for vehicle assemblers from advanced industrial economies as they implement East Asian and global strategies.

In fact, there is a clear trend among vehicle assemblers toward the positioning of Thailand as a production and export base in the ASEAN region (see chapter 4). Since the 1990s, not only Japanese assemblers but also American and European assemblers have regarded Thailand as an appropriate target for massive investment.

We have sought to identify the ways in which the strategies of Japanese assemblers have influenced the Thai automobile industry by dividing assemblers' vehicle production operations into individual processes so that we could determine, through interviews with representatives of Japanese assemblers and parts manufacturers, which processes are carried out in Japan and which in Thailand and what division of roles is planned for the future. (The interviews were conducted in July 2001.) By analyzing the allocation of production functions between parent companies and overseas production sites, it is possible to ascertain an assembler's specific strategy regarding technology transfer. Manufacturers take profitability into account when transferring the technologies that are needed to implement their strategies. Vehicle assemblers do not channel their investment into economies that lack the technology required to realize their strategies at a reasonable cost.

The first half of this chapter consists of an overview of automobile markets and manufacturing infrastructure in East Asia, together with an analysis of the issues confronting Japanese vehicle assemblers, especially from a production perspective, as they implement their new strategies in the ASEAN region. This is followed in the second half by an examination of issues in Thailand, including the allocation of production processes between Japan and Thailand now and in the future, priorities for the acceleration of technology transfers from Japan, and policy priorities for the government.

OVERVIEW OF THE EAST ASIAN AUTOMOBILE INDUSTRY

Most of the world's automobile industries form a four-part structure consisting of Asia, Europe, Japan, and North America.[1] However, the automobile industry is substantially smaller in Asia than in the other three regions, in terms of market scale and volume of production. Moreover, Asia has long been divided into separate national markets, each with many production facilities, a factor that has limited the potential for economies of scale.

In 2000 the combined automobile market of the seven automobile-producing economies in East Asia, not including Japan, amounted to 4.9 million units (table 5.1). China is the biggest market, followed by Korea, Taiwan (China), Malaysia, Indonesia, Thailand, and the Philippines. The total market for the 12 East Asian economies, including those that produce no automobiles, amounts to 5.0 million units. This is substantially smaller than the figures for other regions, which include 18.0 million units for the American market, 16.5 million units for the European Union market, and 6.0 million units for the Japanese market. The biggest single market in the ASEAN region is Malaysia (343,000 units). This is smaller than the total for Brazil (1.3 million units) and Mexico (871,000 units), which are the two biggest automobile-producing economies in Latin America.

In 2000 the seven automobile-producing economies in East Asia had a combined output of 6.5 million units. The fact that this total exceeds the size of the region's market is attributable to the large contribution made by exports in the case of Korea. Korea is the region's biggest producer, followed by China, Taiwan (China), Thailand, Malaysia, and Indonesia. The United States produces 12.8 million units, the European Union produces 17.1 million, and Japan produces 10.1 million.

Notwithstanding its small size, the East Asian market remains important because of its growth potential. While the market for new cars in Japan, the United States, and Western Europe has been stagnant for the past 10 years,

1. Japanese vehicle assemblers divide the world into Asia, Europe, Japan, and North America. From the viewpoint of American and European assemblers, the world consists of three regions, since Japan is part of Asia.

Table 5.1 Sales and Production of Automobiles, 2000 (thousands of units)

Region or economy	Sales			Production		
	Passenger vehicles	Commercial vehicles	Total	Passenger vehicles	Commercial vehicles	Total
Europe						
Western Europe	14,747	2,282	17,030	17,582	2,693	20,275
European Union	14,312	2,206	16,518	14,907	2,236	17,142
Eastern and Central Europe	2,130	305	2,435	2,378	324	2,702
NAFTA	10,309	10,141	20,450	372	9,327	17,699
United States	8,857	9,136	17,993	542	7,258	12,800
Canada	849	737	1,586	1,551	1,413	2,964
Mexico	603	268	871	1,279	656	1,935
South America	1,570	439	2,038	1,646	430	2,076
Argentina	225	82	307	239	101	340
Brazil	1,067	233	1,300	1,348	323	1,671
Middle East	499	277	776	—	—	—
Oceania	617	253	873	324[a]	25[a]	348[a]
East Asia	6,732	4,225	10,955	12,508	4,136	16,643
Japan	4,260	1,703	5,963	8,363	1,782	10,144
Korea, Rep. of	1,058	373	1,431	2,602	513	3,115
China	613	1,475	2,087	605	1,464	2,069
Taiwan, China	263	110	373	265	97	362
Indonesia	47	254	301	257	36	293
Malaysia	297	47	343	259	26	285
Philippines	29	55	84	16	26	42
Thailand	83	179	262	136	190	326
Others	82	29	111	5	2	7
Rest of Asia	2,302	299	2,603	507	287	795
Africa	406	234	640	221[b]	96[b]	317[b]
World total	39,312	18,456	57,800	41,299	16,996	58,296

—Not available.

a. Australia.

b. South Africa.

Source: Compiled from JAMA (2002).

East Asian markets averaged almost 15 percent annual growth rates before 1997. The Asia-Pacific region (excluding Australia and Japan) is projected to account for slightly less than half of incremental global volume between 1998 and 2006, with the largest markets, China and Korea, doubling over the next decade. ASEAN markets are projected to grow at rates between 10 and 20 percent a year.

The automobile industry is currently affected by global overcapacity. East Asia is no exception to this pattern. In the early 1990s predictions of sustained market growth triggered substantial expansion of capacity in ASEAN and Korea. However, the pace of market recovery in the wake of the currency crisis has been slow. Despite this situation, factories planned before the currency crisis are now going into production in Thailand, while foreign assemblers are increasing their capacity in China in parallel with suppliers of parts. In addition, assemblers in ASEAN are undertaking new investment in preparation for the introduction of new models.

Total production value and production value per worker of the East Asian automobile industries are substantially smaller than the corresponding statistics for Japanese industry. As shown in table 5.2, the output of the Japanese automobile industry in 1998 was worth $307.8 billion, which is far higher than China's $50.9 billion, Korea's $46.4 billion, and Thailand's $18.4 billion.[2] Productivity is also low in East Asia. In 1998 production per worker was worth $209,397 in Korea and $171,589 in Thailand, compared with $409,277 in Japan.

Until the first half of the 1980s, there was little need for governments or automobile assemblers to be concerned about international competitiveness. The industry was in the early stages of mass production, and governments used a variety of policy tools, such as tariffs, domestic taxes, import restrictions, and local content regulations, to encourage import substitution production. Governments saw local content ratios as indicators of the development of automobile industries, and their greatest priority for foreign manufacturers was to fulfill local content requirements while coping with a variety of limiting factors.

Between the second half of the 1980s and the middle of the 1990s, governments and automobile assemblers began to envision a future for the automobile industry based on a view of ASEAN as a single economic unit. This led to the introduction of brand-to-brand complementation schemes, whereby parts were supplied reciprocally within the ASEAN region, as a way of maximizing the benefits of mass production at the regional level. In

2. A billion is 1,000 million.

Table 5.2 Overview of Automobile Manufacturing in Asia, 1998

		Automobile manufacturers		Value of automobile production		Employees				
Economy	Year	Number of firms	Percent of all manufacturing firms	Millions of U.S. dollars	Percent of value of all manufacturing	Number (thousands of persons)	Percent of employees in all manufacturing	Average number per firm	Wage rate (U.S. dollars)[b]	Production per employee (U.S. dollars)
Korea, Rep. of	1997	3,083	3.3	46,444	10.2	222	8.5	72	21,094	209,397
China[a]	1998	6,779	4.6	50,876	7.2	3,375	6.8	498	—	15,074
Taiwan, China	1996	2,838	1.8	10,940	4.1	72	2.9	25	17,053	151,599
Indonesia	1998	232	1.1	4,394	1.0	38	0.9	166	965	114,427
Malaysia	1997	317	1.4	4,832	3.4	38	2.9	120	6,508	126,829
Philippines	1997	249	1.7	2,934	6.1	30	2.7	122	3,638	96,513
Thailand	1995	1,095	4.6	18,398	13.2	107	4.4	98	4,776	171,589
Japan	1993	10,436	2.8	307,776	12.6	752	7.6	72	43,007	409,277

—Not available.

Note: Percentages are relative to all manufacturing.

a. Indicates the transportation equipment industry.

b. Indicates annual costs per worker.

Source: Compiled from UNIDO (1999).

1996 agreement was reached on introduction of the ASEAN Industrial Cooperation (AICO) scheme, which contributed to the expansion of intra-regional trade in automobile parts. As of April 2002, a total of 61 AICO approvals had been granted to automobile manufacturers.

The ASEAN automobile industry has experienced dramatic changes since the second half of the 1990s. Assemblers no longer have self-contained ASEAN strategies and are now focusing on the effective use of their production facilities in ASEAN within the framework of their global networks. This shift began with the exporting of commercial vehicles from Thailand to markets throughout the world.

This reflects two environmental changes affecting the Asian region and the automobile industry. First, there has been progress toward trade liberalization within the region. This process will accelerate further in the future. Under the framework of the ASEAN Free Trade Area (AFTA), all of the original ASEAN signatories (Brunei, Indonesia, Malaysia, the Philippines, Singapore, and Thailand) except Malaysia will reduce their tariffs on built-up vehicles imported from within the region to between 0 and 5 percent. (Malaysia moved its deadline for this change back to 2005.)

In addition to these moves to lower tariffs within the ASEAN region, there is also intense interest in a free trade agreement with China. At a summit meeting in November 2000, ASEAN and Chinese leaders agreed to work toward the signing of a free trade agreement by 2010. Since then, consultation has proceeded faster than expected. There is also the possibility of progress toward the negotiation of free trade agreements between Japan and ASEAN and between Korea and ASEAN. Drastic reductions in tariff rates for built-up vehicles and automobile parts would have a major influence on the Asia strategies of automobile manufacturers and on trade within the region. ASEAN members currently impose tariffs of between 30 and 70 percent on built-up vehicles imported from outside the region. A significant reduction in those rates would have a particularly important impact.

Second, leading automobile assemblers began to build strategic partnerships based on capital relationships in 1998 and 1999. Of particular significance were the relationships that formed between Japanese automobile manufacturers and their counterparts in Europe and the United States. These developments would radically change the profiles of automobile manufacturers in Asia.

The formation of strategic alliances between Japanese assemblers, which have numerous facilities in East Asia, especially in ASEAN, and their European and American counterparts, which have production facilities in other regions, such as Eastern Europe, Europe, Latin America, and North America, is starting to have a profound impact on the situation. American and European automobile assemblers have benefited from the powerful brands, procurement networks, and technology of Japanese manufacturers. They have also been able to establish production and sales structures in East Asia almost overnight. Now that their strategies are linked to those of American and European assemblers, Japanese assemblers will need to restructure their business operations from a global perspective.

Each group has production facilities in all or most of the seven economies listed. However, only three groups had formed strategic alliances with Korean manufacturers as of the end of 2001. Those groups are Daimler-Chrysler, which has acquired a shareholding in Hyundai, Renault, which has acquired Samsung Motor, and General Motors, which now owns Daewoo.

Vehicle assemblers have been able to pursue these increasingly complex global strategies thanks to the existence of information and communications technology (ICT) networks. These networks are being used not only to improve the efficiency of internal information flows but also to exchange information with suppliers, sales outlets, and consumers.

This strengthening of links with global networks is not limited to ASEAN. In Korea, the only mass exporter of passenger vehicles in East Asia, local assemblers are starting to link their networks with the global networks of the American and European manufacturers that have become their shareholders. Many automobile assemblers based in the advanced economies are starting to establish facilities in China. In addition, China will have to phase down its tariff rates for imported automobiles as a requirement for membership in the World Trade Organization (WTO). As illustrated by these trends, the rapid implementation of global strategies by automobile assemblers will necessitate a change in the structure of automobile industries in East Asia.

PRODUCTION-RELATED PRIORITIES FOR ASSEMBLERS SEEKING TO ENHANCE THEIR COMPETITIVENESS IN EAST ASIA

Vehicle assemblers have started to clarify the roles of the production sites that they have established in East Asia and to use them in the context of global production networks. This process will require assemblers to confront two issues. First, because they have started to introduce the same models in East Asia as in other regions, assemblers will need to reallocate functions by deciding which operations should remain with the parent company in Japan and which should be transferred to production sites in East Asia. Second, because products are required to meet international standards in terms of quality and price, assemblers need to manufacture products that can compete internationally on price and quality under conditions that are less favorable than in their home country.

Selecting Models to Produce and Allocating Functions

For automobile assemblers, the goal of global management is to maximize profits through the efficient use of management resources scattered throughout the world. To achieve this, companies must decide where to carry out basic research, design, and development, which models to produce at which locations and at what prices and quality standards, and where to sell those models. Multinationals have long faced the task of balancing the benefits of centralization, which allows economies of scale, against localization, which allows products to be tailored to local needs (figure 5.1).[3] By launching the same model on multiple markets, assemblers can reduce research and development (R&D) costs per vehicle. This issue can be paraphrased in the following two-part question: Which models should be sold on East Asian markets, and which parts of the processes leading up to mass production should be carried out in the country concerned?

Model strategies for ASEAN illustrate these changes in corporate strategies. The models sold in ASEAN can be divided into three categories. First, there are vehicles for ASEAN. Second, there are vehicles for emerging markets for sale in developing regions, such Latin America. Third, there are models for sale in both ASEAN markets and the markets of advanced economies. There has been a shift in the models that vehicle assemblers have introduced or are planning to introduce in ASEAN, away from products designed specifically for a single market (so-called "Asia car" models) and toward models that share common features with vehicles designed for other regions, which fall into the second and third categories. One reason for this change is the fact that ASEAN consumers are now demanding newer models. A second reason is that it is becoming possible for vehicle assemblers to concentrate production of specific products and parts within a single country. A characteristic development in relation to passenger cars was the announcement early in 2001 that production of Asia car models would be terminated. First launched onto the market in 1996, these low-priced vehicles were targeted toward the urban middle class, especially in Thailand. To minimize costs, design modifications included the removal

3. Japanese automobile assemblers are pursuing localization strategies. For example, they have launched localized versions of sport utility vehicles not only in East Asia but also in North America.

Figure 5.1 Centralization and Localization in Relation to Global Strategies

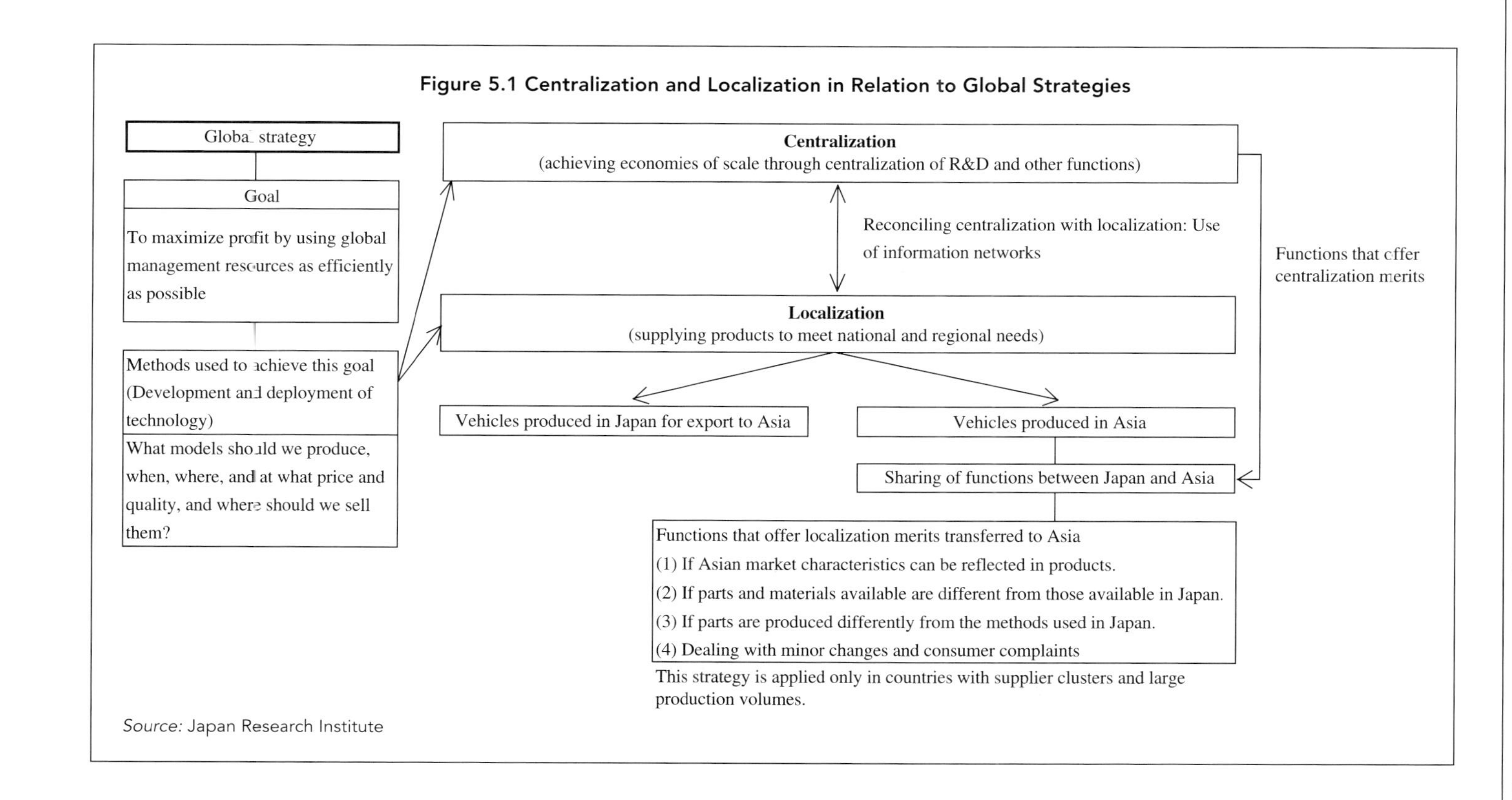

Source: Japan Research Institute

of features not essential for ASEAN economies. Now some groups plan to produce vehicles in specific ASEAN economies for export to neighboring economies or to introduce strategic vehicles for emerging markets. Many assemblers are also planning to introduce "world car" models. These are mostly compact vehicles intended for use in multiple markets throughout the world, including markets in the advanced economies and in the ASEAN region. Thailand has also started to produce models, such as one-ton pickup trucks, that can be exported to many countries.

These changes in the types of vehicles introduced have been accompanied by increased efforts of vehicle assemblers to localize production as part of the sharing of functions between Asia and Japan. The localization strategies of assemblers depend on design and development capabilities in the region for two reasons. First, assemblers must produce a wide range of models in East Asia to suit market needs. Assemblers need to be able to make rapid design changes to reflect local characteristics in East Asian economies, especially when introducing common models in Asia and other regions. To do this, they need to build systems that will allow their plants in ASEAN to turn market information into product plans and to advance those plans in East Asia to mass production without delay. In the case of commercial models that are not produced in the advanced economies, it is quite possible that minor changes that have no impact on the basic performance, frames, dynamic performance, or safety of vehicles will be carried out at plants in East Asia.

Second, the most important requirement in terms of bringing prices in line with international levels is to increase local procurement. To do this, manufacturers will need to establish product engineering capabilities at the East Asian plants and to develop production methods that support the optimized use of locally available materials. Both of these aspects are discussed in detail in the case study of Thailand.

Resolving Two Issues Affecting the Smooth Implementation of Model Strategies

Assemblers will need to resolve two key issues as they implement their model strategies in East Asia: meeting consumer needs and meeting international standards of price and quality.

Meeting consumer needs. The first issue is the need to create a structure capable of supplying a wide range of models to meet consumer needs. East Asian markets have significant numbers of consumers who purchase vehicles configured for local markets (Fujimoto and Sugiyama 2000, p. 406) and manufactured using platforms and core components, such as engines, that are no longer current in the country of origin. At the same time, advances in information technology mean that consumers now have easy access to the latest information from the advanced economies, with the result that it is becoming more difficult to sell outdated models, especially in the passenger car category. The slowdown in sales of Asia car models is attributable in part to this change in consumer behavior. To satisfy the demands of East Asian consumers, manufacturers will increasingly need to shorten the time lag between the launch of new models in the markets of advanced economies and their introduction in East Asia.

Achieving international standards of price and quality. The second issue is that manufacturers are under pressure to measure up to international standards in terms of both price and quality. The major assembler groups are introducing a succession of new models onto Asian markets. As a result, manufacturers must now enhance their price and quality competitiveness. International competitiveness is also crucial for models that are exported to the advanced economies, such as one-ton pickup trucks.

The removal of local content regulations and the reduction of intraregional tariffs under AFTA will force suppliers to compete with imported parts. Moreover, as assemblers switch in earnest to global procurement systems, international prices will become the only prices for parts and materials. Assemblers will first select suppliers on the basis of price information and then consider whether each supplier is able to meet their requirements in terms of quality, delivery dates, and other criteria.

Obviously, cost reduction is an ongoing priority in the automobile industry. It has become a major issue in East Asia over the past few years because the automobile industry, which was previously protected by government policy, has been forced to reduce its costs to international levels over a very short time, despite the inadequacy of its infrastructure. According to an interview survey conducted in Thailand in July 2001, assemblers are targeting cost reductions of around 30 percent over the next three years.[4] In the past, target prices for parts manufactured in Thailand were calculated by adding transportation costs and tariffs to procurement prices in Japan. Recently, however, prices are being based directly on Japanese or international price levels. Moreover, manufacturers previously placed a higher

4. Assemblers are applying uniform cost-reduction targets, not only in Japan but also in their overseas facilities, including those in East Asia. The three-year target of 30 percent is the same as the target for facilities in Japan.

priority on quality than on price in the case of key components because of the emphasis on automobile safety. Today, however, suppliers must meet increasingly high standards in terms of both price and quality.

Three Approaches to the Improvement of Price Competitiveness

Assemblers and suppliers must now achieve substantial improvements in their total competitiveness, based on both price and quality, over a very short time. In particular, manufacturers are using a variety of measures based on regional characteristics in Asia to reduce their costs at the international level. There are three approaches to the improvement of price competitiveness.

Increasing local content ratios. The most important way to enhance price competitiveness is to increase local content. In the case of Japanese assemblers with production facilities in Thailand, parts and raw materials account for about 82 percent of automobile manufacturing costs (cost in 1998 = 100). Depreciation makes up 8 percent, and other manufacturing costs make up 6 percent. Labor accounts for a mere 4 percent of total manufacturing costs (Japan Economic Foundation 1999, p. 18). Similarly, the regional content ratio (the percentage of parts and materials procured in local currencies) in ASEAN was 52 percent in 1998. On a net basis, excluding parts and materials imported for use in the production of parts purchased locally by assemblers, the regional content ratio was around 24 percent.

The major assemblers still regard local procurement as a key issue, despite the abolition of local content requirements and the trend toward globally optimized procurement of parts and materials. This is because they can reduce their exposure to exchange rate risk if they can source local parts that measure up to international standards for price and quality.

To raise local content ratios, it is also necessary to achieve economies of scale. When the number of vehicles produced by each assembler is small, as is the case in East Asia, it is essential to build business relationships that allow suppliers to serve multiple assemblers across corporate groups and to boost production to the required scale through exporting or other means.

Adopting a modular approach. Another way to reduce production costs is to produce parts under modular systems, which increase the unit for production processes to the subassembly level. Modular parts are being introduced throughout the world as part of efforts to cut production costs. In ASEAN there is considerable scope for reducing processes and costs by using similar production methods for parts manufactured using the modular approach in the advanced economies. To achieve these cost savings, multiple suppliers need to cooperate with first-tier suppliers.[5]

Achieving the optimal mix of labor and machinery. The design of efficient production lines is also crucial to attaining the required breakeven point. By deploying workers efficiently, manufacturers can lower the depreciation cost of machinery, thereby achieving cost performance on a par with highly automated production lines in the advanced economies. The percentage of processes automated on assemblers' production lines in ASEAN amounts to a fraction of that in advanced economies. Many processes that are carried out mechanically in the advanced economies are handled by human workers in East Asia. Capital investment and automation tend to be limited to areas in which consistent quality is required. Even then, manufacturers seek to minimize investment costs and do not necessarily install the most up-to-date equipment.

CHANGES IN PRODUCTION STRUCTURE AND TECHNOLOGY TRANSFERS IN THAILAND

How can the East Asian economies take advantage of the global strategies of automobile assemblers to develop their own automobile industries? In this and the following two sections, we use Thailand as the case study for our analysis of this question. We examine the business strategies of multinational corporations, especially Japanese companies, in Thailand and the development of the Thai automobile industry. This analysis focuses on the ways in which manufacturers have positioned their operations in Thailand and on the future role of those operations. We also attempt to predict the future shape of the industry and identify policy priorities for its development.

We focus on Thailand because Thailand is a pertinent example of an economy that has achieved industrial development through FDI and technology transfers from MNCs, especially Japanese companies. A comparison of automobile industries in East Asia (excluding Japan) shows that

5. For reasons of profitability, plants in ASEAN are not used to develop new modular parts that have never been produced at plants in advanced economies. As in the United States, the wage gap between assemblers and suppliers is not large, so only minimal cost benefits would be achieved by allowing suppliers to develop new modular parts. Even where suppliers are able to develop and produce modular parts profitably, there are other issues. For example, the automation of production lines by assemblers is still limited, and neither plants nor production lines are configured for modular parts.

Thailand is the region's third-biggest producer of automobiles after Korea and China. Its industry has grown to the extent that it now occupies an important position in the Asian strategies of automobile assemblers.

Overview of the Thai Automobile Industry

The Thai automobile industry has changed radically since the late 1990s. Until the mid-1990s, the industry produced automobiles only for the domestic market under a policy designed to protect domestic industry. Since 1997, however, there has been a dramatic shift toward a role for Thailand as part of the global strategies of automobile assemblers. The shift reflects environmental changes, including the currency crisis and the subsequent slump in domestic sales, global moves toward trade and investment liberalization, and the formation of strategic alliances among automobile assemblers.

Environmental changes. Table 5.3 provides an overview of automobile production in Thailand. One-ton pickup trucks accounted for 63 percent of all vehicles produced in Thailand in 2001. This type of vehicle makes up a large share of the domestic market because of taxation advantages (also noted in chapter 4). Thailand's exports of one-ton pickup trucks also have expanded in recent years. These trends are reflected in the continuing growth in the share of this type of vehicle in production.

Table 5.3 Production, Sales, and Exports of the Automobile Industry in Thailand

Indicator and year	Passenger vehicles	Commercial vehicles: Total	Commercial vehicles: One-ton pickups	Total vehicles	One-ton pickups as a percentage of total vehicles
Production					
1991	76,938	206,177	169,940	283,115	60.0
1996	138,579	420,849	351,920	559,428	62.9
2001	156,066	303,352	289,349	472,351	63.0
Domestic sales					
1991	66,779	201,781	155,366	268,560	57.9
1996	172,730	416,396	327,663	589,126	55.6
2001	104,502	192,550	168,639	297,052	56.8
Export					
1991	—	—	—	—	—
1996	—	—	—	14,260	—
2001	—	—	—	175,299	—

—Not available.

Source: Thai Automotive Industry Association.

There are 13 automobile assemblers in Thailand, mainly Japanese. Japanese companies first moved into Thailand in the 1960s, and they have adapted to Thailand's local content policies by building production capacity there. They accounted for almost all vehicles produced in Thailand until Ford established a joint venture in Thailand with Mazda in 1998, followed by General Motors in 2000. As shown in table 5.4, Japanese brands still predominate, accounting for 78 percent of total production in Thailand (estimated as the sum of domestic sales and exports in 2001). In addition, many Japanese parts manufacturers have followed Japanese assemblers into Thailand, as noted in chapter 4.

Global move toward trade liberalization. Since the late 1980s, economic growth and the resulting rise in incomes have been paralleled by a steep upward trend in the number of vehicles sold in Thailand (figure 5.2). In the early 1990s, market growth was spurred by falling prices and the introduction of new models. These changes were driven by a gradual easing of policy pertaining to the automobile industry, including the reduction in tariff rates and the removal of restrictions on participation in the industry, and by the escalation of competition as assemblers vied to secure their share of an expanding market. By 1995 manufacturers had increased their production capacity on the strength of forecasts that the domestic market would reach 800,000 units by 2000. It was in this period that Ford and General Motors decided to move into Thailand in order to strengthen their Asia strategies. Investment by assemblers encouraged parts manufacturers to expand into Thailand, leading to formation of the biggest automobile industry cluster in ASEAN.

The onset of the currency crisis in 1997 was followed by a sharp decline in the domestic market. This caused the Thai automobile industry to shift its focus from domestic demand to exports. In 2001 Thailand exported

Table 5.4 Share of Each Nation's Brands in Thailand, 2001

	Number of units			Share of all brands		
Brand	Domestic sales	Export	Total	Domestic sales	Export	Total
Japanese	261,939	105,918	367,857	88.2	60.4	77.9
American	20,196	69,381	89,577	6.8	39.6	19.0
European	11,405	0	11,405	3.8	0.0	2.4
Korean	1,936	0	1,936	0.7	0.0	0.4
Others	1,576	0	1,576	0.5	0.0	0.3
Total	297,052	175,299	472,351	100.0	100.0	100.0

Source: Thai Automotive Industry Association and others.

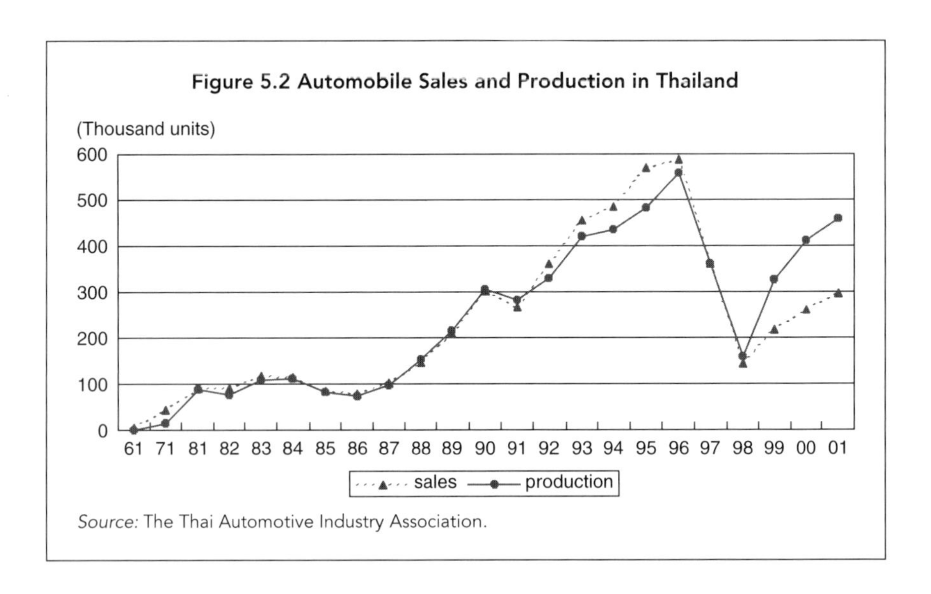

Figure 5.2 Automobile Sales and Production in Thailand

Source: The Thai Automotive Industry Association.

175,299 vehicles, an increase of 15 percent over the previous year. This total was equivalent to 38 percent of total automobile production. Exports thus helped to supplement domestic demand, allowing the industry to maintain production levels during the slow recovery in the domestic market. In addition to Mitsubishi Motors, which has used Thailand as a production base for one-ton pickup trucks since the mid-1990s, Thailand has also become an export base for fully assembled vehicles manufactured by Auto Alliance Thailand, which is a Ford-Mazda joint venture, and General Motors. The presence of these three companies has played a major role in the sustained growth of exports (table 5.5). Together they accounted for 86 percent of total automobile exports in 2001.

Formation of strategic alliances. Thailand needs to improve its international competitiveness. The Thai automobile industry, which produced for the domestic market under protectionist policies, must now make rapid improvements in its ability to compete as an exporter of fully assembled vehicles. In the space of a few years, the benchmarks for the Thai automobile industry reached international levels in terms of price and quality.

Automobile manufacturers are beginning to use their operations in Thailand from a global perspective. This change reflects trade and investment liberalization under the WTO and AFTA and the formation of alliances among American, European, and Japanese companies. For example, some manufacturers are making Thailand their main center for Asian production and export of specific types of vehicles. Alliances between compa-

Table 5.5 Automobile Exports from Thailand, 1997–2001

Brand name	1997	1998	1999	2000	2001		
					Number of units	Share of total	Rate of increase
Mitsubishi Motors	40,072	60,861	60,988	63,540	60,027	34.2	–5.5
General Motors	—	—	—	6,283	48,987	27.9	679.7
Auto Alliance Thailand (Ford-Mazda)		1,213	42,785	49,977	42,077	24.0	–15.8
Toyota	1,563	1,819	12,151	16,031	12,027	6.9	–25.0
Honda	601	2,823	6,682	6,184	6,900	3.9	11.6
Isuzu	—	22	519	5,689	3,683	2.1	–35.3
Nissan	—	—	1,912	4,590	1,206	0.7	–73.7
Others	3	50	665	541	392	0.2	–27.5
Total	42,239	66,788	125,702	152,835	175,299	100.0	14.7

—Not available.

Source: Thai Automotive Industry Association and others.

nies have had a particularly conspicuous effect on the share of production in Thailand (table 5.6). As shown in table 5.4, Japanese brands account for 78 percent of all vehicles produced in Thailand. In terms of corporate groups, however, the General Motors Group leads, followed by Daimler-Chrysler, Ford Group, and Toyota, each with shares of around 20 percent. Today all of the world's leading automobile industry groups own production facilities in Thailand. American and European companies, whose presence in Asia was small in the past, have gained access to facilities in Thailand through strategic alliances and are using those facilities to strengthen their Asia strategies. If Thailand's role in corporate strategies is to grow, the improvement of international competitiveness will be an urgent priority.

Production Processes of Automobile and Parts Manufacturers and the Transfer of Technology to Thailand

The analysis of production technology can be divided into three stages: product development, process engineering (production preparation), and production processes.[6] This subdivision can help to identify which functions have been transferred to Thailand for each process, thereby ascertaining Thailand's level of production technology. The technologies that correspond to each process are design technology for product development, production technology for process engineering, and production control technology at the production stage. If a company carries out a particular process in Japan, it is assumed that the production technology has not been transferred to Thailand. A process that is handled both in Thailand and in Japan is treated as an indication that the technology has been partially transferred, and a process that is carried out entirely in Thailand is seen as indicating that the technology is either in the process of being transferred or has been fully transferred.

Each stage can be divided into individual processes (table 5.7). The product development stage begins with the creation of a concept that will enhance the competitiveness of the new product by meeting market needs. This is followed by product planning, which involves the design of the func-

6. Research for this section included interviews with leading American and Japanese automobile and parts manufacturers. The interviews were conducted in both Japan and Thailand in July of 2001. Although the sample consisted of only four assemblers and seven parts manufacturers, all are representative of the industries in Japan and Thailand. For example, the four assemblers interviewed together accounted for 68 percent of the total number of automobiles produced in Thailand (sum of domestic sales and exports in 2001). The seven parts manufacturers supplied parts to an average of five assemblers each and had steadily expanded their production through business relationships with numerous manufacturers.

Table 5.6 Share of Each Company Group in Thailand, 2001

	Number of units			Share of total volume		
Group	Domestic sales	Exports	Total	Domestic sales	Exports	Total
General Motors Group	73,480	52,670	126,150	24.7	30.0	26.7
Toyota Group	86,172	12,274	98,446	29.0	7.0	20.8
Daimler-Chrysler Group	31,730	60,027	91,757	10.7	34.2	19.4
Ford Group	24,975	42,077	67,052	8.4	24.0	14.2
Honda	38,820	6,900	45,720	13.1	3.9	9.7
Renault-Nissan Group	34,993	1,351	36,344	11.8	0.8	7.7
Others	6,882	0	6,882	2.3	0.0	1.5
Total	297,052	175,299	472,351	100.0	100.0	100.0

Note: Each group consists of the following companies. Daimler-Chrysler Group consists of Daimler-Chrysler, Mitsubishi Motors Corporation, Hyundai Motor Company, and Kia Motors. Toyota Group consists of Toyota Motor Corporation, Hino Motors, and Daihatsu Motor Company. Ford Group consists of Ford Motor Company, Mazda Motor Corporation, and Volvo Car Corporation. General Motors Group consists of General Motors Corporation, Isuzu Motors, Suzuki Motor Corporation, and Fuji Heavy Industries. Renault-Nissan Group consists of Société Anonyme Renault, Nissan Motor Company, and Nissan Diesel Motor Company.

Source: Thai Automotive Industry Association and others.

tional characteristics and basic structure required to implement the concept. Processes at this stage include the use of three-dimensional modeling to create exterior and interior styles, the development of layouts and product specifications, and the establishment of technical formats for core parts. At the product engineering stage, the basic designs are used to create prototypes for use in repeated test cycles, leading to the completion of formal designs for finished products and parts. For models produced in Asia, it is also necessary to modify these formal designs to local specifications. For example, the designs may be adjusted to reflect local regulations concerning exhaust gas or conditions of use.

The process engineering stage begins with process design, which involves the development of basic and detailed designs for production processes. Production facilities, tools, molds, and other requirements are then designed and ordered. Work designs are prepared to provide production line workers with an outline of their tasks, together with skill designs that form the basis for the training of production workers. Other tasks include pilot runs, during which production processes are put under final checks in readiness for mass production.

Table 5.7 Process of Automobile Production

Process stage	Individual process
Product development stage	Concept
	Product planning
	Product engineering including drafting and prototyping
	Engineering change for local specification
Process engineering stage	Process design
	Design and ordering of equipment, tools, and molds
	Work design
	Skill design
	Pilot runs
Production stage	
In-house production management	Production plan
	Process control
	Quality control
	Procurement management
	Cost management
	Equipment maintenance
Supplier guidance	Cost management
	Quality control
	Delivery control
	Productivity improvement, quality control circle activity support

Source: Fujimoto (1999); Imada (1998); various sources.

The deployment of each process, including decisions about whether or not to transfer processes to overseas facilities, depends to a large extent on corporate strategy. Design and development functions are commonly centralized in the parent company's design and development organization. This approach offers a number of benefits, including the merits of scale resulting from the sharing of production development facilities and the ability to coordinate development activities effectively by concentrating product development operations in a single location. At the process engineering stage, specialized technology is required for the design of processes and production facilities. For this reason, many Japanese companies establish specialist divisions at the level of the parent company and support the operations of their overseas plants by sending staff to assist them with the start-up of new products.

The processes that have been transferred to Asia relate mainly to the production stage. In general, however, as local production operations become fully established, companies tend to transfer their design and development activities relating to procurement to local facilities so that they can expand local procurement. This approach is used, for example, when the products are developed for local markets or when the production facility functions as a global supply base for a particular product.

Product development stage. Product development, including planning and design, is generally carried out at the parent companies' development facilities (figure 5.3). Most models produced in Thailand are based on models that have already been developed and produced in Japan. In the case of one-ton pickup trucks, even companies that have terminated production in Japan and centralized production in Thailand still handle development activities for current models in Japan. Three of the four companies surveyed even process engineering changes for the Thai market in Japan. According to the responses from those three companies, the role of facilities in Thailand is limited to the provision of information about the local market. Some companies send staff from their head offices in Japan to carry out market research in Thailand. One company, which carries out engineering changes for the local market in both Japan and Thailand, implements annual minor changes at a development facility in Thailand. The tasks handled in Thailand include changes in interior and exterior design and engine adjustments to meet Thailand's exhaust gas laws.

Parts manufacturers also base most of their product development activities in Japan (figure 5.4). Japanese companies approach the development of a new automobile as a joint effort by the assembler and manufacturers

Figure 5.3 Product Development by Automobile Manufacturers in Japan and Thailand

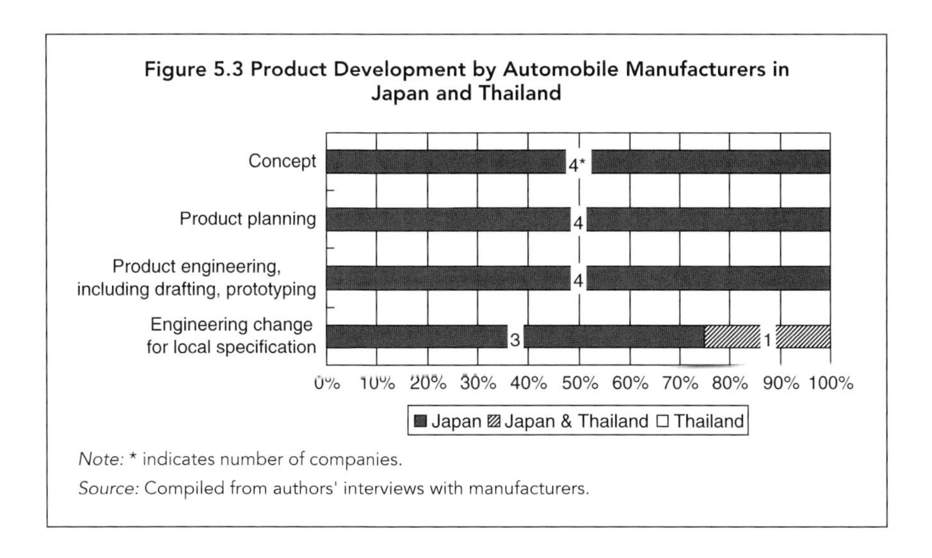

Note: * indicates number of companies.

Source: Compiled from authors' interviews with manufacturers.

Figure 5.4 Product Development by Parts Manufacturers in Japan and Thailand

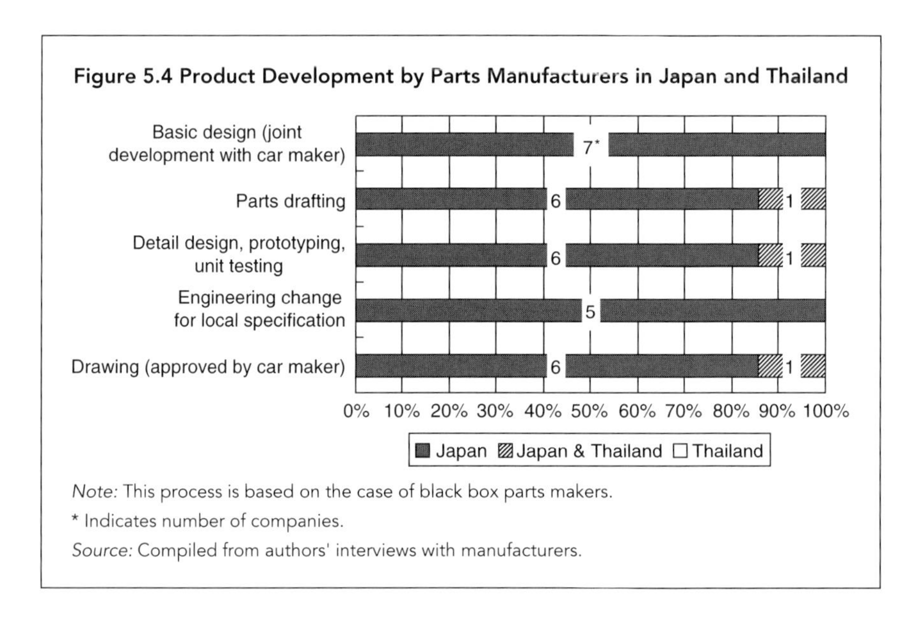

Note: This process is based on the case of black box parts makers.
* Indicates number of companies.
Source: Compiled from authors' interviews with manufacturers.

of core parts. This is called the "design-in" approach. Because product development tasks for automobiles produced by Japanese manufacturers in Asia are carried out in Japan, parts manufacturers also base their design and development activities in Japan, where they share information with assemblers' development facilities. Facilities in Thailand appear to be limited to support activities, such as the provision of information about local parts and materials, the preparation of Thai translations of drawings prepared in Japan, and the rewriting of product numbers on drawings to meet standards in Thai facilities.

Process engineering (production preparation) stage. Thai involvement in process engineering appears to increase in proportion to the proximity of the process to actual production (figure 5.5).

Three companies responded that process design is handled in Japan, five indicated that the task is shared, and two indicated that it is carried out in Thailand. These responses indicate that process design activities are mainly handled in Japan. This is because plant layouts in Thailand are commonly based on the design of Japanese plants. Even the five companies whose responses indicated that the work is split between Japan and Thailand stated that leadership comes mainly from Japan and that the tasks carried out in Thailand are limited to minor changes in layout to suit plants in Thailand.

Three companies responded that the design, development, and ordering of equipment, tools, and molds are handled entirely in Japan, while seven

Figure 5.5 Process Engineering by Automobile and Parts Manufacturers in Japan and Thailand

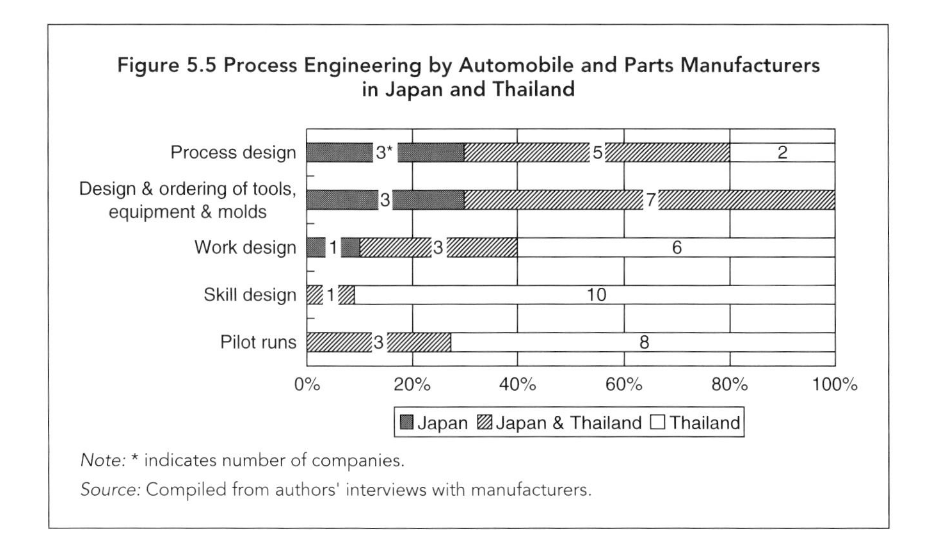

Note: * indicates number of companies.

Source: Compiled from authors' interviews with manufacturers.

said that this work is carried out in both Japan and Thailand. Further analysis of the responses indicates that the typical pattern of the division of labor between Japan and Thailand is based on the design and development of production facilities and complex tools and molds in Japan, with some simple tools and molds produced internally by plants in Thailand. The companies surveyed stated that they plan to increase local production of tools and molds in Thailand in order to reduce costs.

One company responded that work design is handled in Japan, three that the task is divided between Japan and Thailand, and six that it is carried out in Thailand.

Most companies implement skill design and pilot runs in Thailand, although some carry out these tasks in both Japan and Thailand. For new products, instructors are sent from Japan to plants in Thailand to carry out skill training. In some companies, Thai staff are sent to Japan for training so that they can run skill-training programs in Thailand. Most companies implement pilot runs in Thailand. Eight companies responded that they carry out their first pilot runs in Thailand, while three indicated that they transfer operations to Thailand after initial pilot runs at plants in Japan.

Those aspects of process engineering that relate to "upstream" aspects, such as process design, thus either are handled in Japan or are carried out under Japanese leadership with partial support from Thailand. "Downstream" tasks, such as skill design and pilot runs relating to mass production processes, are carried out in Thailand.

All companies responded that process engineering staff from Japan provides guidance and support for process engineering tasks. The practice of sending process engineering personnel from parent companies overseas for the start-up of new plants and products is not specific to Thailand. However, all participating companies thought that the level of reliance on Japan is too high and wanted process engineering tasks to be carried out under the leadership of Thai plants as much as possible.

Production stage. In most companies, day-to-day production management is handled at plants in Thailand, although there is considerable Japanese involvement in cost management (figure 5.6). This tendency is especially conspicuous among automobile assemblers. All three companies whose responses indicated that they handle cost management in Japan or in both Japan and Thailand are assemblers. One automobile assembler stated that day-to-day cost management is under Thai control, but that cost targets at the product development stage are set in Japan. Another stated that personnel are sent to Thailand from Japan to provide cost management support.

Cost management will become increasingly important in the future for two reasons. First, automobile assemblers are working to cut their costs in Thailand. Second, they are expanding their global procurement operations. They are building comprehensive supplier databases. Japanese manufacturers that have formed alliances with European and U.S. manufacturers plan to use their partners' databases and to establish joint procurement systems with them. This means that plants in Thailand will need to upgrade their cost management procedures to conform to these databases.

Figure 5.6 Production Stage by Automobile & Parts Manufacturers

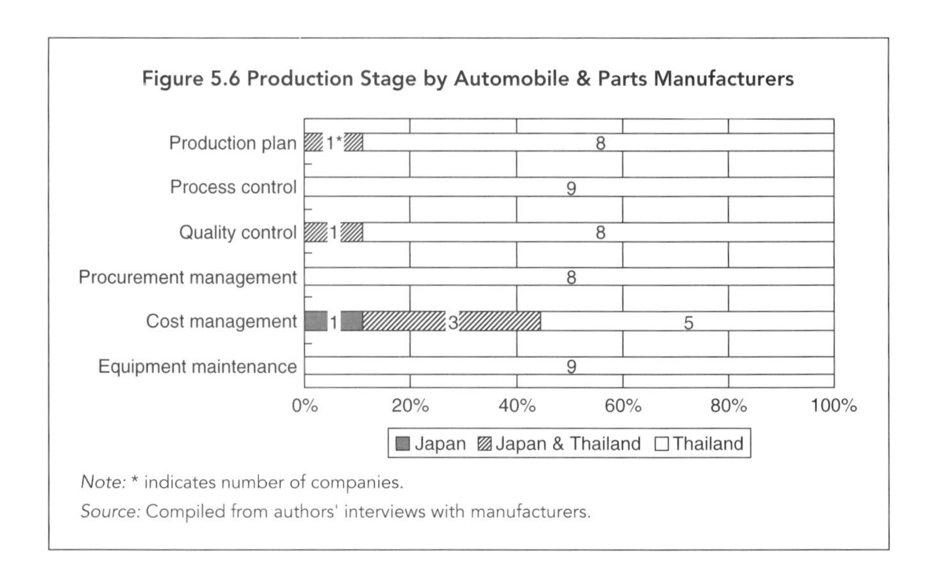

Note: * indicates number of companies.

Source: Compiled from authors' interviews with manufacturers.

All parts manufacturers responded that they handle cost management in Thailand. This appears to indicate not that they are making progress toward localization but rather that their management systems are comparatively loose. If automobile assemblers demand further cost reductions, it is likely that parts manufacturers will need to tighten their cost management systems. This could lead to more intervention from Japanese parent companies, including the transfer of cost management expertise.

Many companies indicated that supplier guidance is provided locally (figure 5.7). However, the Japanese parent companies of automobile assemblers are heavily involved in suppliers' cost management. Japanese-affiliated manufacturers are providing more guidance to help Thai suppliers to improve quality and production efficiency.

Technology transfers in Thailand. As noted, the processes that have been completely transferred to Thailand are the downstream aspects of process engineering, such as skill design, pilot runs, and actual production. The upstream aspects of product development and process engineering, specifically the engineering processes required at all stages from development through the start-up of production, are centralized in Japan. In other words, manufacturers have transferred the production management technology required for Thailand to be a base of production.

We also need to look at the quality of technology transferred to Thailand. The quality of processes handled in Thailand appears to have risen since the 1980s. This reflects a rise in the level of quality that is expected of vehicles made in Thailand. In the 1980s companies transferred the technology needed to produce parts that met local content regulations. In

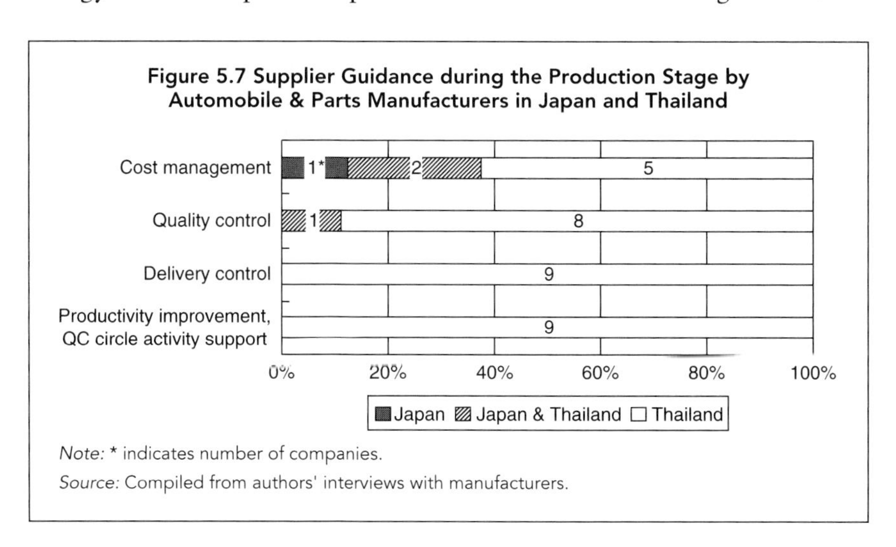

Figure 5.7 Supplier Guidance during the Production Stage by Automobile & Parts Manufacturers in Japan and Thailand

Note: * indicates number of companies.

Source: Compiled from authors' interviews with manufacturers.

the 1990s, with competition among manufacturers escalating, companies made autonomous decisions to introduce enhanced production management technology as a way of improving their competitiveness. The shift to automobile exporting in the late 1990s has created a need for extremely sophisticated production management technology to enable manufacturers to meet export quality standards.

Yet technology has not been fully transferred to Thai personnel. Even where processes have been transferred to Thailand, support from Japanese parent companies or Japanese personnel is still essential. For example, Thai personnel handle day-to-day production management tasks, but Japanese staff stationed in Thailand deal with any problems that may arise. In some cases, support personnel from the Japanese parent company are sent to Thailand. Some companies have expanded their parent company guidance because of the need to cope with the higher standards of quality required by exporting. In terms of the responsibility for technology, the transfer of technology to Thailand and its assimilation there remain incomplete. Real technology transfers will occur only when Thai personnel have acquired management expertise that enables them to improve production.

LONG-TERM OUTLOOK FOR THE THAI AUTOMOBILE INDUSTRY

Given the clustering of automobile and parts manufacturers and the strategies of automobile assemblers in Thailand, Thailand's future as a key producer nation in Southeast Asia seems assured. Manufacturers have positioned Thailand as a major base of production in their Asia strategies, which means that exports of automobiles and parts are likely to expand. Vehicle assemblers plan to allocate the following two functions to Thailand and are expanding their investment for this purpose. The plans of all manufacturers add up to an output of more than 1 million units by the middle of this decade.

First, Thailand will be a supplier of certain types of vehicles, such as one-ton pickup trucks, to the world. Mazda and Mitsubishi Motors have ceased production of one-ton pickup trucks in Japan, and their production operations are now centralized in Thailand. Isuzu has shifted to Thailand production of its one-ton pickup trucks, which used to be produced in Japan for export. Its production target for Thailand is 160,000 units in 2004, which represents a substantial increase from the 72,000 units produced in 2000. Thailand is also positioned as the production base for Toyota's Innovative International Multipurpose Vehicle Project, which will commence in 2004. Toyota also plans to shift production of one-ton pickup trucks that were produced in and exported from Japan. Honda began exporting compact cars from Thailand to Japan at the end of 2002. Because demand for the model in question is relatively small in Japan, Honda's strategy is to achieve optimal use of production facilities in Japan and Thailand by concentrating production of the compact vehicle in Thailand.

European and North American manufacturers, which have few manufacturing facilities in Asia, see their production base in Thailand as a supply base for the ASEAN region or even for the whole of Asia, including Japan. General Motors, for example, produces its Opel minivans at two plants in Germany and Thailand and supplies the world market from Thailand.

Second, Thailand is the core nation for the automobile industry in ASEAN. Since the reduction of regional tariffs under AFTA, vehicle assemblers have started to establish regional divisions of labor that include built-up vehicles. Within these structures, Thailand has become the supply base for cars and pickup trucks, with the result that assemblers supply more models from Thailand than from other nations. Manufacturers also have established regional coordination offices and procurement centers in Thailand as part of their division of labor. In addition to its role as the regional hub for the production of vehicles, Thailand is also assuming a coordinating role in regional divisions of labor.

East Asian governments have commenced negotiations on trade liberalization, including the establishment of free trade agreements. Trade liberalization in ASEAN enhanced Thailand's advantage within the region, but what are the implications of East Asian trade liberalization for Thailand? If liberalization proceeds, vehicle assemblers are likely to build production networks linking countries that possess assets, such as automobile industry clusters and markets, and to establish divisions of labor for built-up vehicles and parts. The bases that are likely to be incorporated into these production networks are Thailand, as the core base in ASEAN, together with China, Japan, and Korea.

Two factors suggest that regional divisions of labor in Asia will be based on cooperative complementation of product ranges for each market. First, there are considerable advantages to be gained by treating ASEAN, China, and Korea as separate markets and by producing vehicles in regions close to those markets. Second, while China is expected to register the biggest growth in automobile production, most of its output will be used to supply the domestic market, and there will be relatively little surplus capacity for exporting in the foreseeable future. Vehicle assemblers can be expected to use production in each base to meet local and regional demand, while establishing divisions of labor based on product differentiation, whereby centralized production will be used to supply the Asian region with models for which demand in individual countries or regions is limited or parts for which economies of scale are needed. Progress toward liberalization is expected to result in the establishment of complementary divisions of labor in the Asian automobile industry. Thailand is expected to occupy an important position in these production networks by virtue of a comparative advantage based on its two roles: as a supply base for one-ton pickup trucks and related parts and as a core production center in ASEAN. The negative implications, such as competition with other economies, are likely to be minimal for Thailand.

Transferring Design, Development, and Procurement Functions to Thailand

What will be the role of production facilities in Thailand in the context of Thailand's role as a production center for Asia? How will the divisions of labor between parent manufacturers and their facilities in Thailand evolve?

First, there is likely to be more local leadership in process engineering and production processes, which are already carried out in Thailand. There is likely to be greater Thai participation in production management, for example. The industry is also expected to move toward the development of self-sufficient production operations in Thailand, including the localization of tool and mold production at the process engineering stage.

Second, automobile assemblers are expected to move toward divisions of labor in which aspects of product development that relate to basic product performance, including concept development and basic design work, will be handled in Japan, while product engineering and subsequent processes will be carried out in Thailand (table 5.8). Specifically, there is a strong possibility that manufacturers will transfer product engineering and engineering change functions for parts to Thailand in order to lift their local procurement ratio for parts and materials.

Evidence for this prediction can be found in the efforts that automobile assemblers are making to increase the percentage of parts procured locally. Japanese automobile assemblers in Thailand have already raised their local content ratio for passenger cars and pickup trucks to 60–75 percent. However, they still rely on imports from Japan for expensive, sophisticated parts and many of the parts and materials that make up "local" parts. Automobile manufacturing costs in Thailand are relatively high. To improve their cost competitiveness in Thailand, automobile assemblers have started to move toward true localization, including sophisticated parts and components for parts. Some companies are aiming for 100 percent ASEAN content.

Table 5.8 Present and Future Division of Labor between Japan and Thailand

Process stage and individual stage	Present	Future
Product development		
Concept generation	Japan	Japan
Product planning	Japan	Japan
Product engineering	Japan	Thailand
Engineering change for local specification	Japan	Thailand
Process engineering	Japan, Thailand	Thailand
Production stage		
In-house production management	Thailand	Thailand
Supplier management	Thailand	Thailand

Source: Compiled from authors' interviews with manufacturers.

There is a strong possibility that automobile assemblers will shift the following three functions to Thailand as part of their efforts to increase local procurement.

First, there is a collection of information for procurement activities. Manufacturers wishing to expand their local procurement will need to gather detailed information about suppliers of parts and raw materials in Thailand.

Second, engineering change functions for local specification and product engineering are likely to be carried out in Thailand. Automobile manufacturers are likely to move engineering change functions to Thailand first. The processes involved include changes to specifications to suit the Thai market and partial changes of drawings to allow the use of parts and materials sourced in Thailand. In the future, product engineering probably will be carried out in Thailand once the product planning stage has been completed. If automobile assemblers shift product engineering of parts to Thailand, suppliers are likely to establish development capabilities at their plants in Thailand. In other words, there will be a design-in structure under which suppliers in Thailand participate in the engineering processes of automobile assemblers.

Third, there is the assessment of parts and materials. Before new parts and materials can be introduced, they must be assessed for compliance with quality requirements and specifications. This task has been carried out in Japan. If it is transferred to Thailand, manufacturers will be able to shorten the lead time from the design and development stage to the start of production. It is unlikely that assessment will be shifted entirely to Thailand because of the high cost of the facilities required. However, assessment systems in Thailand are likely to be developed more fully than in the past. The shift of product engineering functions to Thailand is also likely to result in the relocation of prototyping and unit testing at the product engineering stage.

The fact that it is now possible to relocate procurement functions reflects the tendency of Japanese automobile assemblers to locate engineering functions close to production facilities. Close proximity fosters information sharing and enables automobile assemblers to engineer products that are easy to manufacture.

Considerable benefits can be gained by centralizing functions, such as concept development, product planning, and development of key parts, in a single development organization. In terms of the product engineering function, however, manufacturers have the possibility of choosing another option, which is to locate product engineering close to hub production facilities in each region. For example, proximity siting would provide advantages in the case of vehicles that are produced only in Thailand, such as one-ton pickup trucks. The same would apply to models that are intended to be launched simultaneously onto world markets, such as world cars.

When the product engineering and engineering change functions are implemented in the regional development facility on the basis of product engineering designed by the parent company's development facility, it should be possible to reduce the parent company's development and process engineering costs and to shorten lead times. This would increase the possibility that vehicles can be launched onto different markets at approximately the same time.

Decisions about where to locate engineering and procurement functions will vary according to the strategy of each manufacturer. Some companies choose to centralize global operations for each function in a single location, while others prefer to establish independent structures in each region. However, automobile assemblers in Thailand share the common goal of establishing self-sufficient production operations there in order to raise local procurement ratios and shorten lead times.

Outlook for Parts Manufacturers in Thailand

As Thailand's importance to automobile assemblers increases, the expectations of suppliers will change radically around the criterion of international competitiveness. Suppliers will need to meet four requirements.

First, they must achieve international standards in terms of quality, cost, and delivery (QCD). Thailand's transition from production for a domestic market to production for export over the past few years has been paralleled by a rapid rise in the QCD level that Thai suppliers must attain. To enhance the export competitiveness of finished cars, automobile assemblers will impose tougher QCD standards on their suppliers. Although automobile assemblers will increase their local procurement, they have also adopted global optimized procurement policies and indicated that they will source parts from outside Thailand if better suppliers can be found elsewhere. The QCD levels that Thai suppliers need to meet are now based on global comparisons.

Second, suppliers will need engineering capabilities and expertise in value analysis and value engineering on a level that allows them to participate in product planning and engineering with automobile assemblers. When product engineering and assessment functions are transferred to Thailand, suppliers will need to engage in joint development activities with automobile assemblers. Specifically, they will need to have the capacity to participate in joint development activities, to propose solutions to automobile assemblers, to produce prototypes based on detailed designs, and to carry out unit tests.

Third, suppliers will need to develop global supply structures. Under their global strategies, car manufacturers plan to build identical vehicles, such as compact cars designed for developing countries, in multiple regions at the same time. Suppliers must develop systems capable of supplying all regions. For example, the ability to supply parts to Brazil, Thailand, and the United States was one of the criteria used to select suppliers for a pickup truck jointly developed by General Motors and Isuzu, which went into production in 2003.

Fourth, ICT infrastructure will be essential. Automobile manufacturers are developing a variety of network systems. Suppliers face the costly task of installing multiple system terminals to match the network specifications of each manufacturer with which they deal. After product engineering operations are transferred to Thailand, suppliers will need to link their computer-aided design and manufacturing systems with those of automobile assemblers. They will also need to install high-quality communication systems with sufficient bandwidth to allow the sharing of design drawings.

Parts manufacturers in Thailand will need global resources, including global supply networks, development capabilities, and the ability to invest in assessment facilities and ICT systems. Most major Japanese parts manufacturers have established operations in Thailand. Since the late 1990s, American and European parts manufacturers have also moved into Thailand, following the establishment of plants by Ford and General Motors. Thailand now has an almost complete lineup of parts manufacturers with global capabilities. This has led to a further intensification of competition among first-tier parts manufacturers.

First-tier parts manufacturers will play an increasingly important role in future efforts to enhance the international competitiveness of Thai-made automobiles. Automobile manufacturers may select and restructure their suppliers according to the four criteria cited above. First-tier parts manufacturers will need to structure their production systems and enhance the competitiveness of their products to suit the strategies of automobile assemblers. Moreover, the clustering of the world's leading parts manufacturers in Thailand will allow automobile assemblers to pursue modularization strategies. All automobile assemblers are using modularization techniques to a varying degree. In particular, after forming alliances, Japanese manufacturers and North American or European manufacturers, which have adopted modularization aggressively, are tending to use modular parts in the new models that they have jointly developed. When production of these models begins in Thailand, the first-tier manufacturers that supply modules to assemblers will handle the overall development of modular parts, the coordination of second-tier manufacturers, and the provision of overall quality assurance for modular parts.

Can local parts manufacturers survive? As the world's major parts manufacturers congregate in Thailand, car manufacturers are placing increasingly tough requirements on their suppliers. It is not easy for local parts manufacturers to win orders in this environment.[7] There is no longer time for parts manufacturers to grow through their own efforts while receiving technology from foreign-owned companies. In fact, the growth opportunities for local parts manufacturers are shrinking for a number of reasons.

First, the Thai government eased restrictions on foreign ownership after the currency crisis, allowing foreign companies to establish wholly owned subsidiaries. Second, automobile assemblers have encouraged foreign investment in local manufacturers in order to improve quality. The standard of quality demanded by automobile assemblers has risen with the shift to exporting, and local manufacturers are no longer able to meet the requirements without assistance. For this reason, automobile assemblers now require local parts manufacturers to obtain injections of foreign capital and technology as a condition for the continuation of their business relationships. In addition, many joint ventures between Thai and foreign capital have shifted to majority foreign ownership because of the easing of investment restrictions and the deteriorating financial circumstances of Thai partners since the currency crisis. Foreign-owned companies are taking control of man-

7. A local parts manufacturer is defined as a pure Thai company that is wholly owned by Thai capital and has received no capital from foreign companies, regardless of whether or not it has been given foreign technology.

agement and investing management resources. This pattern is reducing the number of local parts manufacturers and widening the competitiveness gap between foreign-owned and local manufacturers.

Although the situation is difficult, there are ways in which local manufacturers can survive. First, they can specialize in products that allow them to take advantage of their cost competitiveness. Cost reduction is becoming a major focus in Thailand, and local manufacturers will have an edge as suppliers of price-sensitive parts that can be produced with relatively low levels of technology. Another option is to specialize in niche products that foreign-owned companies cannot produce profitability on a stand-alone basis.

Second, local manufacturers can expand their business with foreign-owned first-tier suppliers. Demand from that source is likely to grow as foreign-owned suppliers seek to cut costs by sourcing more parts locally. Demand also will rise as parts suppliers shift from in-house production to outsourcing in order to improve production efficiency.

Whichever path they choose, local manufacturers will survive if they are able to acquire capabilities in world-class management of production and quality. If automobile assemblers move further along the modularization track, parts manufacturers that choose the first path could be positioned as second-tier suppliers. However, competitiveness will be vital even for second-tier suppliers. Local manufacturers are unlikely to survive unless they take steps to enhance their competitiveness.

The Use of ICT Networks

ICT networks will play an important part in enhancing the functionality of production plants in Thailand and linking them into global strategies. What is the current state of development of ICT networks in Thailand, and how will those networks be utilized in the future? In this part we examine the ways in which ICT networks are used, especially by Japanese-owned companies in Thailand.

Automobile manufacturers in Thailand use ICT networks to link (a) with their parent companies and plants in other parts of the world (corporate networks), (b) with automobile dealers in Thailand (dealership networks), and (c) with suppliers in Thailand (supply networks).

Japanese automobile assemblers and major parts manufacturers have been building global corporate networks since the 1980s. Today these corporate networks link parent companies with their operations in Thailand and other parts of the world. Progress in the development of information technology has been extremely rapid in recent years. In the past corporate networks were based mainly on leased lines and value-added networks, but rapid advances in ICT are transforming this pattern. The Internet is expanding and now supports high-speed transmission of large volumes of information. Security has also been enhanced thanks to encryption, virtual private network, and other new technologies. As a result, more and more businesses are switching to Internet-based networks. However, some small and medium Japanese parts manufacturers do not have corporate networks to link them with their overseas operations.

The information exchanged between Thai plants and their parent companies via corporate networks includes production, sales, and accounting data. As described, products manufactured in Thailand are designed and developed by parent companies. For this reason, there is minimal exchange of design drawings with parent companies. Drawings are mailed in CD-ROM format or transmitted through corporate networks as data files.

Within Thailand, the development of corporate networks linking automobile dealerships with manufacturers has accelerated since 1997. Sales stagnated in the wake of the currency crisis, and manufacturers were forced to reduce sales-related costs by improving the efficiency of their order-processing procedures. Networks have been used to achieve a number of goals, including the establishment of production schedules that reflect up-to-date sales trends and the reduction of inventories through the use of dealers' inventory data to shift stock between dealers. Automobile manufacturers are also using networks to improve customer service and marketing, including the development of customer databases.

Automobile manufacturers have also started to establish networks with their major suppliers. At present, information is shared between manufacturers and suppliers mainly by fax, telephone, and e-mail. However, some automobile assemblers have begun to introduce web-EDI (electronic data interchange) systems. Other manufacturers have also announced plans to build ICT networks. Network development is likely to accelerate to accommodate plans for the introduction of new models and the production of world cars in the first half of this decade.

The main types of data shared between automobile assemblers and suppliers are (a) design and development data at the product engineering stage and (b) order data at the mass production stage. Automobile manufacturers in Thailand are starting to use Web-EDI systems and other methods to exchange data generated at the mass production stage. These systems are being used for types of data that do not require large bandwidth, such as orders and accounting data. Companies are progressively upgrading their networks so that they can use them to share information that requires large bandwidth, such as design and development data at the product engineering stage. The advent of the design-in approach in Thailand will necessitate the networking of computer-aided design and manufacturing systems and the sharing of engineering data between automobile assemblers and suppliers.

In addition, Japanese automobile assemblers in Thailand are likely to strengthen their linkages with global production systems via corporate networks. In the future, parent companies' development centers will be able to share product engineering and process engineering data with regional product engineering centers in other parts of the world. It is also possible that manufacturers will build comprehensive databases with information that will be useful when selecting a supplier.

All of these networks are closed networks in which access is limited to automobile assemblers and specific suppliers. In the future, it is possible that open e-markets run by e-commerce intermediaries will also be used in Thailand. One Western car manufacturer, which actively uses the Internet for procurement, has plans to use an e-commerce site to auction parts.

This use of ICT to support the sharing of production and sales information among sellers, automobile assemblers, and suppliers in Thailand will help to create a highly efficient production structure capable of responding more quickly to market trends. Despite the improvement of ICT networks in Thailand, however, it will be difficult for suppliers to reduce their inventories of parts and raw materials. This is because they need to maintain stocks of imported parts and raw materials and because they need to have reserves of finished products to guard against production equipment failures. Thailand does not have sufficient people with the skills needed to repair production facilities. Nor do companies have the technology required to take over production of high-added-value parts when in-house production is interrupted. That is why companies must bring repair staff from Japan to restore facilities to operating condition. Most companies maintain sufficient stocks of finished products to last about one week.

ASSESSING THE THAI AUTOMOBILE INDUSTRY: ISSUES FOR THE FUTURE

The growth of the Thai automobile industry has been driven by investment from MNCs. Today Thailand is the biggest automobile producer in ASEAN and the only ASEAN economy that is achieving sustained growth in its exports of automobiles. Manufacturers of automobiles and parts are likely to enhance their production operations in Thailand to support their future role as production and export bases for specific types of vehicles or as hubs for production activities in ASEAN. For example, it is highly probable that manufacturers will relocate procurement, product engineering, and production of sophisticated parts to Thailand. Automobile production in ASEAN began with the assembly of knockdown units under import substitution strategies. Only in Thailand has the industry grown to the extent that manufacturers are beginning to transfer processes with higher value added.

Why did automobile assemblers identify plants in Thailand as key elements in their global strategies and expand their investment in the 1990s? In other words, what comparative advantage does Thailand have as a target for investment?

First, there is the growth potential of the Thai market. Automobiles have huge numbers of parts, and both parts and assembled automobiles are bulky. It is therefore essential to produce automobiles close to the point of consumption. Merits of scale are also essential in the automobile industry, so the size and potential growth of the domestic market are key considerations. The growth of the Thai economy since the second half of the 1980s has been paralleled by the expansion of the automobile market, which increased from 102,000 units in 1987 to 572,000 in 1995. The expectation that this rapid growth would continue prompted Japanese manufacturers to expand their existing plants. It also led to decisions by American and European manufacturers, such as Ford and General Motors, which did not previously have plants in Thailand, to establish production operations there.

Second, Thailand has the most developed parts and raw materials industries in Southeast Asia. The ability to buy parts and materials cheaply by sourcing them locally is more important to the production of competitively priced automobiles than low labor costs. This is because parts and raw materials make up more than 80 percent of the total production cost of automobiles, while labor costs are relatively small. The expansion of production by Japanese-affiliated automobile assemblers since the second half of the 1980s has prompted an expanding influx of parts manufacturers. By the mid-1990s, Thailand had the biggest clusters of parts and materials industries in ASEAN. By that time Ford and General Motors were also considering expansion into ASEAN. Because one of their goals in establishing production facilities was to produce assembled vehicles for export, the development of parts and materials industries played a major role in decisions to site plants in Thailand.

Third, the industry has also benefited from the sensible policies adopted by the Thai government to encourage investment by MNCs (see chapter 4 of this volume). For example, in the early 1990s the government began a gradual process of liberalization in the automobile sector through measures that included tariff cuts and the removal of restrictions on new participation in the automobile business. Corporate investment naturally gravitates toward economies where there is greater freedom for business activities. In addition to that, the government successfully attracted FDI by developing the infrastructure needed for the automobile industry. To correct the excessive concentration of industry in Bangkok and the surrounding areas, it developed infrastructure and provided investment incentives in the eastern seaboard area located to the southeast of the capital. Government projects included the development of industrial parks, the construction of an expressway network linking Bangkok with the eastern seaboard area, and the establishment of international port facilities capable of accommodating large container vessels at Laem Chabang. The expressway network is vital to procurement and distribution between manufacturers and suppliers and to domestic distribution of finished automobiles, while the international port is a vital access point for imports and exports of finished automobiles and parts (Mori 2001, p. 65).

Investment by parts manufacturers was further stimulated when these three factors prompted a Ford-Mazda joint venture and General Motors decided to establish facilities in the eastern seaboard area in the mid-1990s. This led to formation of the biggest automobile industry cluster in ASEAN, as automobile and parts manufacturers established facilities in an area stretching from the Bangkok metropolitan area to the eastern seaboard. This cluster continues to expand.

In the mid-1990s there was an emerging trend toward global trade and investment liberalization, and companies were restructuring their Asia strategies from a global perspective. Thailand's success was attributable to its comparative advantage in terms of market size (and growth potential) and development of parts and materials industries and to its adoption of investment incentives to assist foreign companies to set up business operations.

POLICY ISSUES

Nevertheless, many issues will need to be tackled before the Thai automobile industry can build closer links with global corporate strategies and take over high value-added functions. Two issues that relate to the aspects discussed in this section are (1) the development of human resources to support advanced production systems and (2) the improvement of ICT infrastructure. As ICT becomes more sophisticated, it will be especially important for the future development of the automobile industry to improve infrastructure and develop human resources in the area of ICT.

Developing Human Resources for an Increasingly Sophisticated Sector

Automobile manufacturers have positioned Thailand as a production base and are starting to transfer more sophisticated production operations there. The development of human resources to work in this field has become an urgent priority. Thailand's evolution as a production base has been accompanied by a rapid rise in the QCD levels that are required. This is reflected in increased efforts to enhance competitiveness, including the improvement of production quality in manufacturing operations.

Moreover, as companies start to transfer design and development functions and improve their ICT networks, they will need personnel trained to new technical standards in new fields. However, Thailand is not able to supply sufficient personnel suitable for work in this changing environment. For example, in 1996 Thailand had only 10,209 research and development personnel (researchers, technicians, and equivalent staff; see table 5.9). With just 103 researchers and 39 technicians per million of population, Thailand falls far short of levels in China and the four newly industrialized economies of Asia. It is also short of science and engineering graduates with the potential to work in research and engineering. For example, the number of natural science graduates in Thailand in 1995 was the lowest among the ASEAN-4 at 23,310 (table 5.10). This is less than one-tenth of China's 1994 output of 364,047 natural science graduates.

For these reasons, Thailand is short of personnel capable of carrying out engineering work in the areas of design and development. In the ICT sector, it has few people with the specialized education needed to build and maintain systems. The diffusion of personal computers in Thailand is low, and few workers have any experience with computers at the time of initial employment.

In other words, human resource development systems in Thailand are unable to respond to rapid qualitative changes in the types of human resources sought by MNCs. Developing countries will need to consider how to shape their human resource development policies to facilitate the development of their automobile industries in a globalized era. The goal should be to supply human resources capable of adapting to the supply chain management systems of MNCs. It is not just in production that personnel are needed. There is also an urgent need to train people with the ability to use ICT networks in various processes, such as design and development,

Table 5.9 Personnel Engaged in Research and Development, by Country

Country	Year	All R&D personnel	Researchers	Technicians and equivalent staff	Researchers per million inhabitants	Technicians per million inhabitants
China	1996	787,000	559,000	228,000	454	200[b]
Indonesia	1985	36,185[a]	21,160	3,888	182[c]	24
Japan	1996	891,783	617,365	—	4,909	827[d]
Korea, Rep. of	1996	135,703	99,433	—	2,193[e]	318[d, e]
Malaysia	1996	4,436	1,893	654	93	32
Philippines	1992	14,578	9,960	1,399	157	22
Singapore	1995	9,497	7,695	997	2,318	301
Thailand	1996	10,209	6,038	2,303	103	39

—Not available.

a. 1984.

b. 1995.

c. 1988.

d. 1994.

e. Full-time plus part-time personnel; does not include the social sciences and humanities.

Note: Data for Indonesia refer to full-time and part-time personnel. Data for Japan and Korea refer to full-time researchers. Data for Singapore exclude the social sciences and humanities.

Source: UNESCO (1999).

Table 5.10 Graduates in the Field of Natural Sciences, by Country

Economy	Year	Total number of graduates	Graduates in the natural sciences	
			Number	Percentage of total
China	1994	1,040,135	364,047	35.0
Indonesia	1996	388,672	106,496	27.4
Japan	1996	1,127,500	261,580	23.2
Korea, Rep. of	1997	380,571	145,759	38.3
Philippines	1995	309,645	86,701	28.0
Singapore	1995	20,603	11,929	57.9
Thailand	1995	130,223	23,310	17.9

Source: UNESCO (1999).

procurement, production, and sales. Apart from personnel for engineering and information technology–related work, ultimately there will also be a need for personnel to work in local operations, from production through to sales, so that the level of support needed from parent companies can be kept to a minimum. There will be demand for people capable of linking local activities to the global production, procurement, and sales strategies of MNCs. In the past, MNCs met this need primarily by assigning personnel to these areas from their home country. However, if foreign companies continue to send the required personnel from their home country, they will lose their cost competitiveness. Without human resources, companies cannot pursue their global strategies or succeed against competition. Human resource development is certain to become an increasingly important consideration in relation to both business and siting strategies.

In addition, manufacturers will need to give proper consideration to shortening the cycle of introducing new technology and new production systems. The technical requirements for workers assigned to existing production lines will change rapidly, and reeducation programs for technical workers will be a crucial consideration. The improvement of training organizations and qualification systems will become even more important. Governments will need to update human resource development programs constantly to keep pace with new technology. This process should be approached in cooperation with private sector companies.

Now that manufacturers are starting to relocate high-value-added functions to Thailand, the continued development of the automobile industry there could depend on Thailand's ability to supply human resources capable of supporting those functions. Multinationals have limited management resources to invest in one country. It is vital in this context that Thai staff are deeply involved in all areas of activity and that there are Thai initiatives to enhance the added value of the automobile industry. If manufacturers decide that their plants in Thailand are not capable of functioning as regional hubs, they could change their strategy.

Human resource development needs include not only production personnel and ICT experts but also people who can cope with operations on a global scale. Although automobile manufacturers have plants and facilities in many countries, it is likely that they will continue to develop core technology and formulate global strategies in their home country. In the future, they are also expected to introduce new technologies and models more or less simultaneously, with some adjustment for market characteristics, at their facilities throughout the world. This will require personnel who can implement head office directives at Asian plants and who can provide head offices with information about Asian operations.

Three types of managers are needed for operations in Asia: managers who can take responsibility for specific areas, such as personnel and human resource development, marketing and sales, and production; managers who can integrate all operations in an entire country; and managers who can develop strategies and maintain efficient logistics across the ASEAN region or throughout Asia.

Governments cannot train managers. Companies will need to have clear policies calling for the extension of localization up to the management level. They will also need to show a strong commitment to the investment of time and money and to the accumulation of knowledge within their organizations.

Improving ICT Infrastructure

As manufacturers are developing global strategies based on the use of ICT networks, governments should respond by developing ICT infrastructure to meet manufacturers' needs. Automobile manufacturers are starting to use ICT networks to share product engineering data among their facilities throughout the world and to support global procurement activities. Thailand will have to develop high-quality, large-bandwidth infrastructure capable of supporting these concepts, together with service options designed to meet manufacturers' needs. To encourage the development and use of corporate networks, it will be necessary to reduce the various charges levied on telecommunications.

Measures are needed to reduce the cost of ICT investment for small- and medium-scale enterprises, especially local parts manufacturers. The use of ICT networks in the automobile industry is increasing. Access to global networks has become a prerequisite for the growth of Thailand's automobile and parts manufacturers. However, ICT investment and the training of personnel with ICT skills are major cost factors for smaller parts manufacturers, and steps should be taken to reduce this burden. Consideration should be given to the development of an industry-standard network to reduce the cost of ICT investment. At present the ICT networks and applications installed by automobile assemblers vary from company to company. As a result, parts manufacturers must install systems and applications for each automobile manufacturer that they supply, which increases the cost of ICT investment. Such a network could be based on standardized industry networks, such as Automotive Network eXchange (ANX) and Japanese Automotive Network eXchange (JNX), which are being developed in the advanced economies to alleviate this problem.

CONCLUDING OBSERVATIONS

The regional or global strategies of Japanese vehicle assemblers hinge on the creation of efficient production networks through the appropriate allocation of functions, with some centralized in Japan and others transferred to overseas sites. These strategies involve higher local procurement ratios and will therefore require efforts to bring East Asian parts manufacturers up to international standards within a short period. Parts manufacturers that are unable to adapt will either fall by the wayside or be repositioned as suppliers to foreign primary parts manufacturers. As described in chapter 4, governments are focusing on their own nation's assets and the strategies of foreign manufacturers as they strive to maximize their nation's advantage within this structure.

East Asian automobile industries will be able to generate more domestic added value if they can develop human resources in a wide range of fields, including design, engineering, production, procurement, sales, and management, albeit under the strong influence of the strategies of foreign manufacturers. Localization is also an important goal for foreign manufacturers in terms of their transition to global management, so this approach offers benefits for both sides.

Over the next 10 years or so, automobile assemblers are likely to move increasingly toward a borderless model of business development. East Asian governments are meanwhile likely to become less committed to the notion that their automobile industries must be self-contained within the national economy. Most of the parts and materials used to make Japanese automobiles will probably be sourced from East Asia. As long as brand images are maintained, consumers can be expected to take a diminishing interest in the country of origin of their vehicles. As ICT networks become more sophisticated, it will become more efficient to carry out design, engineering, procurement, production, and sales activities in East Asia. It is also likely that a steadily increasing percentage of automobiles manufactured in East Asia will be branded as "made in Asia," regardless of the actual country of origin.

REFERENCES

The word *processed* describes informally reproduced works that may not be commonly available through libraries.

Fujimoto, Takahiro. 1999. *The Evolution of a Manufacturing System at Toyota.* Oxford: Oxford University Press.

Fujimoto, Takahiro, and Yasuo Sugiyama. 2000. "Ajia Ka to Gurobaru Senryaku—Gurobaru Rokaru Toredoofu ni Taisuru Dotaiteki na Apurochi [The Asia Car and Global Strategies: A Dynamic Approach to Global-Local Tradeoffs]." In Aoki Masahiko and Juro Teranishi, eds., *Tenkanki no Higasi-Ajia to Nihonkigyo [East Asia in Transition: The Role of Japanese Companies]*, pp. 405–54. Tokyo: Toyo Keizai Shinposha.

Imada, Osamu. 1998. *Gendai Jidosha Kigyo no Gijustu, Kanri, Roudou [Techonology, Management and Labor of Automotive Companies].* Tokyo: Zeimukeirikyoukai.

JAMA (Japan Automobile Manufacturers Association). 2002. *Sekai Jidosha Tokei Nenpo* [*World Motor Statistics*].

Japan Economic Foundation. 1999. "Tonan Ajia no Jidosha Sangyo no Hatten to Ikinai Kyoryoku ni Kakawaru Chosa Kenkyu Hokokusho [Research Report on Automobile Industry Development and Regional Cooperation in Southeast Asia]." Research project commissioned to the Sakura Institute of Research, April. Processed.

Mori, Minako. 2001. "The Formation and Development of the Automobile Industrial Cluster in Thailand." *Pacific Business and Industries RIM* 50(January):54–67.

UNESCO. Various years. *Statistical Yearbook.* Paris.

UNIDO (United Nations Industrial Development Organisation). 1999. *International Yearbook of Industrial Statistics.* Vienna.

CHAPTER 6

CHINESE ENTERPRISE DEVELOPMENT AND THE CHALLENGE OF GLOBAL INTEGRATION

Edward S. Steinfeld

Over the past two decades, China's domestic institutional transformation has been inextricably linked with the nation's broader integration into the global economy (Lardy 2002). At the same time, these processes have interacted with changes in the global economy itself or, more specifically, with changes in the way industrial production is organized internationally. In essence, China's rise as a global manufacturer is inseparable from the emergence more broadly of phenomena like globalized supply chains, outsourcing, and networked production (Gereffi and Korzeniewicz 1994).

Whether for socks or semiconductors, Chinese enterprises have become critical suppliers of goods and components for a host of global production chains. By the 1990s global manufacturing, especially in consumer electronics and household appliances, appeared to be migrating en masse to China, thus leading to cheaper products for consumers globally but also to concerns by Europeans, Japanese, and North Americans that Chinese exports were being "dumped" into their markets. China today ranks as the fourth largest industrial producer in the world (behind the United States, Japan, and Germany), an impressive statistic, but one that only hints at a widely held sense in global industrial circles that China represents a force to be reckoned with (Leggett and Wonacott 2002).

At the same time, many of China's individual producers, rather than celebrating, find themselves locked in a downward spiral of intense cost-based competition, primarily with one another. Anecdotes abound of cutthroat competition across a number of product areas, what in China have been termed the "television wars," "refrigerator wars," and "air conditioner wars," just to name a few. Regardless of product area, the Chinese environment seems to be characterized by a proliferation of relatively small-scale firms producing standardized products and competing on the basis of price. In recent years, this has proven to be as true for the myriad, essentially nameless, subcontractors manufacturing for globally branded multinationals as it has been for some of China's better-known companies—Haier in home appliances, Legend in personal computers, and Changhong and Konka in home electronics.

The question is why this is so. Why have Chinese firms been able to master extremely complex manufacturing processes, but not apparently the functions that can move them away from nondifferentiated commodity production? Why have they been able to enter global supply chains, but not control them? Why do even the firms producing branded products face such difficulties in achieving sustainable competitive advantage? What does achieving sustainable competitive advantage, particularly for developing-country firms, even mean in an era of globally networked production?

As this chapter argues, Chinese firms are structured in a fashion that allows them to compete on the basis of low cost in relatively low-value—albeit often highly sophisticated—manufacturing activities. The problem is that this structure does not easily allow them to move upward into nonsubstitutable, higher-return activities: control over brands, provision of unique services, or development of proprietary knowledge. To the extent that Chinese firms remain shut out of these activities and stuck in basic manufacturing, they will have no choice but to compete on the basis of cost, thus eroding their profit margins and further inhibiting efforts to upgrade.

Had it not been for China's economic reforms, Chinese firms could never have entered the global economy. However, current bottlenecks in the institutional reform process, combined with certain aspects of national industrial policy, encourage an enterprise-level corporate structure that severely limits the manner in which Chinese firms compete globally, the extent to which they can upgrade, and the likelihood that they will challenge the multinational firms currently exerting the greatest control over globally networked production.

THE SHIFTING ARCHITECTURE OF GLOBAL PRODUCTION

Technological advances in the management of information, particularly digitization, have dramatically altered the architecture of production processes globally. Digitization, by facilitating the management and transmission of vast amounts of information, has allowed the codification of highly sophisticated processes of production.[1] Once codified, processes can be split into discrete steps—modules, in effect—and standards ensuring their connectivity can be established.[2] Modularization, in turn, has permitted activities that once had to be co-located and managed within the confines of a single firm to be spread across great geographic and organizational expanses.

The issue is not that any activity can be done anywhere or that all manufacturing has become completely modularized, but rather that a number of new options for structuring activities now exist.[3] For some processes—or parts of processes—individual steps have become completely modularized, such that the rules of connectivity with upstream and downstream steps are fully codified and stable. At any given stage in such cases, that which is produced stands alone both functionally and structurally, thus allowing it to be easily "plugged in" to other modules in the overall production process (or even in multiple, seemingly unrelated, production processes). At the other extreme are processes whose component steps cannot easily be codified and disaggregated. They may be separated geographically and organizationally, but their integration into a final product requires extensive coordination and communication among the producing parties. That which is produced in each step has no stand-alone capability, but instead must be carefully tailored to, coordinated with, and often co-designed and developed with specific upstream and downstream components. This sort of "integral" production architecture may be pursued as a matter of choice by a lead firm (that is, a vertically integrated organization), but it also may be dictated by the state of technology.[4]

Taken together, these emergent architectures of production have had a profound impact on China's development (as has China's development on them). Chinese enterprises have skillfully exploited the opportunities of modularization, aggressively upgrading their manufacturing skills so as to meet outsourcing demands by leading global players. In some cases, Chinese firms have autonomously pushed the replacement of traditional integral architectures of production with more modularized, open forms, thus forcing the commodification—and outsourcing to China—of certain activities, regardless of the preferences of overseas lead firms.

1. Roughly, the idea is that "recipe books" can be written for processes that in the past could be successfully run only with tremendous prior experience, vast amounts of accumulated tacit knowledge, and a certain degree of "art."

2. On modularization, see Aoki and Takizawa (2002); Baldwin and Clark (2000).

3. For a comprehensive taxonomy, see Fujimoto (1999, 2002).

4. The term "integral" comes from Fujimoto (2002).

Yet, while modularization affords new opportunities, it also creates major vulnerabilities, whether for Chinese producers or for those from other developing countries. Fully modularized, codified, open production architectures—virtually by definition—entail the manufacturing of standardized, nondifferentiated products. Firms focusing on such activities have little choice but to compete on the basis of low cost and high volume. Moreover, they continually run the risk of being unseated by the next low-cost entrant, particularly since fully modularized products are easily substitutable from the consumer's perspective. That Chinese firms have mastered modularized production accounts for China's emergence as the globe's shop floor. It also accounts for the fact that Chinese firms across a variety of sectors today find themselves locked in mutually destructive price competition.

The lesson is that once new entrants commence modularized production, they rapidly face pressures to upgrade, not so much in terms of the complexity of their manufacturing activities, but rather in terms of the source of their competitiveness. Several options exist. The modularized producer can attempt to control the supply chain by actively setting rather than passively accepting the rules of connectivity in the upstream and downstream directions.[5] In other words, the producer creates the modularized product that everybody else must design around (for example, an Intel processor). Alternatively, the producer might elect to shift away from modularization and move back toward more integral processes, ones that must be coordinated and co-designed with upstream and downstream partners in the network. Finally, as is done by many leading global players, the firm may compete by providing key services—overall product definition, branding, and marketing—that shape the entire supply chain and command the bulk of the final product's value.

All these options require innovation in some sense, a daunting challenge for even the most sophisticated commodity manufacturers. Again, the global shift toward more modularized, networked forms of production offers both opportunities and pitfalls in this area. To the extent a modular manufacturer is engaged in multiple supply chains—that is, produces a stand-alone component that can be plugged into a variety of downstream products—the manufacturer's fate is tied to no single final product in particular.[6] Hence, the manufacturer is free to innovate in ways that not only

5. Examples include Microsoft on personal computer operating software and Intel on processor designs. Controlling the rules means that information can be dispensed selectively to downstream producers, customers can be locked in, and premiums can be charged on new products.

6. The challenge here becomes to move into an integral component or module, one around which upstream and downstream producers must tailor their products, but one that is not easily copied by others. Again, examples, however fleeting, are Intel processors or Microsoft operating software.

incrementally improve existing downstream products ("sustaining" innovation) but also in ways that unseat such products by facilitating new substitutes ("disruptive" innovation).[7] Similarly, open, modularized supply chains permit the rule makers—those determining the rules of connectivity—to shift the standards and thus force the rule takers to scramble in compliance (Aoki and Takizawa 2002). Such freedom undoubtedly contributes to the extremely rapid product cycles and dizzying pace of innovation characteristic of high-tech industry today.[8] Yet it also creates major vulnerabilities for the rule takers, the followers, the commodity producers, and all the rest of the supply chain participants that must respond to innovative lead firms.

THE PHENOMENON IN CONTEMPORARY CHINA: PRICE WARS AND CORPORATE STRUCTURE

As indicated, much of Chinese industry today consists of small-scale firms competing intensely on the basis of discounting. In theory, this could be understood as a prelude to industry-wide shakeouts that should eliminate smaller firms and consolidate activities into a few larger producers, presumably the sorts that might engage in industrial upgrading. Evidence of such progression, however, remains sparse. Undoubtedly, thousands of firms—particularly in the state sector—were either liquidated or substantially restructured in the 1990s, often with significant societal ramifications (Hu 1999; Tenev and Zhang 2002). Similarly, the private sector burgeoned, with enterprises entering and exiting at high rates (Gregory, Tenev, and Wagle 2000). That said, there is relatively little evidence of rationalization, whether in terms of coherent consolidation across given sectors or movement into higher-value proprietary activities by individual firms.

Instead, a pattern of corporate organization has persisted that sets Chinese firms apart from many of their global counterparts and certainly from the lead firms in global supply chains.[9] First, and not surprising given China's relatively recent emergence, Chinese firms tend to be both newer and smaller in scale than their global counterparts. In the World Bank's 2001 survey of 1,500 higher-technology enterprises in China[10], firms averaged just over 600 employees (table 6.1—although the average for firms in

7. The distinction between sustaining and disruptive innovation (and innovators) comes from Aoki and Takizawa (2002); Christensen (1997).

8. This, in a sense, is what Fine (1998) refers to as "clockspeed."

9. Similar attributes are described with great clarity in Nolan (2001).

10. The data can be obtained from the William Davidson Institute. [Davidson Data Center and Network (DDCN)].

Table 6.1 Average Number of Employees of Firms in China, by Location, 1995–2000

Year	All firms	Beijing	Shanghai	Tianjin	Guangzhou	Chengdu
1995						
Average number of employees per firm	798	1,152	1,203	382	582	681
Number of observations	1,057					
1998						
Average number of employees per firm	686	985	956	353	568	572
Number of observations	1,312					
1999						
Average number of employees per firm	635	931	819	339	552	532
Number of observations	1,424					
2000						
Average number of employees per firm	611.9	872	798	352	539	499
Number of observations	1,500					

Source: World Bank survey conducted in early 2001.

Beijing and Shanghai is higher) and generally had been in existence for only 10 to 15 years.[11] Even China's more famous firms—those with known brands and national, if not global, status—tend toward the smaller side. China's premier computer and information technology firm, the Legend Group, had revenues in 2001 of approximately $3.5 billion (Legend Group 2001).[12] Its nearest multinational counterpart, IBM, had revenues that year of $85.9 billion (IBM 2001). The Chinese state-owned petroleum company Sinopec had revenues of $34.0 billion in 1998, compared with $182.3 for ExxonMobil in 1997 (Nolan 2001, p. 166). Capital Iron and Steel had revenues of $2.2 bil-

11. The World Bank's survey involved 500 firms in each of the following cities: Beijing, Chengdu, Guangzhou, Shanghai, and Tianjin.

12. A billion is 1,000 million.

lion in 1998, compared with Nippon Steel's $21.6 billion. Haier—China's premier home appliance manufacturer, one of the best-known Chinese brands internationally, and the fifth largest producer of white goods worldwide in terms of market share—is dwarfed by its global competitors: Bosch-Siemens, Electrolux, General Electric, and Whirlpool.[13]

Second, although their output often ends up either in foreign hands or in overseas markets, Chinese firms tend to be extremely localized in terms of their actual operations. In the World Bank's 2001 survey, 41 percent of manufacturing firms in the sample reported producing to specifications set by foreign firms.[14] Of the firms surveyed, 21 percent reported directly producing parts for foreign firms, while 25 percent reported producing final products for such customers. Indicative of China's liberal policies toward FDI, 25 percent of all firms in the survey reported having foreign equity partners, with the foreign ownership stake on average hovering just over 50 percent.[15] Despite all this foreign interaction, however, the firms' upstream supply network and downstream customer base tend to be confined geographically. The 2001 survey suggested that, on average, more than 50 percent of upstream suppliers are located in the respondent's own city.[16] Approximately 75 percent of the supply network on average is located within China. Downstream, the survey indicated that, for the average Chinese firm, approximately half of the customer base is located within the firm's own municipality.[17] Approximately 15 percent of the customer base on average was reported to be overseas. Whether for upstream or downstream interactions, rather traditional means prevail: communication is conducted primarily by phone and fax, while goods themselves move primarily via surface transportation.[18]

---

13. In the U.S. market, where Haier recently established a refrigerator manufacturing facility, Electrolux (Frigidaire), General Electric, Maytag, and Whirlpool account for 98 percent of the refrigerators sold. See Sprague (2002).

14. There were 995 manufacturing firms in the sample.

15. There were 1,500 firms in the total sample.

16. Of the five cities surveyed (Beijing, Chengdu, Guangzhou, Shanghai, and Tianjin), Guangzhou proved to be somewhat of an exception to this point, with respondents reporting higher levels of overseas suppliers and less concentration of intracity supply networks.

17. Again, Guangzhou was the exception, with respondents reporting 38 percent of the customer base within the city and 29 percent located overseas.

18. Downstream, 67 percent of goods are moved by surface transportation and 84 percent of all communications are conducted either face to face, by phone, or by fax. Upstream, 77 percent of goods are delivered via surface transportation and 87 percent of communications are either face to face, by phone, or by fax.

The localized nature of Chinese commercial networks leads to a third point, the relative shallowness with which Chinese firms integrate into global supply chains. Despite high levels of foreign ownership, only 15 percent of the manufacturing firms surveyed by the World Bank in 2001 reported designing parts for foreign customers, a sign that the respondents are essentially "rule takers" in open, fully modularized production processes. Only 7 percent reported providing customers with research and development (R&D) or other specialized services. The figures are surprisingly low given that the sample specifically targeted higher-tech sectors, the very ones in which we should expect high degrees of innovation, networking, and development of firm-specific proprietary knowledge.

The firms are failing not only to design for downstream customers but also to develop deep relationships of any kind with such customers, again, a sign of open, modularized production. Among survey respondents, 69 percent reported using trading companies to handle interactions with the broader customer base, thus suggesting essentially arm's-length rather than deeply enmeshed customer relations.

In terms of identifying the factors inhibiting greater exports, respondents focused on the difficulties of meeting foreign product standards, the high costs of meeting such standards, and particularly the intense cost competition they face (table 6.2). Managers apparently prefer to produce for export markets, and few claimed that targeting the domestic market offers better financial gains, but managers perceived that their firms lack the capabilities needed to meet foreign standards in a cost-effective manner. At the same time, they perceived themselves to be in an intensely cost-competitive environment, with pressures bearing down from both domestic and foreign counterparts. Presumably, one reason why managers find markets so price driven is that they neither sell directly to, nor design directly for, downstream counterparts. If a firm does not sell directly, it cannot develop the sort of service-based, specifically tailored relationships or products that lead to customer loyalty. If the firm—a component producer, perhaps—does not design directly for the downstream user, it is not likely to have deep interactions with that user and hence likely will forgo the important learning opportunities (whether in terms of product or process technology or even of marketing skills and service provision) that might come from dealing with potentially more sophisticated customers.[19] Nor is the firm likely to be able to exert any leverage if the user decides to source the upstream component elsewhere.

---

19. A positive example would be the Taiwan Semiconductor Manufacturing Corporation. See Burns (2000).

Table 6.2 Main Inhibitors of Export Growth in China, by Location (percent)

Inhibitor	All firms	Beijing	Shanghai	Tianjin	Guangzhou	Chengdu
Shipping and transport costs	9	6	8	7	15	8
Cost of meeting foreign legal and product standards	15	13	11	13	18	16
Inability to produce to clients' standards, specifications, and schedule	15	15	12	13	19	13
Inability to match prices of domestic competitors that export	11	13	10	14	8	9
Inability to match prices of foreign competitors	12	11	17	11	4	9
Inability to meet demands by foreign clients for product upgrades and changes in specifications	6	4	7	9	7	5
Difficulty of recovering payments from abroad	3	4	2	5	4	2
Higher profitability of supplying the domestic market	7	7	8	6	4	8
High costs of establishing a foreign distribution network	17	16	18	12	13	22
Domestic content requirements	4	11	7	9	8	8

Source: World Bank survey conducted in early 2001.

That, then, leads to a fourth and final point regarding innovative capacity. Chinese enterprises today face great pressure to upgrade their technological capabilities, and managers—as they did in the 2001 survey—routinely report high levels of what they at least perceive to be innovative activity. The pressures are understandable. Modern production, whether for low-end goods like textiles or high-end goods like semiconductors, virtually by definition entails the management of complex processes, complex and often capital-intensive machinery, highly refined product-specific materials, expectations of high quality on the part of customers, and generally rapid turnaround time. Simply to be involved in global production networks today, even at the relatively low end, and even just to produce modularized products that can be "thrown over the wall," new entrants must climb exceedingly steep managerial and technological learning curves, and they must do so rapidly. As indicated earlier, while modularized, networked production has opened up new doors for new entrants, the bar to basic participation has been raised quite high. That Chinese firms are involved so extensively in production for overseas markets represents a major achievement and indicates extremely impressive degrees of learning on their part.

It would be incorrect, however, to assume that such learning actually constitutes—or necessarily leads to—innovation. In other words, it is not at all clear that these firms are developing intellectual assets, production skills, modes of serving customers, or actual products that can be understood as in any way proprietary—things that cannot be duplicated by hundreds or thousands of other firms in their immediate environment. In the 2001 survey, nearly half of all firms reported innovations in shop-floor production processes, and another 46 percent reported innovations in managerial techniques, all measures that cut costs. What few, if any, of the firms reported were innovations that allow the firm to charge a higher margin rather than a lower one—in other words, innovations that encourage customers to pay a premium for the product. Moreover, given the prevalence of product "wars" and cutthroat discounting among the proliferation of small producers in China, it appears that nobody has discovered the sort of proprietary cost-cutting solutions that afford competitive advantage over a reasonable period of time. Instead, the cost-cutting measures get duplicated from firm to firm, and the margins continue to erode.

The response to this dilemma often entails another activity that survey respondents termed innovation: the introduction of new products or entirely new lines of business.[20] Commodity producers end up chasing one surplus market after another, a pattern true even for China's more advanced branded companies. The nation's leading television manufacturers, including Konka—facing declining profits, rising inventories, and overseas import quotas—have moved aggressively, and en masse, into mobile phone manufacturing (Leggett 2001).[21] Similarly, Guangdong Galanz Enterprise Group, a major producer of low-end microwave ovens, has coped with declining profits by jump-

20. In the 2001 survey, 36 percent of the respondents reported introducing new products within existing lines, and 21 percent reported introducing new products in entirely new lines of business.

21. Konka, China's second largest television producer, lost $84.5 million in 2001. TCL International reported a 32 percent drop in profits that same year and, like Konka, has moved into other appliance manufacturing. See Stevens (2002).

ing into the air conditioner market, a market already suffering from high inventories (Lee 2002). Galanz's strategy, not surprisingly, has been to pursue extensive price discounting. Even the most established firms cope with increasing competition by aggressively discounting and expanding sales volume on existing products, by entering new product areas in which they can compete again only on the basis of discounting and razor-thin margins, or, finally, by trying to export their way out of trouble by pursuing overseas markets.[22]

In essence, firms focus on activities with low barriers to entry. Once the cost pressures become too intense, rather than moving upward into higher-end activities or taking the time to develop proprietary skills, the firms diversify into other low-entry-barrier markets. The products themselves—be they televisions, personal computers, refrigerators, or data routers—are standardized. In the local context, therefore, innovation becomes associated with cost reduction and flexibility, the ease and rapidity with which a firm can jump from one saturated product market to the next. In this same local context (let alone globally), neither of these capabilities has proven defendable over extended periods.

EXPLANATIONS: REFORM STYLE, GOVERNMENT CAPACITY, AND INDUSTRIAL POLICY

The pattern described above stems in large part from the interaction between three factors: style of governmental reform, state capacity, and industrial policy. This interaction has at once permitted the integration of Chinese firms into the global economy and substantially constrained the extent of that integration.

Reform Style

Since the dawn of reform, China's approach to market transition has been characterized by informality, experimentation, and decentralization. Central leaders have set the overall policy aim (economic growth) and the basic constraint (the maintenance, in the vaguest terms, of socialism). Local

22. Haier, China's largest refrigerator and air conditioner manufacturer—and an emerging global brand—suffered a 45 percent year-on-year drop in net profit in the first half of 2002. The company's response was to increase exports of low-end air conditioners and further ramp up domestic production of refrigerators and air conditioners. See "China's Haier First Half Profit" (2002).

officials, then, have been granted broad leeway to engage in policy experiments, virtually all of which have involved elements of market economics. Socialism is maintained simply to the extent that the experiments remain informal. When experiments prove successful, the center encourages their implementation—again on informal terms—nationally. If success continues, the experiments stand to be adopted post hoc as official government policy. Finally, in some—but not all—cases, the center formalizes the outcomes with new institutional rules, many of which directly challenge the initial condition of "maintaining socialism." Through a certain element of linguistic legerdemain, that which began as an experimental alternative to socialism (and hence its explicitly informal status) gets legitimized as socialism itself, albeit socialism "with Chinese characteristics."

The approach has proven brilliant in many respects. Without it, China's transition to what much of the world terms capitalism could never have proceeded smoothly.[23] It also explains how private enterprise—anathema just 20 years ago in China—now constitutes the predominant form of ownership in Chinese industry and has gained official status in the Chinese constitution.[24]

That said, there are negative ramifications of informality and decentralization as well. Entrepreneurial firms can thrive under such circumstances, and they can engage in international commerce, but their property rights tend to remain either undefined or—to the extent they tuck themselves under the auspices of a government bureau or state-owned firm—inaccurately defined. Without clear property or formal title to assets, these firms—as they would in any commercialized economy—face limited financing options. Borrowing from a bank becomes virtually out of the question. Instead, they have little choice but to self-finance, a situation that may ensure hard budgets, but one that also tends to limit enterprise growth.[25]

In a pattern consistent with that of virtually all firms in China save for larger state-owned enterprises, firms in the World Bank's 2001 sample reported relying primarily on retained earnings as their main source of financing (table 6.3). Firms consistently reported that upward of 50 percent of all

23. For a useful definition of what such a transformation means, see Kornai (2000). See also Steinfeld (2002).

24. A change in attitude was signaled in 1997 when the Communist Party recognized private firms as an important part of the Chinese economy. That change was followed up in 1999 with an amendment to the Chinese constitution officially recognizing private enterprise. Finally, in 2001 China's president and Communist Party general secretary, Jiang Zemin, called for welcoming private entrepreneurs—capitalists, in effect—into the Communist Party.

25. On the ramifications of informality, see de Soto (2000); Gregory, Tenev, and Wagle (2000); Mackenzie (2001).

Table 6.3 Sources of Financing among Chinese Firms, by Location

Source of financing	All firms	Beijing	Shanghai	Tianjin	Guangzhou	Chengdu
Retained earnings or internal funds						
Percent of all financing	51.5	51.7	51.2	49.2	50.1	55.1
Letter of credit						
Percent of all financing	0.8	1.1	0.6	0.7	0.4	1.0
Supplier credit						
Percent of all financing	3.3	3.1	4.2	3.0	4.0	2.8
Bank loans						
Percent of all financing	18.9	17.0	19.3	14.9	18.7	24.8
Other financial institutions						
Percent of all financing	1.6	1.2	2.0	1.1	0.7	3.1
Parent or partner company						
Percent of all financing	8.4	7.6	11.9	8.6	8.9	5.1
Equity finance						
Percent of all financing	0.6	0.1	1.1	1.0	0.1	0.6
Personal, family, and friends						
Percent of all financing	8.6	7.8	3.2	17.2	8.6	6.2
Other sources						
Percent of all financing	6.3	10.3	6.6	4.4	8.5	1.9

Note: Number of observations is 1,486.

Source: World Bank survey conducted in early 2001.

financing comes from retained earnings. Bank loans amount to 19 percent of total financing on average, although the figures are somewhat lower in Tianjin (15 percent) and somewhat higher in Chengdu (25 percent). Equity financing, not surprisingly given governmental quota restrictions on stock market listings, is low across the board (averaging 0.6 percent across the sample). Personal loans from family and friends constitute an important source of financing, averaging 8.6 percent of total financing for firms in the sample.

Limited financing options, of course, frequently lead to tight liquidity constraints. The response of enterprises often involves operating on a cash basis, but that leads them to forgo transactions that in more formalized systems allow for greater enterprise expansion.[26] Furthermore, rather than investing in existing business lines and developing specialized skills, cash-starved firms jump to alternative businesses simply to maintain cash flow

26. Or they engage in informal channels of financial intermediation, credits that tend to be high in price and small in scale. See Tsai (2002).

(that is, if low-end television production is not generating revenue today, they move to low-end mobile phone manufacturing). Such diversification addresses liquidity issues, but it does not encourage the development of firm-specific proprietary assets or skills. Instead, firms remain stuck in activities with low barriers to entry.

Informality, to the extent it dilutes the firm's legal status, also limits the firm's geographic reach. Without legal standing, the firm must engage predominantly in trust-based transactions (both regarding with whom it chooses to deal and who chooses to deal with it). The surest way to ensure trust is to stay local, essentially by buying from known local suppliers (or, better yet, backward integrating to ensure reliable supplies) and selling to reliable local customers (so as to ensure payment). When dealing with international markets, the main option becomes to sell to a local trading company.

For foreign companies dealing with such informal organizations, the strategy of choice often entails either buying from a more formalized state trading company or actually buying equity in the local producer itself. Indeed, foreign direct investment, to the extent that it places the recipient into the special regulatory category of "foreign owned," constitutes a formalization mechanism, one that benefits provider and recipient alike. In some cases, Chinese firms sell their assets to foreign firms at a discount, but in so doing they achieve a degree of formality that affords them access to credit and insulation from arbitrary governmental policy (Huang 2003).[27]

Like informality, governmental decentralization leaves its own mark on entrepreneurial organizations. Many local governments in the reform period have eagerly promoted economic development and, as part of that goal, have frequently promoted local entrepreneurship (see Oi 1999). They have been less eager, however, to facilitate development that benefits areas beyond the locality. Early in the reform era, this reluctance manifested itself in regional trade wars and overt barriers to interprovincial trade.[28] More recently, in the 1990s, given central crackdowns on overt protectionism, localities used more subtle methods: selective enforcement of product standards, more rigorous registration and licensing requirements for outsiders, and prejudicial application of health codes, just to name a few (Institute of Industrial Economics 1998, p. 294; also cited in Naughton 1999, pp. 20–21).

27. 2003. The implication is that China has an inordinately high demand for FDI.

28. For alternative perspectives on the problem, see Naughton (1999); Young (2000). On interprovincial trade, see also Hu and Wang (1999).

The more positive interpretation of Chinese-style decentralization is that localities have been permitted to build institutions that best suit local circumstances and best support local development.[29] The more negative view, one not wholly irreconcilable with the first, is that institutional ingenuity today has come to mean local content and "buy local" rules for local producers ("local" defined as "within the municipality"), discriminatory regulatory enforcement for outsiders trying to enter the local market, and restrictions on enterprise mergers and acquisitions.[30] Managers on a day-to-day basis may not be spending the bulk of their time on governmental matters, but it is fairly clear that administrative factors impinge on, and often restrict, commercial activities.

Similar conclusions can be drawn regarding sectoral and geographic rationalization in industry. Whereas rational mergers and acquisitions are frequently blocked through various administrative interventions, commercially *irrational* mergers are often imposed by local administrative fiat. Particularly in the state sector, financially sound firms have been forced, often under considerable duress, to assume ownership of insolvent organizations simply to stave off bankruptcies.[31] That the acquiring firm is sometimes accorded preferential policy treatment as a sort of quid pro quo further distorts budget constraints and incentives for productive growth.[32]

More generally, when firms are forced to merge with failing local neighbors or to source only from local counterparts, they are often indirectly prevented from interacting with the best, most advanced suppliers. In an era of networked production, when innovation is understood as emanating as much from interfirm learning across the supply chain as from isolated activities within the individual firm, linking up with the best upstream and downstream partners becomes a key component of upgrading. Administratively imposed restrictions on such linkages, particularly restrictions that limit the linkages to a given municipality, prevent Chinese firms from ac-

29. For more on the regional nature of development and institutional reach, see Segal and Thun (2001).

30. Decentralization was perhaps politically necessary—and sound from a policy sense—to allow firms to start up and avoid policy rigidity. The problem is that once those firms are established and need to grow, local idiosyncratic institutions and local discrimination inhibit growth. For extensive discussion, see Wedeman (2003).

31. A prime example involves the forced merger between the Chongqing Iron and Steel Group and the deeply troubled Chongqing Special Steel Company. (Author's interviews, Chongqing, 2000.) See also Pomfret (2001).

32. For example, Shanghai's Baoshan Iron and Steel Group, after being forced to take on a financially insolvent local producer as a subsidiary, was subsequently able to have the subsidy's debts forgiven through a debt-equity swap with state lenders. (Author's interviews, Beijing, 1999 and 2000.)

cessing not only the best global suppliers, but even the best national ones. Deprived of high-quality components and important learning opportunities, many Chinese firms are pushed further down the road of low-end manufacturing and cost-based competition. Moreover, when localities try to keep the firm local, the firm's problems of small scale and limited financial resources simply deepen.[33]

State Capacity

By the later 1990s the architects of Chinese reform began to tackle many of the problems discussed above. It became understood that informality and decentralization, while critical initially for achieving local acceptance of reform (and for easing the ideological problems associated with what ultimately became a wholesale adoption of market economics), had run their useful course and were now inhibiting further growth. Efforts to formalize China's market system across a variety of areas rose to the top of the policy agenda. The problem, however, is that these imperatives collided with the reality of limited state capacity in China.

The issue of capacity manifests itself in at least two respects: the ability of the center to coordinate policy across the government's administrative hierarchy and the ability of the government as a whole to regulate commercial activity in the civil sphere. The first problem, discussed in the previous section, has arguably receded in recent years. The second, however, has proven more vexing. As might occur in any developing economy, the Chinese system has experienced a dramatic increase in the complexity and density of interactions between economic actors, most of which are no longer under the direct administrative control of the state. Across the board—whether in financial relationships, contracts, issues of corporate control, or intellectual property rights—demand within the civil sphere has increased for both objective rules and reliable enforcement (Peerenboom 2001). Understandably, given the rapidity of China's economic growth, the demand for market governance has outpaced the ability of the state to provide governance-related public goods. Courts are overwhelmed with cases, judges are often inadequately trained, and enforcement mechanisms are generally weak at best (see Alford 2000; Lubman 1999).

It is widely recognized in China today that rule of law is essential for sustained growth, but it is far less clear how rule of law can be achieved or even

33. Some scholars have argued that Chinese private firms, rather than innovating, spend most of their energy cultivating clientalistic ties with political patrons. See Wank (1999). Others have argued that clientalism has receded in recent years, as institutionalization has increased. See Guthrie (1999).

exactly what it entails (see Peerenboom 2002). Meanwhile, the absence of effective legal institutions encourages the sorts of rent-seeking behavior that further erode trust in commercial transactions and society more broadly. In the financial area, for example, we have witnessed the emergence of what some Chinese describe as a "nonpayment" economy (Zhou 1999, p. 6). Commercial buyers make purchases and then refuse to pay. Borrowers take out loans and then default. Banks accept deposits and then squander them in ill-advised lending. In each case, the victim is left with little recourse. As the saying in China goes, "You sue, but the court won't accept your case; the court accepts your case but won't begin the trial; the court begins the trial but won't issue a judgment; the court issues a judgment but then doesn't enforce it [*qi gao bu shou li, shou li bu kai ting, kai ting bu xuan pan, xuan pan bu zhi xing*]" (Zhou 1999, p. 6).

What results is neither utter lawlessness nor an absence of growth. Instead, there exists a subtle pattern of unclear rules, low levels of trust, and frequent efforts to skirt the boundaries of legal strictures, conditions that—as indicated earlier—all affect the organizational structure and global competitiveness of Chinese firms. At the very least, the environment impinges on both the capacity and inclination of firms to innovate.

In more concrete terms, low state capacity impedes restructuring of a key economic chokepoint, the bank-dominated national financial system. For the first 15 years of reform, the state-owned banking system was employed as a quasi-fiscal mechanism to channel funds into state-owned firms (Lardy 1998; Steinfeld 1998). Over time, that pattern resulted in a systemic buildup of nonperforming loans and a situation of overall technical insolvency (for recent data, see Ma and Fung 2002). To its credit, the central government, particularly since the Asian financial crisis, has acknowledged the problem and has undertaken a series of measures to encourage both the recapitalization and commercialization of the financial system (Steinfeld 2000). Yet, while it has proven able by fiat to freeze credit provision, the government in the past five years has proven consistently unable to produce modern financial regulation or to achieve anything resembling healthy commercialization. The banking system remains mired in conditions of moral hazard: bank balance sheets remain awash in red ink, large borrowers have not been shut down, real loan write-offs have been deferred, and bankers have then been told to loan money on commercial terms.

Under such conditions, it is no wonder that nonperforming loans continue to accumulate (Kynge 2002).[34] In 1999 the Chinese government established four asset management companies to take on nonperforming

34. Kynge (2002). Ma and Fung (2002) place the number at 35 percent of outstanding loans.

loans from the four main state banks and begin recovery efforts. Since the inception of this debt-equity swap program, distressed assets valued at approximately RMB 1.4 trillion, some 20 percent of the value of the banking system's entire loan portfolio, have been transferred to asset management companies (Ma and Fung 2002). Even after these massive transfers, though, state banks have been reporting high levels of nonperforming loans. Officials of the China Construction Bank reported that, as of the end of 1999, nonperforming loans still constituted at least 30 percent of the loan portfolio (Fang 2001). In 2001 the president of the Bank of China reported levels of nonperforming loans at 26 percent of outstanding loans, and this was again after some 20 percent of the bank's total loan portfolio had been transferred to an asset management company. Along similar lines, analysts at the Bank for International Settlements have recently estimated that, at the end of 2001, China's nonperforming loans amounted to 35 percent of GDP (Ma and Fung 2002, p. 3). The overall sense is not only that the stock of nonperforming loans is large but also that the flow is still increasing. Under such conditions, it is no wonder that Chinese industrial producers—particularly the newer, more dynamic firms that have no established relationship with state banks—face tight liquidity constraints.

Industrial Policy

Lurking behind these capacity issues, of course, is the issue of ultimate government aims. China throughout the 1990s pursued institutional reforms that encouraged market deepening and a level playing field among all participants in the economy. Nonetheless, it would be incorrect to suggest that the Chinese government has abandoned either the notion of state ownership or the ambition of building key state firms into vertically integrated, globally competitive "national champions." In other words, traditional industrial policy in the style of Japan and the Republic of Korea, at least in some respects, still enjoys considerable appeal in China.

Industrial policy as the unleashing of comparative advantage. Contemporary Chinese industrial policy appears schizophrenic in many ways. On one side, the government has pursued what it describes as a "comparative advantage" strategy of development (Chen 2002). In what amounts to an essentially Heckscher-Olin-Samuelson approach, the strategy is premised on the idea that development and industrialization are natural byproducts of the convergence of factor prices across nations. The country's relative factor endowments at any particular time are taken as given (in China's case, surplus labor and scarce capital), and then development is understood as unfolding as the country specializes in the production and export of goods intensive in the use of the abundant factor. As long as external trade and internal markets are opened up—conditions that become central goals of this sort of industrial policy—a dynamic international division of labor should ensue. Poorer countries like China begin by exporting labor-intensive products, but as capital flows in (seeking higher marginal returns in the capital-scarce environment), the developing country presumably begins spontaneously climbing the ladder of industrial sophistication and capital intensity. Over an extended period, provided that markets remain liberalized, industrialization should proceed in pace, and factor prices should ultimately equalize across countries (for an alternative view, see Amsden 2001).

Policymakers in Beijing, at least on this particular side of industrial policy, have followed the theory's prescriptions, albeit with some modifications. Reform, since its very inception, has been promoted as a process of "opening up," and opening up specifically to foreign trade, knowledge, and technology. China throughout the 1980s, and particularly after 1992, dramatically lowered statutory import tariff rates (Lardy 2002, p. 37). Since 1997 the government has substantially expanded policy initiatives that exempt certain domestic firms and institutions from paying the import duties that formally do exist (Lardy 2002, p. 36). Finally, in 2001, China formally became a member of the World Trade Organization (WTO), binding itself to an accession protocol more expansive, in terms of both market access and permissible trade practices, than that faced by any other developing nation in history (Lardy 2002, pp. 63–105). Moving from informality to formality, the guiding assumption for this face of industrial policy is that the surest way to stimulate development is to expose domestic firms to international competition.

Equally important, on the other side, reformers have coupled this with an aggressive courting of FDI, pursuing what has amounted to the most liberal policy on FDI of any Asian developing economy. Here, a bit of practicality has tempered devotion to theory. Heckscher-Olin-Samuelson theories assume knowledge to be perfect. That is, as long as capital and labor are allowed to flow freely, prices should equalize across countries, and productivity should equalize across firms. The actual knowledge of *how* to produce is presumed to be trivial: it simply flows across the ether like a television program washing across the eyes of a passive viewer or a library book changing hands freely from borrower to borrower (this point is critiqued in Amsden 2001, pp. 2–8). As long as the prices are right, the firm is presumed capable of producing. Lack of competitiveness, therefore, can be attributed primarily to bad policy: government distortion of prices, excessively high wages, and illiberal trade regimes (Amsden 2001, p. 3).

Policymakers in Beijing, though, have hedged their bets on this front. Instead of waiting passively for "natural" transfers of knowledge and technology, they have chosen proactively to build a vector: FDI through industrial joint ventures. The idea is that for Chinese firms to upgrade, they must work hand in hand with leaders from the developed countries. In exchange for transferring technology and know-how to Chinese counterparts, outsiders are then granted privileged access to the Chinese domestic market or preferential treatment on other grounds.[35]

The efficacy of such policies can, of course, be debated. Some scholars, by pointing out the discounts at which foreigners have been able to purchase industrial assets, have suggested that China has been inadequately compensated for such transfers (Huang 2003). Others have suggested that traditional Chinese industrial firms devote far too much attention to hard technology—physical assets—and in the process severely neglect the training and expertise needed to absorb those technologies (see Gilboy 2003).

Nonetheless, it is eminently clear that over two decades, China's policies on FDI—not to mention its liberal policies toward emigration—have led to a monumental scaling up of managerial expertise in the country (on emigration issues, see Zweig 1997). Whether in foreign or domestic firms, an essentially world-class population of managers has been created at the highest tiers of the economy. Increasingly, this population has begun to flow back and forth between employment in foreign and domestic companies and between employment within China and outside.

Of course, the question is whether, particularly in domestic firms, these managers can operate in an institutional environment conducive to enterprise success.[36] Skilled domestic managers now exist, but can physical assets really flow? Will commercially moribund firms—legacies of the prior era for the most part—be allowed to go under and entrepreneurial firms be permitted to rise from their ashes?

Again, at least on this side of the industrial policy ledger, major strides forward have occurred. Between 1994 and 2000, with the government's policy of *zhua da fang xiao* ("grasping the large and releasing the small"), almost 60,000 small- to medium-size state-owned enterprises have been "restruc-

35. Examples include tariff exemptions for imported raw materials used in the processing of exported final products.

36. Indeed, one of the arguments of those who do not like FDI policy is that such policies provide foreign firms a privileged domestic institutional environment and thus encourage domestic assets to flow to them.

tured," a term that generally signifies outright liquidation, privatization, or transfer to employee ownership (Tenev and Zhang 2002, p. 30). At the same time, the private sector has been permitted to burgeon and now constitutes the single largest form of ownership in Chinese industry. In the past, enterprise reform in China meant measures to improve performance in existing state-owned firms. Today, enterprise reform has increasingly come to mean measures for *eliminating* poor performers.

Conceptually, then, this particular guise of Chinese industrial policy—the exposure of firms to foreign competition, the encouragement of FDI and knowledge transfer, and the ruthless downsizing of poor performers—can be understood as a "creative destruction"–centered vision of development (a view similar to that outlined in Schumpeter 1968 [1911]). Industrialization becomes the progeny of market forces, and those forces themselves are understood as the mechanism for winnowing winners from losers. The continual composition and decomposition of constellations of assets—in other words, the rise and fall of firms—are treated as a good unto itself, one that outweighs the intrinsic value of any given firm. Innovation, the driver of development, is not envisioned as the product of a steady accumulation of tacit knowledge and internal experience within long-lived corporate organizations, the kinds that must be protected and nourished by governmental policy. Rather, innovation grows out of the maelstrom of intense competition among firms, the continual overtaking of conservative incumbents by radical newcomers, and the wild dynamism of organizational destruction and re-creation.

One could argue that this vision fits the contemporary era of rapid deverticalization and modularization in global industry. To the extent one believes a true technological and organizational revolution occurred in the late 20th century, one would expect innovation to fall mostly within the "disruptive" rather than the "sustaining" category. Traditional, vertically integrated incumbents—the firms with the greatest stake in the products and processes of the prior era—are the least likely to remain on the cutting edge of technology. They have accumulated knowledge and experience, but not the kind relevant for a changed world. Moreover, to the extent one believes that the production of goods and services truly has become modularized, one would anticipate within any given module rising incentives for rapid and disruptive innovation. The condition itself of being a modular producer implies that one is no longer captive to single final products downstream. One simply innovates within the module, regardless of the degree of disruption that others may suffer (or added sustainability that others might enjoy). As waves of modular innovation cascade up and down the production chain, winners in any given activity presumably emerge from communities of flexible, highly specialized entrepreneurial entrants, entrants that compete among themselves at the same time as they cross-pollinate.

Industrial policy then becomes focused on creating environments, not firms. It focuses on creating Silicon Valleys and innovative clusters rather than industrial conglomerates and vertically integrated giants. Moreover, to the extent that supply chains overlap and individual modularized producers somehow exist at the nexus, it becomes pointless to think about unified "industries" in the traditional sense. The goal becomes not to build a national electronics, auto, or aerospace industry per se, but rather to develop capabilities that accord control and high returns in the international supply chains crisscrossing all those industries.

Industrial policy and the building of national champions. What makes Chinese industrial policy so difficult to comprehend is that, for all its focus on market-based approaches and comparative advantage, it also has an entirely different side, one that embodies assumptions of heavily statist Japanese and Korean models of the past. Policymakers in Beijing may employ all the mechanisms associated with comparative advantage strategies, but the ultimate aim of such policies remains the creation of "national champion" firms in self-reliant, vertically integrated pillar industries (Nolan 2001, p. 16). This, after all, is what the "grasping the large" side of the *zhua da fang xiao* enterprise restructuring policy is all about. It is about creating exactly the type of organizations associated with the Japanese and Korean models of yore: large, vertically integrated business groups that encompass entire industries from upstream to down, operate at the cutting edge of technology, and dominate global markets from their home base in China. Yet this is a story that involves more than just new techniques for achieving old industrial ambitions. Rather, it is a story about a government claiming as its ultimate policy aim precisely the type of firms that its highest-profile restructuring (and trade) policies militate against. In essence, the government is seeking to create the very firms that comparative advantage, not to mention global technological change, select against.

Of course, as some policymakers in Beijing are inclined to admit, China's effort to build national champion conglomerates must differ from earlier Japanese and Korean efforts in a few respects. First, the Chinese economy today is much larger and more diversified than were the Japanese and Korean systems at the height of their respective experiments with dirigiste industrial policy (Perkins 2001). In Park Chung Hee's Korea, more than half of all manufacturing was accounted for by fewer than 200 large industrial groups. In China, one would have to turn to tens of thousands, or even hundreds of thousands, of firms to account for a similar percentage of industrial activity. The point is that simply to exert the same degree of control associated with the Korean model, Chinese policymakers would have to deal with an exponentially larger task and exponentially more complex information flows than anything experienced in 1970s Korea.

Second, the government of China, in no small part because of the reformist legacies of decentralization and informality, operates in a less unified manner than that of Japan or Korea decades ago. Whether by design or by default, policymakers in Beijing today implement most national policy through local agents.[37] This has certainly proven true in the effort to build "national" pillar industries, a task that has been essentially farmed out to individual provinces and municipalities. Consequently, while China's industrial planners proclaim the need for national steel, auto, or machine-building firms, what results is the duplication of such entities in virtually every province and large municipality. Traditional Japanese and Korean industrial policy may have been all about focusing limited national resources on the development of just a handful of industrial giants. China's industrial policy, however, is about signaling regional governments to focus resources on what ends up being thousands of firms and tens of locally duplicated industrial structures.

Third, Japanese and Korean developmental efforts were premised on the idea that, at least in their home market, key industrial conglomerates would be granted sweeping protection. They would be held to international standards and encouraged to compete head to head with foreign firms in foreign markets, but on the home front, they would be showered (selectively) with subsidies and sheltered from outside competition. As signified by the terms of China's accession to the WTO, though, the world today is not that of the 1960s and 1970s, in no small part because the world's wealthiest countries—though hardly paragons of free trade—simply do not tolerate the sorts of protectionism they once did with regard to Asian developers. Nor, somewhat ironically, are they inclined to tolerate the sorts of export flows previously generated by Asian "national" firms.

Whether or not the Korean-style industrial policy was effective on its own terms and in its own era is a major question, but one not immediately relevant to this chapter. What *is* relevant, however, is the basic reality that, while China may seek to build the kinds of firms associated with such models, it has at its disposal few of the policy instruments and external conditions enjoyed by industrializers decades ago.

37. The 1994 national tax reforms and 1999 national banking reforms all attempted to move away from this approach, but only partially succeeded.

That said, even if the model were feasible on practical policy terms, it still would be difficult to reconcile on logical terms with China's "other" industrial policy, the comparative advantage approach. Indeed, practical issues of implementation aside, how can Chinese policymakers square the circle between the highly divergent conceptions represented by each of these approaches? Decisionmakers may presume that, to the extent they get industrial policy "right," the resulting national champion, pillar industry organizations will be globally competitive and hence sustainable after WTO-mandated market liberalization takes place.

Yet that really begs the question of how the divergent premises of comparative advantage and national champion can be reconciled. After all, one view stresses the primacy of churning and market selection—creative destruction—as the driver of innovation and growth. The other stresses virtually the opposite: the degree to which innovation occurs through the evolution and sustenance of established incumbents, corporate repositories of knowledge, and experience.[38] One view emphasizes the market's role as a selection mechanism, a ruthless judge of winners and losers. The other emphasizes the market's role as an incentive mechanism, a treatment that, when applied to preexisting organizations, encourages efficiency. One view says that firm-level incentives are inseparable from, and indeed can be understood only as emanating from, the system-wide process of creative destruction. The other suggests that market incentives, by encouraging existing firms to maximize efficiency, obviate—or at least reduce the likelihood of—such destruction. Indeed, in this latter view, if selection begins spontaneously to operate—if losers start to appear, particularly on a grand scale—then something must have interfered with the proper operation of the market, be it politicization, insufficient liberalization, or "bad policy" in any of its other guises. One view, in essence, understands the firm as a by-product of the market. The other takes the firm, particularly the modern industrial conglomerate, as the linchpin and driver of the market.[39]

Policymakers could try, as is done in China, to hedge by operating on both sets of premises simultaneously.[40] In so doing, however, they frequently adopt policies that function at cross-purposes. For example, the quest for a "national team" has led to persistent governmental distortions of financial markets.[41] Those distortions, though, by withholding capital from China's

38. Origins of the view can be seen in Schumpeter (1976 [1942]). Note the differences between Schumpeter's 1942 perspective and his 1911 views cited earlier.

39. For a history of this perspective, see Chandler (1990); Chandler and Hikino (1997).

40. A number of nations attempt this. See Steinfeld (n.d.).

41. As evidenced by the disproportionate representation of state-owned enterprises on Chinese equity and debt markets.

most dynamic market-oriented firms—its private enterprises—limit the ability of these firms to respond to competitive pressures being induced by comparative advantage market liberalization measures. In essence, the distortions aimed at building the national team undercut the global (and domestic) competitiveness of a huge swath of Chinese industry.

Along similar lines, policymakers encourage the development of vertically integrated pillar industry firms but then pass the actual developmental task indiscriminately to localities. What results is neither the verticality nor overall scale that traditional Korean-style industrial policy calls for. National champion firms end up in reality as little more than local or regional players. At the same time, the focus on verticality encourages localities to think not in terms of cluster economies, innovative communities, or crosscutting supply chains—the sorts of environments from which effective comparative advantage competitors are likely to emerge today—but instead in terms of self-contained industrial units, units that may co-exist, but not interact. Firms end up with locally focused captive supply chains, a worst-of-all-worlds situation even if one agrees with the goal of building integrated national conglomerates. To the extent the supply chain is held captive, it should at least be permitted to extend broadly in geographic terms (so as to incorporate "best in class" suppliers nationally). Keeping it local almost guarantees that the firm will fail to access the best suppliers and hence will fail to produce world-class products. At the other extreme, to the extent one believes that firms should focus on modular activities and then link into upstream and downstream activities on a global basis (in line with the comparative advantage approach), administratively enforced captive supply chains should disappear altogether.

More generally, by merging essentially irreconcilable visions for industrial development, policymakers end up achieving the aims of neither. Localization and geographic duplication undermine the scale and supply chain quality that might, under the theory's own assumptions, produce globally competitive conglomerates. At the same time, the institutional distortions induced to achieve national champions (local as they may be) undercut the ability of private sector firms to compete effectively on purely market terms. The firms shielded from creative destruction remain weak, while the distortions behind that shielding leave everybody else handicapped in the face of creative destruction. That many in the latter group have survived is testament more to their fortitude than to the brilliance of industrial policy per se. Unfortunately, such survival, achieved primarily through commodity production and cutthroat discounting, is hardly the basis for extended success in the future, whether at the enterprise or the national level.

CHINA'S EMERGING MULTINATIONALS AND THE QUESTION OF CATCH-UP

The preceding sections have considered the impact of institutional factors on the overall population of enterprises in China. Worth noting, however, are the Chinese firms at the far end of the spectrum that have emerged as important national and even global players. Such firms have managed to do exactly what their peers have not: develop powerful and clearly identifiable brands, operate beyond local and regional boundaries, and achieve significant market share in their respective product areas. Within the Chinese domestic market, they have been able to match—or even surpass—the positions of powerful multinational players. Some of these firms have gone a step further by reaching out to major global markets, whether through the export of branded products, the establishment of overseas R&D and representative offices, the acquisition of overseas subsidiaries, or even the opening of overseas manufacturing facilities.

In the telecommunications equipment sector (switches and routers), at least two Chinese firms—Huawei and ZTE—have emerged as plausible competitors to global leaders like Alcatel, Cisco, and Nortel (Einhorn, Elgin, and Reinhardt 2002). Both Huawei and ZTE have achieved significant share in the home market, one that until just a few years ago was completely dominated by foreign producers and, in telecom fixed equipment, is now the second largest globally. Huawei produces low-cost, reasonably high-quality versions of devices sold by industry leaders like Cisco. Whether the products are knockoffs—mere copies—and whether they represent violations of Cisco's intellectual property are questions open to debate. What seems clear is that, while Huawei products lag somewhat behind those of industry leaders technologically, they sell for as little as 60 percent of the cost of state-of-the-art devices (Einhorn, Elgin, and Reinhardt 2002).

On at least some dimensions, the approach appears to be working. Huawei in 2003 had total sales of $3.87 billion and recently won major equipment contracts with China Mobile and China Telecom, the nation's largest telecom service providers.[42] Huawei has also begun making inroads into foreign markets, achieving in 2001 a 156 percent increase in year-on-year international sales. The Shenzhen-based firm has branch offices in 32 nations and research institutes in India, Russia, Sweden, and the United States. Overseas sales still represent only 10 percent of the firm's business, but the company aims to expand that figure (Einhorn, Elgin, and Reinhardt 2002).

42. www.huawei.com.

Huawei has recently won equipment contracts in Brazil, Ecuador, Egypt, Germany, Kenya, and Russia. The company has also opened sales offices in San Jose, Northern Virginia, and Dallas.[43] In terms of international research collaborations, in 2002 Huawei agreed with Microsoft to set up a joint lab in Shenzhen and with NEC to set up a lab in Shanghai.

An analogous trajectory has been followed by Haier in home appliances and Legend in personal computers. In 2003 Haier had total sales of RMB 80 billion ($9.75 billion) and by its own estimates enjoyed 30 percent domestic market share for refrigerators, freezers, air conditioners, and washing machines.[44] The firm has a highly recognizable brand within China and increasing brand recognition overseas as well. The company today commands half the U.S. niche market in compact refrigerators and has begun carving out a position producing electric wine cellars (Sprague 2002). While the U.S. market for standard home appliances is still dominated by global giants like General Electric, Maytag, and Whirlpool, Haier—in a unique move for a Chinese firm—recently attempted to gain ground by acquiring manufacturing facilities in South Carolina for full-size refrigerators. In another high-profile move, Haier America purchased the historic Landmark Building in downtown Manhattan as its new corporate headquarters. Haier products are today marketed under the company's own brand in 12 of Europe's top 15 and eight of the United States' top 10 chain store retailers.[45]

Like Haier, Legend—China's premier information technology company—has developed a powerful domestic brand and has matched that brand with significant market share. For personal computers, through the middle of 2002, Legend accounted for 28 percent of the Chinese market and 12 percent of the Asia-Pacific (excluding Japan) market.[46] By the mid-1990s Legend had become one of the world's top five suppliers of computer motherboards and add-on cards. The company's initial growth was fueled by innovation in Chinese word-processing technology, namely the development and commercialization of a word-processing add-on card for IBM-compatible personal computers (for an extensive description, see Lu 2000). The firm, in effect, offered an innovative and much-needed method for tailoring personal computers to the Chinese market. In the late 1980s Legend elected to tailor its add-on card to the AST personal computer and be-

43. www.huawei.com; Einhorn, Elgin, and Reinhardt (2002).

44. www.haier.com.

45. www.haier.com.

46. www.legend-holdings.com. Total sales in 2003 were $2.59 billion.

came the sole distributor of AST machines in China. As AST became the leading brand in China and Legend's revenues soared, the Chinese company's expansion—not to mention rapid climb up the manufacturing learning curve—began in earnest. Through a series of Hong Kong (China) acquisitions, Legend quickly moved into the design and manufacturing of motherboards, shipping a steady volume of 3,500 units per month by the fall of 1989. The next step involved combining the firm's Hong Kong (China)–based manufacturing capabilities for motherboards with its personal computer distribution network in China (developed for AST) to introduce Legend-branded personal computers into the home market. This occurred precisely when the market for personal computers in China began to boom. By the end of the 1990s Legend personal computers were the top brand in China, and the firm was publicly listed on the Hong Kong (China) stock exchange. The firm also was pursuing a strategy of vertical integration, both backward (from motherboard and add-on card to printed circuit board manufacturing) and forward into system integration services and software. Today, Legend is still primarily a manufacturing company, with products ranging from personal computers, hand-held computers, laptops, and mobile phones all the way to basic hardware components. By its own account, the firm aims to extend its reach into high-margin information technology services and applications.

The final illustration involves not a single firm, but rather an entire industry—the Chinese motorcycle sector. What is interesting here is not so much the rise of a major branded player—numerous ones exist in China today—but rather the growing footprint of China's producers as a group both domestically and abroad. China today has hundreds of motorcycle manufacturers and almost as many local brands. Many of these manufacturers during the 1990s—and some still today—engaged in joint ventures with leading Japanese firms, companies like Honda, Suzuki, and Yamaha. The Japanese entered these arrangements in the early 1990s for obvious reasons. They sought access to the Chinese motorcycle market (in terms of sheer numbers of units, the largest in the world), and they found that the Chinese could manufacture a motorcycle at a fraction of the cost of Japanese companies (Zaun and Leggett 2001). The Japanese supplied the brand and design, and the Chinese produced the motorcycle. Over time, Chinese manufacturers—those with Japanese joint ventures and even those without—proved increasingly able to produce motorcycles independently under their own brands. The designs are often still Japanese—essentially Chinese knockoffs—and the Chinese firms often produce under a brand that either closely resembles that of the Japanese competitor (that is, Yameha, Suzaki, or Honea) or somehow directly incorporates a Japanese firm's name (that is, Taizhou Yamaha or Nihon Yamaha, firms that have no relationship with Yamaha in Japan and, indeed, firms that the Japanese protest should be shut down; Zaun and Leggett 2001). The result has been that, while the Chinese motorcycle market experiences rapid growth, Japanese producers are witnessing an erosion of their market share.

A similar phenomenon, albeit with a more effective Japanese response, has been under way in Southeast Asia, particularly Vietnam and Indonesia (countries that rank, respectively, as the third- and fourth-largest motorcycle markets in the world). Through much of the 1990s, Japanese firms—namely Honda, Suzuki, and Yamaha—produced motorcycles in Indonesia and accounted for most of the motorcycles sold in the country.[47] In 1999 Indonesia lifted import restrictions, and low-priced Chinese kits representing nearly 60 brands flooded in. Chinese motorcycles quickly came to occupy 20 percent of the market, although the Japanese responded effectively with low-priced, high-quality alternatives. The same sort of competition between Chinese and Japanese producers obtains in the Vietnamese market. There again, the Japanese have focused on localized joint venture production, while the Chinese have pushed imported kits. In 2001, of the 2.5 million kits imported to Vietnam, Chinese brands accounted for 78 percent, a nearly 30 percent increase from 1999 (Cohen 2002). Again, the Japanese responded with a heavily marketed, low-cost, high-quality product—the Honda Wave Alpha. The Vietnamese government complicated matters by imposing "emergency" tariffs on imported motorcycle kits, ostensibly to address traffic congestion, infrastructure development, and safety issues. Meanwhile, the intense Sino-Japanese competitive dynamic continues in this market, with each side employing political strategies to overcome market access problems.

Several characteristics are common across these examples. First, the firms' success to date, at least in terms of market share, has for the most part been restricted to the Chinese home market. Even within that market, the emerging Chinese branded competitors still tend to occupy the lower-end segments. Their products compete mainly on the basis of low cost, and customers tend to view them as somewhat lower in quality, technological sophistication, and style than foreign alternatives. The same, of course, can be said for overseas markets, although Chinese branded firms have, for the most part, yet to achieve significant market share.

The point is not that these firms necessarily produce low quality products or even direct knockoffs. They have mastered extremely complex production

---

47. Story recounted in Dhume (2001).

processes, and they are producing goods that are in some sense viewed as substitutes for those of leading foreign firms. Rather, the point is that the Chinese are still at the stage in which their prime source of competitive advantage, even in their home market, is still low cost.[48]

Second, Chinese firms have shown little ability to engage in product innovation or overall product definition, areas that remain the domain of leading foreign firms, whether globally or even within the Chinese market. For example, in China's mobile phone handset market, companies like Ericsson, Motorola, and Nokia still set the overall design trends and occupy the high end of the market, while Chinese entrants like Ningbo Bird follow in step with lower-cost alternatives. Even Legend, whose early successes came through innovative customization for the Chinese home market, appears to have ceded higher-value product innovation to foreign partners. It has teamed with Microsoft to develop a tablet personal computer, it incorporates IBM storage technology in its own storage products (while also producing as an original equipment manufacturer for IBM), it employs IBM voice recognition software in its pocket computers, and, of course, like most personal computer manufacturers, it uses Intel Pentium processors. Again, the point is not that Legend or companies like it are backward, but rather that they are primarily commodity manufacturing enterprises. The higher-value, proprietary aspects of their manufactured hardware are usually still controlled by foreign leading firms. Companies like Legend, in effect, are still in the position of being design takers rather than design makers.

The third issue is that, while Chinese firms innovate in manufacturing, they do not do so in a manner that easily translates into sustainable competitive advantage. Both branded and nonbranded manufacturers in China, almost regardless of the level of complexity at which they operate, have exhibited an uncanny ability to push modularization. They take manufacturing processes that in the past were performed by foreign firms in an integral, uncodified manner and aggressively reconstitute those processes in a codified, standardized form. The motorcycle example is particularly relevant here. Japanese assemblers, even at present, for the most part design their motorcycles to accept only customized, product-specific parts. Because Honda motorcycles take only Honda parts or Yamaha motorcycles take only Yamaha parts, the leading assemblers organize production around captive supply chains, whether in Japan or in their overseas joint ventures. Chinese manufacturers, while they may copy the overall design of Japanese products,

48. Moreover, since they are operating in their home market, it becomes hard to rule out explanations based on issues of privileged access, particularly when government agencies and firms are primary customers.

shift production over to standardized, interchangeable parts. The product's architecture becomes "open" rather than proprietary, and the manufacturing process becomes modular rather than integral (integral in the sense that each step must be customized and coordinated with the next). Whether quality is sacrificed is debatable, but what is eminently clear is that the standardized manufacturing process becomes much cheaper.

The opening up of production architecture may be quite advantageous to consumers. It explains why high-quality appliances, personal computers, home electronics, and other goods are available today at a fraction of the cost of just a decade ago, more often than not with a "made in China" label. It is also advantageous to lead firms that control the definition, design, and marketing of products but seek to outsource much of the manufacturing. Similarly, it is threatening to lead firms that resist outsourcing and instead attempt to hold onto internal supply chains even in the face of commodification.[49]

Yet it is important to remember why open product architectures lower prices for customers—modularization and codification lower entry barriers for manufacturers. That is the heart of the problem for Chinese firms pushing this trend. They innovatively codify a process, but, in so doing, they provide entrée for a host of competitors in their immediate environment. In its darker variant, open product architecture leads to phenomena like the countless Japanese knockoffs—bearing dubiously familiar Japanese-sounding brand names—in China's motorcycle market. In its more positive face, though, we see the production of extremely complex, high-quality standardized products: personal computers, routers, motherboards, digital cameras, and cell phones. Chinese producers are able to achieve high enough quality and low enough cost to make these manufacturing activities quite unattractive to overseas lead firms. Hence the overseas firms shed the activities. The problem, however, is that, to the extent they are commodified, the activities are not terribly attractive over the long run to the Chinese manufacturers themselves. For all the reasons discussed before, the margins remain razor thin and the competition vicious.

The problems that Chinese motorcycle makers have faced in Southeast Asia are illustrative. Many Chinese producers can underprice the Japanese, and some 50 Chinese brands flooded into Indonesia in 1999. Yet, given their narrow margins and liquidity constraints, as well as lack of experience, no Chinese firms have been able to respond to Honda's marketing counterattack.

49. Japanese motorcycle and auto manufacturers are an example, as is Philips in electronics.

China's best firms have to some extent compensated by integrating markets across China and investing more seriously in the building of brands. Yet even among the best firms, the focus remains on standardized manufacturing (and products), vertical integration, and diversification. In other words, much like their less famous counterparts, they cope with narrow margins and nonproprietary products by diversifying and by integrating both forward and backward in the supply chain. For example, from its initial base in add-on card and motherboard manufacturing, Legend moved forward into personal computers and routers and backward into printed circuit boards and semiconductors.[50] Similarly, at a different level of complexity, Chinese motorcycle manufacturers, even competitors within single cities, in many cases still try to integrate their supply chains.[51]

Vertical integration may be a reasonable response to various institutional features of the Chinese business environment. Yet it is not an effective response to the technological and organizational change signified by modularization. Chinese firms modularize processes and then try to push them together through vertical integration. There is a difference, however, between a firm's decision to integrate certain processes and the degree to which processes are somehow integral in a technological or product architecture sense. The distinction becomes apparent in the Legend case. Legend may try to own entire supply chains from printed circuit boards right through personal computers. Yet the technological reality remains that each one of these devices today is a standardized, stand-alone component, one that can—and often is—produced in fully substitutable form by other competitors. Legend may choose to integrate, but it is far from clear how vertical integration could lead to any advantage over competitors specializing in any single piece of the modularized chain. Vertical integration *would* make sense, though, if the firm could somehow reverse the standardization process, converting an open architecture product back into an integral architecture. In other words, if a firm like Legend could create a new product that depended on high degrees of coordination between customized components, vertical integration might prove an effective strategy. Alternatively, a firm like Legend might command value if it could control the rules of connectivity across various modularized steps. At present, however, the fact that Legend products still have "Intel inside," "Microsoft inside," and "IBM inside" suggests that the overseas firms still control the rules of con-

50. Although the semiconductor side of the business remains for the most part an aspiration rather than a reality.

51. Author's interviews, Chongqing, 2000.

nectivity, while Legend has been left to focus on low-value, commodified hardware.

The preceding discussion still leaves open the question of catch-up—the question of whether China's lead firms, for all their problems today, may just be in the first stages of catching and ultimately surpassing their foreign rivals. In other words, might we be witnessing the opening stages of a situation analogous to the Japanese auto industry's rise vis-à-vis American auto companies in the 1970s? As firms like Haier or Huawei enter North American and European markets, are we simply seeing a replay of the Japanese and Korean story? Is it the same story of new competitors figuring out how to produce products inexpensively, introducing those products overseas first into lowest-end market segments, gradually and quietly building market share, and eventually becoming dominant in high-value products?

In answering these questions, it is worth considering the conditions under which Japanese and Korean industrial firms rose decades ago. In that era, industries could still in a meaningful sense be understood as separate, self-contained entities and often self-contained in national terms. We could refer to the American steel or the French auto industry, and we could contemplate whether rising industrializers like Korea would develop strength in a particular industrial sector. Moreover, in these relatively autonomous industries, product innovation occurred in incremental terms, and manufacturing processes tended to be integral. The various steps in the process, while perhaps understood in broad terms in these stable industries, were uncodified (and given the state of information technology at the time, probably uncodifiable). As such, they could not be pulled organizationally apart from one another, and they tended to be particular to each firm or each firm's captive supply chain. Challengers, to the extent they could amass the resources needed to enter these capital-intensive industries, could compete on the basis of process innovation, the ability to produce the same products as incumbents, but at significantly lower cost (see Amsden 1989). Because manufacturing processes remained uncodified and integral within the firm, shop-floor innovations were truly proprietary. They were, in effect, a form of art or craftsmanship that neither incumbents nor other entrants could easily copy. There was no open recipe to follow.

What that description suggests is that rising Chinese firms today are operating in a dramatically transformed era, one in which the methods of the past are not directly applicable. First, activities, not whole industries, are what move today from developed to rising nations. What has moved to China en masse, whether at the bequest of leading global companies or through pressures from Chinese firms themselves, are the manufacturing-intensive segments of particular value chains. More precisely, it is the codified, commodified, non-integral manufacturing activities that move. Competing in these areas, while hardly trivial, often does involve mastering open processes rather than developing proprietary ones. That is the reason, in part, that we see so many new entrants from China in manufacturing rather than the handful of firms that entered from Japan and Korea in previous decades.

Second, when Japanese and Korean competitors emerged, they were rising up against relatively stable incumbents, incumbents whose focus was still on manufacturing. As such, the incumbents were essentially stationary targets whose products could be substituted by lower-cost alternatives. Today, the situation is quite different. In large part because of modularization, the incumbents—global lead firms—are hardly stationary and in many cases have completely transformed themselves. Firms like Ningbo Bird, Haier, Huawei, and Legend may be rising on the basis of their low-cost manufacturing expertise. At the same time, most lead firms—whether Cisco, Dell, Electrolux, IBM, Motorola, or many others—are moving away from manufacturing entirely. Instead, they are increasingly focusing on what may be broadly termed the "service" side of production: overall product definition, design, marketing, and supply chain management. Legend may manufacture and sell personal computers, but its nearest American counterpart, IBM, has effectively become a business services and software company. Similarly, Dell, the leading American seller of personal computers, engages in very little manufacturing and assembly but instead has built a business on distribution.

Rising firms may be capturing manufacturing activities, but the former manufacturers have increasingly specialized in high-value nonmanufacturing activities. In so doing, they either retain control of the supply chain's rules of connectivity—perhaps by controlling the key operating software or state-of-the-art processor designs in electronics—or retain control of activities that remain truly integral and proprietary. Examples of the latter often involve efforts by lead firms to embed services in the higher-end products they either manufacture or outsource. The point is not so much that Chinese firms are performing poorly in the developmental process—quite to the contrary. Rather, the point is that Chinese entrants and developed-country lead firms are jointly responding to major transformations in the organization of production. Their joint response is better understood as complementary rather than as substitutable or somehow "competitive" in head-to-head terms.

That leads to a third point about the way the terms "industry" and "national industry" are understood today. In previous decades, it made sense—with a certain degree of simplification—to conceive of industries as distinct silos. Particular nations, then, could be mapped over one or more of those silos. In the current era of modularization, certain activities within industries have been separated from one another and moved across national boundaries. Hence, in any given silo, we may see companies or countries occupying certain activities, but not others. Yet modularization is really about much more than that. It is not just that activities within discrete industries have been split apart, but rather that these independent, highly specialized activities now cut across multiple industries. What were once distinct industry supply chains now overlap, intersect, and interact in myriad forms. As such, it becomes increasingly difficult to say exactly which "industry" a given firm or nation specializes in. Is a semiconductor foundry in the electronics industry or, since its chips go into mobile phones, in the telecommunications sector? Is the "fab-less" semiconductor design house mapping out chips for automobiles—along with semiconductors for a host of other applications—in the auto industry? Perhaps the firm is also integral to aerospace, telecommunications, or home appliances. IBM today is basically a business solutions and services company. What industry, in the traditional sense, does it belong to?

The specialization associated with modularization has led to the blurring of boundaries between industries and the growing interaction across them. Today, it may make more sense to think of matrices and webs of specialized activities than discrete, stand-alone industrial sectors. Among other things, such organizational change leads to the phenomenon of modularized innovation and ripple effects across formerly unrelated industries. The fab-less chip design house, in its efforts to design a telecommunications application, may come up with a new capability applicable to aerospace. Then again, perhaps the new chip design will have its greatest impact on "smart" home appliances. For the chip innovator, the ultimate downstream application may be irrelevant, so long as the design gets purchased in great quantity. Yet the downstream application certainly is not irrelevant to those who are competing in the downstream activities, particularly when the new application may lead to downstream substitutes. Similarly, a firm like Microsoft may keep churning out operating software for personal computers, but it also focuses on enabling the sorts of products—hand-held computers, digital writing tablets, Web-capable mobile phones—that may undercut or otherwise replace personal computers. One can begin to see how, in the modularized world, specialized innovations lead to unpredictable outcomes.

One can also begin to see the challenge for contemporary industrial policy. It is not just that the pace of change is faster now than in the heyday of Japanese or Korean industrialization. More important, the organizational mechanism of change—particularly the extent to which change is spread across ostensibly unrelated firms and "industries"—is completely new. For a nation to be strong in autos, aerospace, or telecommunications, what fundamentally does it need? Software companies? Semiconductor design houses? Handset manufacturers? Steel firms? Marketing firms?

That it is hard to say underscores the risks entailed in forcing the vertical integration of industries. From the perspective of product architecture, it may be impossible to determine the exact boundaries of a given industry. Yet Chinese industrial policy, by selecting pillar industries, does precisely this in an artificial sense. It operates under the idea that a country can, from upstream to down, "build" a steel or auto or aerospace sector. Similarly, for various institutional reasons, individual Chinese companies may themselves elect to integrate their activities vertically. Whether through institutional default or conscious policy, they end up forcing the integration—whether under a single company roof or within a single national geography—of activities that are not in any technological sense "integral." In effect, they push together within a given organizational boundary activities that could just as easily stand apart from one another. In so doing, as such activities are held captive within single "industry" supply chains, policymakers and corporate strategists limit the extent to which modular innovation and cross-fertilization can occur. It is not surprising, therefore, that China perceives itself, probably correctly, as lagging behind India, let alone developed countries, in industries like software. Similarly, it is not surprising that China lags in high-end semiconductor design capabilities. Legend may put motherboard and printed circuit board manufacturing under one organizational roof, but that does not change the fact that these are essentially modular activities. Nor does it change the fact that Legend's competitors, firms like IBM, are shedding precisely those standardized activities so as to focus on the higher-value activities that, for technological reasons, truly are integral in the supply chain.

The point is not that single lead companies can necessarily control a supply chain. After all, who has more economic power in the personal computer supply chain: Dell, Intel, Legend, or Microsoft? The question not only is difficult to answer but also may be irrelevant. Given the degree of specialization and narrow innovation permitted by modularization, perhaps there is value to be achieved in any activity. Achieving that value, however, requires the firm not simply to integrate activities, but rather to focus on integral activities—the functions, products, or services that are absolutely critical to upstream and downstream counterparts and that competitors have difficulty duplicating. Of course, the companies that achieve this today—whether Intel, Microsoft, or anybody else—may not be around tomorrow. Nothing guarantees that the firm setting particular standards and rules of connectivity today will last into tomorrow. Yet, from the national perspective, the role of industrial policy should be to ensure that the home environment at least can spawn a series of such firms. In China today, neither the goal of creating national pillar industries nor the institutional problems that independently encourage vertical integration fulfill that purpose. Instead, they isolate even the best Chinese enterprises from state-of-the-art technology, reduce the likelihood that Chinese firms will set rules of connectivity globally, and facilitate specialization among foreign lead firms. What results is not so much catch-up as a greater division of labor, one that arguably widens the gap between overseas lead firms and Chinese follow-on producers.

CONCLUSIONS

This chapter has argued that the innovative capacity of Chinese firms and their ability to upgrade within global supply chains are impeded by legacies of the Chinese style of reform, current bottlenecks in the institutional reform process, and inconsistencies in governmental industrial policy. To be sure, progress has been made on a number of fronts. The Chinese government has moved aggressively in administrative terms to tackle issues of market fragmentation, local protectionism, and regulatory inconsistency. Unfortunately, a number of these issues extend beyond the administrative and into the political. Their resolution, at least in part, depends on the willingness of the state and ruling party apparatus to subject itself fully—at any jurisdictional level—to the rules and regulations of the system.

At the same time, the government must come to terms not just with the benefits of market economics but also the limits. The goal of building nationally autonomous industries may be justifiable on societal or national security grounds. Yet the goal is not consistent with the sorts of corporate organizations and production architectures that in today's world achieve commercial sustainability. Efforts to employ market liberalization to achieve "national industries" will, therefore, likely produce two equally undesirable, albeit related, results: the industries themselves will fail (and the resources that went into building those firms will have been wasted) or markets will get administratively distorted to ensure the industries' "success." Unfortunately, both outcomes are likely to inhibit the further integration and upgrading of Chinese firms in global production networks.

Chinese policymakers, with their vision of national champion producers, may operate on entirely inaccurate notions of what a globally competitive commercial organization is or should be. Their confusion, though, is understandable given the uncertainty surrounding the nature of competitiveness more broadly. Even given the fact of globalized production networks, the assumptions we make about the nature of those networks can lead to highly divergent implications for policy. To the extent one understands technology as relatively stable, product cycles as fairly long, and production networks as consisting of fully modularized, discrete processes, one could imagine that innovation would fall primarily within the category of sustaining technologies and that it would occur within the confines of the incumbent firm. The goal of industrial policy then might be to create the kinds of large, self-contained organizations that could dominate a particular piece of the supply chain. These organizations would diverge significantly from the ideas of vertical integration popular in China today, but so too would they diverge from the vision of small start-ups conjured by creative destruction.

Yet, if one understands technology as highly unstable, product cycles as extremely short, and production networks as defined by extensive coordination between upstream and downstream producers in integral processes (in other words, if the need for coordination undercuts full modularization), then innovation would fall primarily in the category of disruptive technologies and might be understood as occurring primarily through interaction between firms. Under such circumstances, the policy goal would be to create not particular kinds of companies, but rather particular kinds of communities (à la Silicon Valley). Chinese industrial policy may be creating neither of these, but one can sympathize with the dilemmas policymakers face.

Second, there is uncertainty about what the relevant comparison (or comparator) for China really is. Chinese firms in most sectors are locked in intense competition, competition for which the dominant strategy still seems to involve deep discounting rather than specialization and innovation. By virtually any measure, Chinese firms are not as innovative as global leaders—namely, multinationals producing branded products—in any given supply chain. But are the global leaders really the relevant comparison? China's per capita income in 2001 was $890, roughly one-fortieth of that of Japan or the United States.[52] It is perhaps not surprising that Chinese firms are failing to unseat incumbents from these far richer countries. Yet, in terms of positioning in global supply chains, can we say that Chinese firms are performing poorly relative to Mexican firms, Malaysian firms, or Thai firms (firms hailing from countries with per capita incomes, respectively, six, four, and two times that of China)? The point is that Chinese firms may not be innovating relative to one another

52. Income data (Atlas methodology) from World Bank (2002).

and relative to globally branded leaders. Yet they are out-competing rivals from far wealthier developing countries, and they are doing so by rapidly developing competence in increasingly complex manufacturing processes. Simply to remain in the game—simply to compete even on the basis of cost—firms in the contemporary era must upgrade rapidly, and that is precisely what Chinese enterprises have proven able to do. They may not be innovating in the traditional sense, but they are keeping pace with a dynamically evolving system of global production, an achievement that appears to elude many of their developing-country counterparts. Although that may not fit the Chinese goal of catching up, keeping pace represents an achievement worth celebrating—and understanding analytically—in its own right.

REFERENCES

The word *processed* describes informally reproduced works that may not be commonly available through libraries.

Alford, William. 2000. "The More Law, the More . . . ? Measuring Legal Reform in the People's Republic of China." Working Paper 59. Stanford University, Center for Research on Economic Development and Policy Reform, Palo Alto, Calif., August. Processed.

Amsden, Alice H. 1989. *Asia's Next Giant: South Korea and Late Industrialization.* Oxford: Oxford University Press.

———. 2001. *The Rise of the Rest: Challenges to the West from Late-Industrializing Economies.* Oxford: Oxford University Press.

Aoki, Masahiko, and Hirokazu Takizawa. 2002. "Modularity: Its Relevance to Industrial Architecture." Paper prepared for the conference on innovation within firms, Saint-Gobain Centre for Economic Research, Paris, November. Processed.

Baldwin, Carliss Y., and Kim B. Clark. 2000. *Design Rules: The Power of Modularity.* Cambridge, Mass.: Massachusetts Institute of Technology.

Burns, Simon. 2000. "I-Sight: Everyone's a Winner." *Far Eastern Economic Review,* July 20.

Chandler, Alfred D. 1990. *Scale and Scope: The Dynamics of Industrial Capitalism.* Cambridge, Mass.: Harvard University Press.

Chandler, Alfred D., and Takashi Hikino. 1997. "The Large Industrial Enterprise and the Dynamics of Modern Economic Growth." In Alfred D. Chandler, Franco Amatori, and Takashi Hikino, eds., *Big Business and the Wealth of Nations.* Cambridge, U.K.: Cambridge University Press.

Chen, Qingtai. 2002. "Promoting Regional Economic Cooperation in Asia amidst Economic Globalization." Paper presented to the conference on Asian economic integration, Research Institute of Economy, Trade, and Industry, Tokyo, April 22–23. Processed.

"China's Haier First Half Profit Down 45 Percent on Low-Priced Exports." 2002. *Dow Jones International News,* July 29.

Christensen, Clayton M. 1997. *The Innovator's Dilemma: When New Technologies Cause Great Firms to Fail.* Cambridge, Mass.: Harvard University Press.

Cohen, Margot. 2002. "Biker Wars: Chinese and Japanese Manufacturers Battle for Vietnam's Fast-Growing Motorcycle Market." *Far Eastern Economic Review*, November 7.

de Soto, Hernando. 2000. *The Mystery of Capital: Why Capitalism Succeeds in the West and Fails Everywhere Else.* New York: Basic Books.

Dhume, Sadanand. 2001. "Road Warriors: In Indonesia, Chinese Motorcycle Manufacturers Are Challenging the Japanese for a Slice of the Market." *Far Eastern Economic Review*, December 27.

Einhorn, Bruce, Ben Elgin, and Andy Reinhardt. 2002. "The Well-Heeled Upstart on Cisco's Tail." *Business Week*, October 28.

Fang, Xinghai. 2001. "Reconstruction of the Micro-Foundation of China's Financial Sector." Paper presented at the conference on financial sector reform in China, Harvard University, September. Processed.

Fine, Charles H. 1998. *Clockspeed: Winning Industry Control in the Age of Temporary Advantage.* Reading, Mass.: Perseus.

Fujimoto, Takahiro. 1999. *The Evolution of a Manufacturing System at Toyota.* Oxford: Oxford University Press.

———. 2002. "Organizational Diversity and Corporate Performance." Paper prepared for the conference on innovation within firms, Saint-Gobain Centre for Economic Research, Paris, November. Processed.

Gereffi, Gary, and Miguel Korzeniewicz, eds. 1994. *Commodity Chains and Global Capitalism.* Westport, Conn.: Praeger.

Gilboy, George J. 2003. *Nodes without Roads: Pockets of Success, Networks of Failure in Chinese Industrial Technology Development.* Ph.D. diss., Massachusetts Institute of Technology, Department of Political Science.

Gregory, Neil, Stoyan Tenev, and Dileep Wagle. 2000. *China's Emerging Private Enterprises: Prospects for the New Century.* Washington, D.C.: International Finance Corporation.

Guthrie, Doug. 1999. *Dragon in a Three-Piece Suit: The Emergence of Capitalism in China.* Princeton, N.J.: Princeton University Press.

Hu, Angang. 1999. "The Greatest Challenge of the New Century: China Enters the Stage of High Unemployment [Kuaru xinshiji de zui da tiaozhan: wo guo jinru gaoshiye jieduan]." Chinese Academy of Sciences and Tsinghua University, Joint Center for Chinese Studies. Processed.

Hu, Angang, and Wang Shaoguang. 1999. *The Political Economy of Uneven Development: The Case of China.* Armonk, N.Y.: M. E. Sharpe.

Huang, Yasheng. 2003. *Selling China: Foreign Direct Investment in the Reform Era.* New York: Cambridge University Press.

IBM (International Business Machines). 2001. *Annual Report 2001.* Available at www.ibm.com.

Institute of Industrial Economics. 1998. *Zhongguo gongye fazhan baogao.* Beijing: Jingji Guanli Chubanshe.

Kornai, Janos. 2000. "What the Change of System from Socialism to Capitalism Does and Does Not Mean." *Journal of Economic Perspectives* 14(1, Winter):27–42.

Kynge, James. 2002. "Creaking Economy Needs Stronger Foundations." *Financial Times*, October 29.

Lardy, Nicholas R. 1998. *China's Unfinished Economic Revolution.* Washington, D.C.: Brookings Institution Press.

———. 2002. *Integrating China into the Global Economy.* Washington, D.C.: Brookings Institution Press.

Lee, Jane Lanhee. 2002. "China's Price Wars Build Pressure on State Companies to Reform." *Dow Jones International News*, April 17.

Legend Group. 2001. *Annual Report 2001.* Available at www.legend-holdings.com.

Leggett, Karby. 2001. "Konka's Loss Shows Effects of Price Wars." *Asian Wall Street Journal*, August 30.

Leggett, Karby, and Peter Wonacott. 2002. "Burying the Competition." *Far Eastern Economic Review*, October 17.

Lu, Qiwen. 2000. *China's Leap into the Information Age: Innovation and Organization in the Computer Industry.* Oxford: Oxford University Press.

Lubman, Stanley B. 1999. *Bird in a Cage: Legal Reform in China after Mao.* Palo Alto, Calif.: Stanford University Press.

Ma, Guonan, and Ben S. C. Fung. 2002. "China's Asset Management Corporations." BIS Working Paper 115. Bank for International Settlements, Basel, August. Processed.

Mackenzie, Davin. 2001. "A Healthy Financial Sector Requires Enterprises That Deserve Financing." Paper presented at the conference on financial sector reform in China, Harvard University, September. Processed.

Naughton, Barry. 1999. "How Much Can Regional Integration Do to Unify China's Markets?" Paper prepared for the conference on policy reform in China, Stanford University, Palo Alto, Calif., November. Processed.

Nolan, Peter. 2001. *China and the Global Economy: National Champions, Industrial Policy and the Big Business Revolution.* New York: Palgrave.

Oi, Jean C. 1999. *Rural China Takes Off.* Berkeley: University of California Press.

Peerenboom, Randall. 2001. "Seek Truth from Facts: An Empirical Study of Enforcement of Arbitral Awards in the PRC." Available at www.law.ucla.edu/erg/pubs/arbitral_awards.pdf. Processed.

———. 2002. "Let One Hundred Flowers Bloom, One Hundred Schools Contend: Debating Rule of Law in China." *Michigan Journal of International Law* 23(3, Spring): 471–544.

Perkins, Dwight H. 2001. "Industrial and Financial Policy in China and Vietnam." In Joseph E. Stiglitz and Shahid Yusuf, eds., *Rethinking the East Asian Miracle*, pp. 247–94. Washington, D.C.: World Bank.

Pomfret, John. 2001. "Legacy of Socialism Keeps China's State Firms in Red." *Washington Post*, June 20.

Schumpeter, Joseph A. 1968 [1911]. *The Theory of Economic Development: An Inquiry into Profits, Capital, Credit, Interest, and Business Cycles.* Cambridge, Mass.: Harvard University Press.

———. 1976 [1942]. *Capitalism, Socialism, and Democracy.* New York: Harper and Row.

Segal, Adam, and Eric Thun. 2001. "Thinking Globally and Acting Locally: Local Governments, Industrial Sectors, and Development in China." *Politics and Society* 29 (4, December):557–88.

Sprague, Jonathan. 2002. "China's Manufacturing Beachhead." *Fortune*, October 28.

Steinfeld, Edward S. n.d. "Market Visions: The Interplay of Ideas and Institutions in Chinese Financial Restructuring." Unpublished mss. Massachusetts Institute of Technology, Department of Political Science. Processed.

———. 1998. *Forging Reform in China: The Fate of State-Owned Industry*. New York: Cambridge University Press.

———. 2000. "Free Lunch of Last Supper: China's Debt-Equity Swaps in Context." *China Business Review* 27(4, July–August):22–27.

———. 2002. "Moving beyond Transition in China: Financial Reform and the Political Economy of Declining Growth." *Comparative Politics* 34(4, July):379–98.

Stevens, Clint. 2002. "Domestic Chinese Appliance OEMs on the Ropes." *Appliance*, June 1.

Tenev, Stoyan, and Chunlin Zhang. 2002. *Corporate Governance and Enterprise Reform in China*. Washington, D.C.: World Bank and International Finance Corporation.

Tsai, Kellee. 2002. *Back-Alley Banking: Private Entrepreneurs in China*. Ithaca, N.Y.: Cornell University Press.

Wank, David L. 1999. *Commodifying Communism*. New York: Cambridge University Press.

Wedeman, Andrew. 2003. *From Mao to Market: Local Protectionism, Rent-Seeking, and the Marketization of China, 1984–1992*. New York: Cambridge University Press.

World Bank. 2002. World Development Indicators. Washington, D.C., August.

Young, Alwyn. 2000. "The Razor's Edge: Distortions and Incremental Reform in the People's Republic of China." *Quarterly Journal of Economics* 115(4, November):1091–135.

Zaun, Todd, and Karby Leggett. 2001. "Road Warriors: Motorcycle Makers from Japan Discover Piracy Made in China." *Wall Street Journal*, July 25.

Zhou, Xiaochuan, ed. 1999. *Chongjian yu zaisheng*. Beijing: China Financial Press.

Zweig, David. 1997. "To Return or Not to Return? Politics versus Economics in China's Brain Drain to the U.S." *Studies in Comparative International Development* 33(1, Spring): 92–125.

CHAPTER 7

LOGISTICS IN EAST ASIA

Trevor D. Heaver

The provision of logistics services adequate to the needs of industry in East Asia faces many challenges. They arise, in part, from the rapid expansion and growing sophistication of the manufacturing industry, which is placing increasing demands on the logistics industry. Simon Hsu, chairman and chief executive of E-Commerce Logistics of Hong Kong (China), argues that the efficiency of manufacturing has outgrown the efficiency of the logistics industry (Wong 2001). The gap results also, in part, from the changing nature of product markets, which is requiring manufacturers to seek more sophisticated logistics services. As a consequence, improved logistics capability is becoming an important competitive tool in the rivalry among firms and manufacturing regions.

Producers require logistics services of comparable standards to those available to their competitors elsewhere. In the context of this chapter, these standards may be taken as the generally expected attributes of efficient logistics services in the foreign trade of North America and Western Europe. To trade with those regions, the East Asian economies must compete with economies in other parts of the world, which are often closer to those markets. Their success in that competition is influenced by the efficiency of the regional economies, including the logistics services within and among East Asian economies, and by the efficiency of the logistics services to distant regions.

Most countries face challenges in ensuring that the right level of facilities and services, appropriate policies, and sufficient labor force skills are in place. For example, in Europe the quality of rail freight and intermodal services is inadequate. In the United States the ability of ports to handle large ships is a constraining factor that may become more serious. Growing congestion on the roads serving ports is a concern. Also the level and extent to which information technology is speeding processes such as customs clearance may lag behind that found in some countries, for example, the Netherlands. Deficiencies in the attributes needed for efficient logistics to meet national needs are greater in most East Asian economies than in Western Europe and North America.

There are many reasons for the lag in adequate conditions in East Asia, and they vary in importance among economies of the region. The problems are least in the open economies of Singapore and Hong Kong (China). They are greatest where the economies have been open to modern trade for the least time or have experienced the least effects of economic development. As developing economies, the economies started off with serious deficiencies in infrastructure, policies, and the competence of organizations and the work force. For example, ports in Vietnam are only now opening new facilities (Gooley 2001). Furthermore, most economies have experienced very rapid growth in economic activity and foreign trade as they have become more open. It is not surprising, then, that there is a serious lag between logistics capabilities and the desirable level of service.

Improvements in logistics capabilities are a prominent objective in private and public policies. In China many public authorities are pushing logistics as important to economic development. For example, Guangzhou Province, which is facing more competition from lower-cost areas of manufacturing elsewhere in China, is seeking to enhance its logistics capabilities ("Guangdong Seeks Foreign Logistics" 2001). In the business community of Hong Kong (China), there is also recognition that enhanced logistics capabilities are important as the community faces new competition from locations within the Pearl River Delta and Shanghai, which have become continental gateways. In 2001 the Hong Kong Port and Maritime Board responded to the challenge by initiating a study to strengthen the role of Hong Kong (China) as the preferred international and regional transportation and logistics hub. As part of the process, a Committee on Logistics Services Development was established to interact with consultants conducting the study. The result was an overarching master plan to create a more productive logistics environment (Hong Kong Port and Maritime Board 2001).

Similar concerns and diverse initiatives are found throughout East Asia. Indeed, they are present in countries around the world, as exemplified by the TRILOG (Trilateral Logistics) Project of the Organisation for Economic Co-operation and Development (OECD), covering logistics in Asia-Pacific, Europe, and North America (Nemoto and Kawashima 2000). Governments in Australia have been promoting logistics studies and meetings involving Asian economies. In part, this has been done to improve logistics performance in Australia's domestic and foreign trade, and in part it has been to create business opportunities for Australian providers of logistics services (Bangkok Freight Logistics Workshop 2001). A common conclusion of the studies and the meetings is the need for multifaceted and cooperative initiatives. This chapter outlines the elements of such initiatives.

PURPOSE, OUTLINE, AND LIMITATIONS

This chapter has two purposes that tie it closely to earlier chapters dealing with global value chains. The first is to elaborate on the changing influence of logistics requirements on the location of industrial development in East Asia. The second is to use three national case studies to ascertain key strategies that could assist regional economies to improve their level of logistics service and enhance the competitiveness of firms.

Since "logistics in East Asia" is a very wide subject, two approaches are used to bound and structure the chapter. First, specific conditions are examined in only three economies. Second, logistics performance is viewed as dependent on a range of community attributes. This is consistent with the multifaceted approach recommended in other studies and with the approach of successful gateway communities such as the Netherlands and Singapore. Although the chapter is not structured to deal individually with them, those attributes provide a logic for the investigation of conditions in the region.

The community attributes may be grouped in various ways. Here, five key pillars of efficient logistics services are recognized. They are similar to the target activities developed in the logistics master plan of Hong Kong (China):

1. Provide efficient infrastructure and services in transportation
2. Provide efficient infrastructure and services in communications
3. Develop appropriate public policies and institutional practices
4. Create effective private sector logistical organizations
5. Create a work force knowledgeable of logistics skills and concepts.

The chapter is divided in four further parts. The third section defines logistics and supply chain management and examines the role of logistics in supply chain management by using examples of practices in selected industries, focusing on the implications for firms in East Asia. It also describes the structural changes that are taking place globally in the logistics service industry, as these developments, too, have ramifications for the logistics industry in East Asia.

The fourth section is devoted to the need within the region for investments in physical infrastructure to accommodate the growth in trade. Emphasis is given to investments in the hub facilities of ports and airports. The need for and implications of investments in information technology are noted, but not dealt with at length. The examination of infrastructure investments is not intended to be comprehensive. Little is said about the need for regional investments in rail and road infrastructure. The treatment of infrastructure recognizes the huge past and current needs for additional infrastructure in East Asia. However, as such, they are frequently the subjects of large and specialized studies.

The fifth section examines logistics developments in three countries within the context of their economic and policy regimes. The three countries are China, Singapore, and Thailand. Consideration of these three countries covers several dimensions of diversity in the logistical conditions and challenges in East Asia arising from the varying degrees of openness and country size. In each dimension Thailand occupies a middle position between Singapore and China. It is not surprising that the level of accomplishment in logistics services is quite different among the countries. Conclusions are presented in the final section.

Research for this chapter relies on three sources. The first source consists of published materials on logistics and supply chain management in general as well as in East Asia. The second source is the Internet, especially for information on current developments and for government regulations. The third source consists of shippers and logistics service companies. Interviews were held with shippers and logistics service companies in person in Vancouver, Canada, and by phone and e-mail with those elsewhere in North America. Phone interviews were conducted and e-mail exchanged with logistics service companies in Asia.

The research does not provide quantified measures of logistics performance for East Asian economies. Not only are there conceptual problems about what to measure in logistics performance, but there also is a fundamental lack of accessible data even on transportation costs and service levels. Keeping rates confidential may improve market performance, but it does not aid rate monitoring. Further, the implications of logistics costs for the level of activity in a particular location are often more dependent on their level relative to competing locations than on their absolute level. Therefore, assessing the quantitative effects of logistics costs on industry in the region would have required a larger data set than was available.

Unfortunately, reliance on information from the users and providers of logistics services has some limitations. First, the shippers contacted were dominantly North American importers of goods from Asia. This creates an unfortunate bias in the requirements of foreign trade (especially Asian exports) to the detriment of the perspectives of domestic shippers. However, as export trade is likely the "first mover" for new logistics requirements, this minimizes the negative effects of the bias. Second, interviews were held during 2001 with 41 individuals in 25 companies. This is not a large number, but the consistency of the views and their match with the published literature provide confidence in the representativeness of the opinions. Transportation service providers in East Asia were not contacted directly in this research. Consequently, specific concerns related to port terminals, airports, and inland water, truck, and rail transport may not have been identified in this work. However, information was obtained from persons directly familiar with those services.

ROLE OF LOGISTICS IN SUPPLY CHAIN MANAGEMENT

The primary means by which firms compete successfully has changed from one period in history to another. Henry Ford's success with the mass production of cars is used as a marker for the importance of a production focus on economies of scale. It was not until the 1930s—and then after World War II—that the benefits of placing greater emphasis on modern marketing were recognized; the contributions of logistics to corporate performance have been widely recognized only recently.

In this section of the chapter, four topics related to the role of logistics in supply chain management are examined.

Evolution of Logistics Management

Prior to and immediately after World War II, companies seeking to control their transport costs focused on reducing the costs of shipping. Alternative modes were available only to a limited extent. The development of trucking and air freight made it possible to forgo more expensive transportation in return for inventory savings. Marketing-dominated firms attempted to manage physical distribution (for example, consumer goods), while production-dominated firms attempted to manage materials. In reality, most firms had to integrate inbound and outbound flows, so the term common in the military—logistics—was adopted. Integration was sought in the context of decisions affecting logistics first within the plant, then within the firm, and then among the firms involved in the design, creation, and delivery of goods and services. The span over which integration was recognized as desirable extended to all firms that were a part of the process.

Porter's concept of the value chain (1985) gave currency to the emerging term "supply chain" for all of the functions and firms involved in the creation and delivery of goods and services to consumers. Supply chain management is much broader than logistics. It reflects a management philosophy that spans traditional silos. It is about managing the business in the broad view, highlighting the importance of managing a range of external relationships. These include where to source, whether to make or buy, and what sort of relationship to have with buyers and customers. Many of the decisions affect logistics, but only matters immediately related to product flows are a part of logistics.

This approach is consistent with the current definition of logistics by the U.S. Council of Logistics Management. It recognizes that logistics management is only one way in which businesses organize and manage their affairs involving external as well as internal relationships. Logistics is defined as that part of the supply chain process that plans, implements, and controls the efficient, effective flow and storage of goods, services, and related information from the point of origin to the point of consumption in order to meet customers' requirements. Unfortunately, some confusion does arise from the varied contexts in which the term supply chain management is used. It may be used as synonymous with logistics. For example, in many firms the supply chain manager was previously the logistics manager; if his responsibilities have changed at all, it is only because they involve relationships with more outsourced services than was previously the case. In this chapter, logistics is used consistent with the definition above, and supply chain management is used to refer to the management of all business functions in the supply chain (Mentzer and others 2001).

Businesses today are viewed as supply chains in recognition of the complex systems of which individual firms are a part. In effect, firms compete as parts of supply chains. Competition is among chains. The management practices of firms in some supply chains are well integrated. This does not imply that all relationships are long term and have "partnership" characteristics. It does imply that all parties recognize the importance of their contributions to the supply chain and that the success of the supply chain and of their own business has much in common. Firms have a supply chain perspective in their dealings. In other supply chains, integration may be lacking and performance poor. If some participants in a region are inefficient, all members of the chain are disadvantaged. Hence logistics services in East Asia influence the competitiveness of firms in the region.

Demands on Logistics from Supply Chain Management

Supply chains are designed to meet consumers' needs while producing a profit for suppliers. Logistics services are a part of meeting those needs. Pressures to redesign supply chains work concurrently in two different directions. The first pressure is the desire to reduce costs, often driven by the pressure of competition. The second pressure is the need to meet changing consumer preferences, independent of or because of suppliers' initiatives. The main direction of this change has been to offer enhanced logistics services at a competitive price. In each case, the demands imposed on logistics services have changed substantially over the last two decades.

Global improvements in logistics performance have contributed to and been required by heightened competition in product markets. The reduction of tariff and other trade barriers, improvements in the efficiency of transport services, and the higher value and lower weight of many products have all contributed to the ability of products from distant locations to compete locally. Multinational corporations that used to be organized with regional marketing and production divisions have switched to product-based structures that source and market globally. As a result, competition has intensified in most markets. Spatial competition (the competition in common markets for products from far away) is more important now than ever before.

The pressures for change are multifaceted and vary in importance continually. They have worked dynamically over time. This section focuses sequentially on the pressures to reduce costs and to improve service, although they work concurrently. Examples are used to illustrate the working of particular forces and their dynamics and changing importance over time. The examples chosen have special relevance to East Asia.

Cost reduction. Cost reduction—always important to increase profits—becomes a heightened need in the face of increased competition. Four cost reduction strategies provide examples of the central role of logistics. They are sourcing in low-cost locations, supply chain visibility, just-in-time delivery, and postponement. Although these strategies are not strictly separate, they are useful to illustrate the role of logistics in reducing costs.

Industries have always made tradeoffs among locations based on the benefits of locating close to low-cost resource inputs compared with the benefits of locating close to markets. Typically, industries in which labor costs are high and that require low skill levels gravitate to low-wage and low-cost locations. The ability of firms to do this depends on the existence of efficient logistics services that allow them to get products to where they are needed.

Ship breaking (demolition) in South Asia is an extreme example of the pull of low-cost labor. More typical for East Asia is the shift of textile, footwear, automobile, electronic products, and toy manufacturing to the region. The relocation of product manufacture is complex, reflecting the variety of products in each of the categories mentioned. It is also dynamic because the skill and wage levels of workers in locations change and because the logistics needs of products change as the products move through their development cycle. Only detailed studies of industries can reveal the complexities. McKendrick, Doner, and Haggard (2000) examine the American hard disk drive industry.

Within the hard disk drive industry, production competencies and efficient logistics services in other parts of the world have enabled much production to shift away from North America. Few production-related activities have remained tied to or close to (within the country of) the head office. These include research and development, the production of some components with exceptional production requirements, and the development of production processes. Through the development of production processes, firms can conveniently modify the product in the light of production development experience, evolve processes to ensure quality control, and train production managers from elsewhere. Adequate skills elsewhere enable mass production to take place offshore. In the hard disk drive industry, this occurred first as manufacturing migrated from the United States to Singapore in response to good logistics services. Developments in Singapore, made possible by public policies favoring education, improvements in logistics, and industry agglomeration, led to growth in electronics manufacturing. However, success and growth resulted in rising costs and the migration of processes suitable for less skilled and less costly labor to Malaysia, Thailand, and, most recently, China and the Philippines. The extent of migration was influenced by the need for effective communication and rapid fulfillment at certain stages of manufacture. The ability of plants to deliver products daily and quickly to Singapore was vital. Thus the fragmentation of the hard disk drive industry from America to Southeast Asia describes the development of an industry supply chain involving multiple parts and various processes tied together by an efficient logistics system. The evolution of the supply chain was driven by competitive conditions and corporate visions. McKendrick, Doner, and Haggard (2000) highlight the importance of understanding the vision and attitudes of firms by contrasting the mobility of American firms with the slowness of Japanese firms to follow similar strategies. This contrast is related to differences in the purchasing strategies of American and Japanese firms at the time (McCann 1998).

The lessons of firms in the hard disk drive industry are mirrored in other industries. The case of the computer manufacturer Acer is a further example (Nemoto and Kawashima 2000). Initially, the company's headquarters and the plant in which it made products and from which it shipped worldwide were located in Taiwan (China). In the second stage of operations, the company added joint venture facilities in China and Malaysia. These plants were introduced to reduce the costs of low-technology parts, while leaving the strategic core of technology and assembly in Taiwan (China). Because it held a minority position in the parts plants, Acer did not use those plants for core production processes. After the subsidiary plants were in operation, the company acquired related firms in the United States to augment the company's access to improved technologies and to increase its exposure to U.S. market conditions. However, U.S. orders were channeled to Taiwan (China), with parts ordered from China or Malaysia as needed.

Subsequently, the company responded to the need for shorter lead times in its supply chain by adopting a more decentralized approach. The U.S. subsidiary now has an assembly capability and may order parts directly from Chinese or Malaysian plants or from elsewhere. This quick-response model allows the U.S. subsidiary to decide whether to follow the higher-cost but shorter lead-time strategy of shipping components directly from plants or using the slower but lower-cost strategy of sourcing from the high-volume assembly facility in Taiwan (China).

Acer's strategy is consistent with the research findings of a survey by Professor Cheng-Min Feng and his colleagues at the National Chiao Tung University (Feng, Chang, and Chia 2001). Among firms listed on the Taiwan Stock Exchange, 65 reported an increase in offshore factories from five to 92 between 1985 and 1999. While in 61 percent of the cases the main reason for creating the offshore facility was to reduce costs, proximity to customers was the main reason in 24 percent of the cases and access to new technologies was the main reason in 7 percent. Low-cost locations were China, Malaysia, the Philippines, and Thailand. Proximity to market was the main reason for locating in Japan, Singapore, the United Kingdom, and the United States. Singapore was chosen because of its excellent position as a hub for Asian markets and good connections to global markets. Their research also explored the relationship of the product life cycle with the production and logistics strategy.[1] The researchers found shifts as a product moved to the high-volume, more stable, mature stage of its life. Production shifted to speculation—the production of batches on the basis of

1. The product cycle covers the stages of start-up, growth, maturity, and decline in a product's life.

forecasts—while logistics shifted to postponement—the holding of inventory centrally for fast distribution later. The rationale appears to be that, for electronics, both shifts are cost-reducing strategies. Planned production runs result in lower costs and centralized inventory more than lower costs of responsive distribution. The implication is that excellent information technology and delivery are vital.

Firms are striving to achieve better visibility along their supply chains. Forward information reveals point-of-sale information from bar-code scanning and shared real-time or, frequently, automatic electronic transmission with suppliers. This allows members of supply chains to make quick and consistent decisions (with different levels of sophistication among supply chains) about replenishment based on common sales information. It enables the use of cost-effective methods such as manufacturing resource planning or continuous replenishment planning and avoids the violent fluctuations of inventories associated with sequential and delayed information provided along disjointed supply chains (Forrester 1958). Visibility backward along supply chains enables receivers to be fully aware of the status of orders so that they can take early actions to minimize costs associated with delays in supplies or with changes needed to meet new sales requirements.

The visibility of the status of purchase orders from the time they are placed until goods are delivered is one of the competitive services offered by major logistics service companies. It also lies at the heart of increasingly sophisticated supply chain software, for example, the enterprise resource planning software of California-based QAD. Such software is widely used in large companies. For example, Ford adopted QAD's software in 1995 to manage component manufacturing, vehicle assembly, and distribution operations in new markets such as Thailand (QAD 1998).

The successful use of the *kanban* system in the Japanese automobile industry has had a major influence on all industries around the world as they shift to new delivery systems. Just-in-time supply is but one approach to reducing inventories in supply chains through frequent, small-quantity, and highly reliable deliveries. These deliveries are timed to respond to the immediate needs of users; the products are pulled through the supply chain by demand. This is a major difference from the time when goods were produced in long, low-cost runs based on forecasts well ahead of demand.

The strategy of postponement involves the delay of processes so that undifferentiated products can be held in a centralized inventory, thereby reducing the total inventory held and the costs built into it. As information is obtained about demands, differentiating processes can be performed: final manufacture of a product, labeling the product to meet national market requirements, or simply shipping the product to one market rather than another.

The results of new approaches are that products move to users more reliably and responsively at less total cost. When manufacturers are the immediate recipients of goods, the term lean manufacturing is used to describe the new environment. When the recipients are retailers, the term lean retailing is used.

One of the reasons for the agglomeration or co-location of electronic firms and auto parts makers in concentrated locations in East Asia has been the desirability of proximity to ensure responsive availability of products. In the hard disk drive industry, proximity is especially important during periods of product innovation (McKendrick, Doner, and Haggard 2000). The Japanese car industry developed parts suppliers in Thailand as a means to reduce costs by taking advantage of the country's low wages and by shortening the supply chain for vehicles assembled there, as described in chapter 5 (see also Kittiprapas and McCann 1999).

True just-in-time delivery is dependent on the proximity between provider and user so that very small numbers of units can be delivered several times a day for placement on the production line as needed. High product reliability is necessary, so the practice may not be appropriate for new local suppliers in developing economies (Jongsuwanrak, Prasad, and Babbar 2001). However, simply shortening chains or, rather, the time to delivery can substantially reduce total logistics costs. Shifts in location to be closer to the customer and the use of sophisticated information technology, solutions software, and efficient transportation services can reduce inventory by shortening lead times and enabling more frequent, more reliable, and better-scaled deliveries to meet short-term needs. Saving time to reduce costs is especially desirable in industries such as fashion clothing (Abernathy and others 2001) and fad toys (Johnson 2001) that are characterized by highly uncertain demand and highly seasonal products.

Abernathy and others (2001) attribute the rapid growth of Mexican and Caribbean apparel exports to the United States between 1991 and 1999 to a constellation of factors.[2] They include the role of lean retailing practices in increasing the importance of shorter and more reliable times for logistics functions. The relative growth of exports was almost as high in goods that were not constrained by quotas in China and Hong Kong (China) as in goods that were constrained by such quotas. The greatest growth was in

2. In 1991 Mexican and Caribbean apparel exports to the United States were equivalent to one-third of the value of those from the ASEAN-4 (China, Hong Kong [China], Taiwan [China], and Republic of Korea); in 1999 they surpassed them.

apparel items for which frequent replenishment was expected. Moreover, developing the textile and apparel industries in close proximity entailed short supply chains and thus enabled apparel manufacturers to shorten the order cycle time (the time from placement of the order to delivery of the product).[3] Competitive conditions also propel manufacturers to greater use of information technology: for example, apparel manufacturer Esprit de Corps is using the Web-based platform of Qiva iQ-Logistics to manage suppliers and the flow of goods from Asia (Cottrill 2001).

The toy industry, too, is faced with great (arguably, greater than in most industries) challenges of uncertain demand for many products and very high seasonal demand, in spite of various mitigating strategies.[4] Toys are "ideal products to chase cheap labor," and their manufacture has moved progressively to countries with low wages (Johnson 2001). In recent years changes in toy retailing, especially in North America, have heightened pressures on prices and, therefore, on costs, accelerating the shift overseas. The major toy manufacturers mitigate the problems of managing the flow of toys from Asia by integrating the management of suppliers and logistics.

Three strategies stand out. First, companies manage the risks of supply—whether caused by production, logistics, currency, or political factors—by sourcing from different countries. Second, the large toy companies outsource production to multiple suppliers. In the case of Mattel, Vendor Operations Hong Kong manages relations with about 30 suppliers. Selected on the basis of price, quality, and flexibility, the suppliers are registered companies in Hong Kong (China) and finance and manage factories largely in southern China. The use of multiple suppliers increases Mattel's flexibility in sourcing to meet surges in demand. As suppliers produce for a number of buyers, they enjoy some protection against dependence on a few toys of a major buyer should expected demand fail to materialize. Finally, companies improve logistics performance by making increasing use of information technology. This enables them to capture sales information quickly, to place orders to manufacturers later when better information on demand is available, to better match the choice of carrier and mode of transportation with the urgency of the shipment, and to use more centralized warehousing to postpone shipment to markets until demand is better known.

3. The United Kingdom's Marks & Spencer traditionally was very successful in managing short European supply chains. As its competitive position began to weaken in the 1990s, one of its cost-cutting strategies was to source more from distant, lower-cost locations, including Asia. It did not manage the longer and slower supply chains well, which compounded its problems.

4. Various strategies are used to reduce seasonality, for example, employing diverse marketing channels (adding sales at gas stations or giveaways at McDonald's) and delivering a rolling mix of products to extend the duration of buyer interest.

Service enhancement. Competition among products is multifaceted, including price, product quality, and availability. The relative importance of these dimensions varies among products. Availability is more important for frequently purchased items than for goods purchased periodically. However, availability has grown in importance for most products. Many goods that were regarded as highly differentiated among producers have become more like commodities. As a consequence, availability is one of a number of service attributes that have become a competitive feature. Electronic goods fall into this category; availability is now more important in the competition among computer manufacturers. To counter commoditization (see chapter 1), product customization is being used to redefine the nature of the product being sold. This involves postponing final production or assembly until an order has been placed, not to reduce costs but to achieve product and marketing differentiation.

The characteristics and consequences of time-based competition are varied. Examples follow of the need for more responsive logistics systems in the hard disk drive and auto industries.

The time required to complete functions is important in a number of aspects of producing and marketing hard disk drive products (McKendrick, Doner, and Haggard 2000). Being the first to market with a product is a considerable advantage. This time-to-market is dependent more on product development and manufacturing setup than on logistics. The ability to introduce a product first is dependent on the time required to ramp up production to achieve low costs. This is dependent on effective manufacturing in conjunction with effective suppliers and inbound and outbound logistics. When a product is announced, a company needs to be able to deliver it at a competitive price in volume. Once a product is in production, the order cycle time becomes important. This is the time from placement of an order to the availability of the product for sale or use. The order cycle time is affected by the characteristics of the supplier's supply chain, its production scheduling and processes, and its delivery system. A supplier that is only able to have long production runs severely limits the logistics of the supply chain. Such a supplier forces the chain to receive only occasional orders and makes it difficult to meet uncertain market demands without holding excess inventory. Finally, transit time is the most easily measured aspect of time in logistics. It is usually measured from the departure of goods from the factory to their availability for use at the destination. (Narrower definitions that focus on the time in transit with one or several carriers are common.) As the time involved in other processes is reduced, transit time becomes a more important consideration. The practice of cross-docking operations rather than warehousing products calls for excellent information systems as well as specially designed terminal buildings and site layouts.

In all of these time dimensions, reliability is very important. A slow but certain transit time adds to the amount of inventory in transit and to the amount of safety stock needed to cover the added days. An uncertain lead time requires firms to carry safety stock to cover uncertain demand during potential delays in transit. It is not surprising that reliability is one of the top priorities of most shippers.

Logistics are also important in automobile manufacturing. Car manufacturers no longer simply sell the cars available; they now offer the vehicle the consumer wants. In order to provide individual car buyers with the style, color, and options they desire within ever-shorter times and at reasonable cost, the industry has to offer options without holding excessive inventories throughout the retail system. Consequently, the logistics system must be more responsive throughout the supply chain. As with cost-saving measures, they must redesign and even relocate activities (see chapters 4 and 5).

In the auto industry, seats are usually assembled close to auto assembly plants, for the obvious reason that their bulk makes them expensive to transport. However, the manufacture of components for seats may be distributed more widely. For example, the manufacture of seat covers can be labor intensive. As a result, production is found in Asia, often using imported materials. The finished covers are shipped back by sea container. Products of good quality have been produced in systems with long, though precise, roundtrip cycles. However, the pressures to give car buyers choice and fast delivery, at reasonable cost, require changes to the system. Meeting variable demands without keeping excessive inventory close to auto sales outlets requires a much more flexible logistics system. In the absence of forecasts of demand long enough in advance, some seat covers, at least, have to be sent by airfreight. The difference in costs between shipping by sea and by air threatens the viability of exports from Asia. The challenge increases the competitiveness of alternate locations with low labor costs closer to the major markets, for example, Eastern Europe for Western Europe and Mexico or the Caribbean for North America. If effective supply solutions are not found, the range of options available to consumers in a short time will become limited.

General implications of the need for change. Two changes in logistics are occurring in response to the pressures for cost reduction and service enhancement. First, more attention is being given to speed in the design of supply chains and the management of logistics. The reduction in time is becoming more common and more important as a strategy to reduce costs and improve customer service. Hummels (2001) estimates that, for manufactured goods imported into the United States, each day of travel is worth an average of 0.8 percent of the value of the good and that each day in transit reduces the probability that a country will be a source by 1.5 percent. Second, more attention is being given to the cost of disruptions, whether variations in demand, logistics, or supplier performance (Levy 1995). The effects of disruptions are particularly great when lead times are long, even when information systems are good. Greater recognition of the costs of disruptions encourages the design of shorter, faster supply chains. This encourages local sourcing, the use of improved information technology, and the use of expedited logistics services.

These examples show that many forces are heightening the value of speed and illustrate the role of logistics in the wide context of supply chain management. Many dimensions of logistics and product availability affect procurement and marketing policies. The implications are important at various levels.

The need for better integration across functions applies from the operating to the executive levels in firms and has relevance for community leaders. For example, at the operating level, ongoing efforts are usually needed to ensure effective communication between product buyers and logistics managers. At the executive level, the adoption of supply chain management practices has been made necessary by the dynamics of cost and competitive conditions. Executives must establish close working relationships with other participants in the supply chain. They must devote more attention to external relationships. Supply chain management is now a widely followed practice, not just a philosophy preached by a few firms. Finally, community leaders have begun to recognize the importance of logistics services. Community services and policies either hinder or enable firms to deliver products competitively and supply chains to function well, and community leaders and others in East Asia need to be aware of the rapid structural changes taking place in the logistics service industry.

Logistics as a Factor in Economic Development Models

The progressive shift of management strategy to supply chain management has been matched by greater attention in the literature to the role of logistics services in industrial location and economic development. Although models are still based on the tradeoffs of performance at alternate locations, reflecting the relative costs and benefits of proximity to markets or inputs, the variables to be taken into account have widened.

Two aspects of the change warrant recognition here. They are the need to recognize a range of costs related to choice of location and the need to recognize the compound efficiencies available from enhancing the inputs to logistics services.

Range of logistics costs in choice of location. At their simplest, traditional models of industrial location compared the total costs at alternate locations by summing the processing and transfer costs associated with the acquisition of inputs and distribution of products. The literature dealing with the economics of agglomeration recognized that distance is associated with various costs of "friction." However, models did not reflect the fact that firms are part of supply chains in which inventory costs are one element in the choice of location. The inclusion of inventory costs in a formal location model is recent (McCann 1993, 1998), making the logistics-costs approach a more realistic framework in which to structure location analysis.

Once researchers began paying more attention to the effects of transport quality on inventory levels and on market penetration, they also began paying more attention to the effects of transport services and logistics on industrial location. This is exemplified by the studies of McKendrick, Doner, and Haggard (2000) of the hard disk drive industry and of Abernathy and others (2001) of the clothing industry. The location of industries is recognized now as reflecting the relative ability of communities to meet the diverse needs of firms, including logistics services. Logistics performance affects costs in the supply chain and the quality of service provided to customers.

Multiplicative efficiencies from enhancing inputs to logistics services. Transport has long been recognized as a necessary, but not a sufficient, condition for economic development to take place. Improvements in transport have been expected to stimulate economic development. However, in measuring the effects of investments in transport, attention has focused on the direct effects measurable through changes in the costs and volumes of traffic. The potential for firms to realize additional benefits by achieving economies of scale as a result of better accessibility has not been measured (Mohring and Williamson 1969). However, the recognition of significant benefits in other logistics costs as a result of improved transport has led to new efforts to identify and quantify wider benefits. (The research was undertaken during the 1990s in the midst of renewed interest in the macroeconomic relationship between infrastructure investments and economic growth and productivity. For a paper linking the two lines of research, see Weisbrod and Treyz 1998.)

The U.S. Federal Highway Administration has played an important role in studies to improve the assessment of transport investments. David Lewis (1991) outlines the need for evaluations to incorporate changes in logistics structures and practices. Subsequent studies have elaborated on the rationale for benefits associated with logistics and economies of scale and have set out frameworks for their assessment (HLB Decision Economics 2001; HLB Decision Economics, ICF Consulting, and Louis Berger Group 2002; ICF Consulting and HLB Decision Economics 2002; Lakshmanan and Anderson 2002; Weisbrod, Vary, and Treyz 2001).

Lakshmanan and Anderson (2002) suggest that the importance of logistics considerations for the location of industries gives rise to a "new economic geography" as a result of shifts in activities as firms pursue logistics cost savings. But new patterns may emerge as firms take advantage of innovations resulting in endogenous changes in products and in production and logistics technology. Benefits can be gained not just from doing traditional things more efficiently (holding smaller safety stocks, for example) but also from doing different things (for example, adopting pull rather than push inventory systems or shifting to mass customization strategies). An example of endogenous innovation is to remove customs as the collector of import duties and to use computer information systems instead (see Heaver 1992).

The magnitude of the benefits for logistics from transport investments depends on the environment in which the investments take place. The environment may be thought of as the physical hinterland. It may also be thought of as the conditions in that hinterland, in particular, the aggregate characteristics of programs that change capabilities and encourage innovation. In this context, multifaceted approaches to the development of logistics that include developments in transport, information technology, education, and public policies create multiple bases for innovation.

Structural Changes in the Logistics Service Industry

It is not surprising that the changing needs of shippers have been associated with the changing structure of the logistics service industry. It has long been customary for firms with small volumes of international traffic destined for or originating in a foreign country to use a freight forwarder. Freight forwarders have long been specialists in arranging the transportation, storage, and handling of goods along with the processing of documentation between and within countries.[5] They manage the activities through their own offices and through those of partners. They may act as agents in arranging transportation or assume responsibility for transport and issue their own bill of lading. In shipping, they do this by acting as a non-vessel-owning common carrier coloading freight onto a shipping line's vessel. As their volume of airfreight increases, they may also operate some air cargo services (for example, Panalpina's ASB Air). In the trade of countries with highly specialized logistics and transportation conditions, even large firms with a substantial volume of trade into the country have traditionally sought the assistance of such specialists.

The higher demand for logistics services in the last 15 years has resulted in substantial changes in the freight-forwarding industry. The development of the logistics services industry has been marked by four trends: the provision of more sophisticated services, the development of global networks, the development of domestic and contract logistics capabilities, and the integration of transportation and logistics services.

Provision of more sophisticated and comprehensive services. Major freight forwarders have generally lagged behind new logistics service companies in the use of information technologies. However, increasingly, all logistics companies are introducing Web-based capabilities so that shippers can perform activities on-line and can access information about the location and status of their shipments. Initially, this information was based on container location. Now information can be provided by purchase order through proprietary systems and shared portals. The trade literature is replete with examples of logistics companies adding to their information technology capabilities, often through the acquisition of or partnerships with specialized software companies.

The better data on current sales, production, and inventory in facilities and in transit are in keeping with the needs of shippers for more sophisticated information technology capabilities for improved decisionmaking. Indeed, many logistics service providers are shifting away from their traditional asset-base to becoming knowledge-based companies using advanced management decision tools such as dynamic routing and network optimization models. They are prepared to design and manage a logistics system, not just be responsible for its administration; in words typical of their

5. Freight forwarders formed to serve growing 19th-century trade among European countries faced with the problems of many border crossings. The leading European forwarders went on to become the major international firms, for example, Danzas, Kuehne & Nagel, Panalpina, and Schenker.

advertisements, they are "providers of customized integrated logistics management solutions." Three examples relevant to East Asia follow. Schenker has entered into a contract to move the electrical products of the French firm Legrand from France to the Singapore Logistics Center and to handle distribution through the Asia-Pacific region. This role is facilitated by Schenker's extensive physical presence in the region and its Web-based information system, which gives transparency to the whole supply chain ("Distribution Hub" 2001). Similarly, starting in 2002, Panalpina began handling the movement of IBM computer components from Asia to Europe, where it will operate a vendor-managed inventory hub in Hungary for delivery to factories on a just-in-time basis. End products will be received by Panalpina ex-factory and exported back to the Pacific from Japan to New Zealand (Panalpina 2001). Finally, UPS and Samsung have set up a joint task force to streamline the manufacturer's supply chain. Samsung expects outsourcing to improve its competitiveness (Parker 2000).

The progressive shift to more sophisticated and comprehensive logistics services has been associated with changes in terminology. Outsourcing responsibility for providing a wider range of logistics functions gave rise to recognition of third-party logistics service providers, known as 3PLs. Then knowledge-based companies assumed responsibility for more of the planning and management of logistics services provided by others, giving rise to recognition of so-called 4PLs. However, in international logistics, shippers almost always negotiate ocean rates and shares of traffic among lines (Heaver 2002). Contracts for the planning and management of systems dependent on airfreight go to companies with significant assets in distribution or significant airfreight capacity, as these examples have shown.

Development of global networks. The major freight forwarders had a wide international presence for most of the 20th century. They were present in most continents, but their presence varied in intensity from country to country; in some countries they relied on partners. However, in the last two decades they expanded the number of countries in which they had a direct presence and intensified their presence in most countries. They did this largely through acquisitions and mergers. Examples are the acquisition of Royal Cargo Corporation of the Philippines by Danzas (2001) and the strategic alliance between the Swiss-based Kuehne & Nagel and SembCorp Logistics of Singapore. Under the agreement, the companies are cross-purchasing equity (SembCorp Logistics 2000).

The rationale for the strategic alliance was for the partners to be able to provide customers with a better global presence. Kuehne & Nagel sought to have a stronger position in Asia. SembCorp Logistics recognized that the integrated logistics industry in Asia should no longer be regional but needed to be a part of global business. Further, international logistics was being driven by scale in operations, which involves the integration of distribution networks, systems, technology, and a "seamless flow of products and information." In its press release, Sembcorp Logistics (2000) notes,

> Recent market trends have shown that the logistics industry in Asia is experiencing rising competition, particularly from companies in the U.S. and Europe. In light of global competition and the need for change in the definition and scope of its integrated logistics business, SembLog is adopting a global strategy that is driven by scale, integration, reconstruction, and e-commerce. . . . SembLog's strategic alliance with KNI [Kuehne & Nagel] . . . forms an important part of SembLog's strategy to become a global logistics player, dominant in Asia, with a global logistics network that can extend its reach to new and existing multinational customers. Via this strategic alliance, SembLog can take greater advantage of the fast-growing third-party (outsourcing) global logistics market.

The alliance created a partnership in which Kuehne & Nagel is SembCorp Logistics' global freight-forwarding partner and SembCorp Logistics is Kuehne & Nagel's exclusive partner focusing on supply chain management in the Asia-Pacific region. More recently, Kuehne & Nagel acquired the privately owned USCO Logistics for approximately $300 million plus an earn-out based on performance. Through its management contracts, USCO is one of the largest providers of warehouse space in North America. It also acquired a majority interest in Virtual Integration Associates of Canada, a leading supplier of supply chain management and vendor-managed inventory programs for the electronics industry. These are a part of Kuehne & Nagel's strategy to become a global player in contract logistics.

Development of domestic and contract logistics capabilities. The development of logistics services has been marked by the expansion of services to deal with the challenges of international trade and the needs of more sophisticated supply chain managers. Contract logistics suppliers have developed to take advantage of the interests of some large companies in outsourcing logistics functions. The best examples of these new third-party logistics providers in the United States are Ryder Logistics and Schneider Logistics, both of which expanded into logistics from trucking operations. In the United States the contract logistics business is estimated to have grown more than 20 percent in one year to a total of $56.4 billion in 2000 (Armstrong and Associates 2001).[6]

6. A billion is 1,000 million.

European contract logistics firms, such as EXEL, have gone international more aggressively than firms in the United States. However, Ryder acquired Ascent Logistics in October 2000 as a part of its strategy to expand its capabilities globally. Ascent Logistics was incorporated in Singapore in 1996 and in Taiwan (China) in 1998. Similarly, Schneider acquired Hellmann Worldwide because of its international scope ("Schneider Logistics" 2001).

Integration of transportation and logistics services. A major shift in the logistics services industry has been the entry of carriers responding to the demands of shippers seeking to deal with fewer suppliers and to outsource logistics activities.[7] In many developed countries, truckload and less-than-truckload trucking lines have added logistics services. In China such an extension of business has been made impractical by the restriction of licenses to particular types of business. Internationally, shipping lines and air carriers have become important players in the international logistics business, and their expansion has contributed significantly to the growing presence of U.S. firms in international logistics.

The North American trade in manufactured goods with Asia has created a specialized need, arising from the particular challenges of doing business in Asia, including distinctive business practices and attitudes as well as specific types of transportation challenges that are compounded greatly by the wide range of public policy and regulatory constraints on doing business. American importers of Asian goods felt a strong need in the early 1990s for better assistance in managing the flow of imports. The value of imports from Asia was increasing, and U.S. shipping lines were interested in extending their range of services to customers.

The result was the development of consolidation services, most notably initially by APL, known as American Consolidation Services, and by Sea-Land, known as Buyers. The services differed in some important respects from those offered by freight forwarders. They were more focused on the needs of U.S. importers for monitoring the movement of goods to consolidation points, managing the consolidation of goods, and shipping according to the specifications of the buyers. The companies were aided in the development of consolidation services by their shipping services, which familiarized them with the buyers' needs. They were able to offer consolidation services with high visibility by using their existing links with the documentation processes of shipping lines.

---

7. Although they seek logistics firms offering wide solutions, most shippers prefer to retain control of their choice of shipping lines. This is less true with other modes of transport.

The companies effectively act as control agents, ensuring that shipments from many suppliers are executed efficiently, economically, on time, and according to each customer's logistics strategy. In the process, they can capture relevant shipment data in their information systems and give the customer complete access to status reports from the time of booking to the time of delivery. The companies have emphasized the development of leading-edge information technology services. In addition to being able to track their shipments, importers can use the resulting database to plan, budget, forecast, negotiate, and manage their businesses.

The development of these specialized firms has mushroomed as Asian trade has expanded, the outsourcing of logistics has increased, and transportation companies have increased their presence in the logistics service activities. The logistics service companies such as Maersk Logistics and APL Logistics have become global. Other companies have strong logistics service activities in Asia such as Century Distribution Services of "K" Line, MOL Logistics of Mitsui O.S.K. Line, and the logistics services of OOCL (Orient Overseas [International] Limited). Although the logistics enterprises adopt the brand name of the shipping line with which they share common ownership, they are managed as arm's-length businesses. Shippers expect the logistics services to be independent and not feed business to the shipping line (Heaver 2002).

In airfreight, passenger airlines, such as Lufthansa, offer logistics services, but because they are carriers for freight forwarders and are constrained by passenger schedules, they are not dynamic players. The dynamic players are the integrated courier services that have gone global mainly since the 1980s and that, more recently, have been forming logistics divisions. For example, UPS Logistics and FedEx Logistics are offering wider services through UPS Supply Chain Solutions and FedEx Supply Chain Services, respectively. The reduction of airfreight costs and the higher value placed on speed in logistics systems account for the very rapid growth of these services. They are now significant drivers of change, facilitated by the relaxation of restrictions on airfreight services.

Implications of the Changing Structure and Role of Logistics for East Asia

The increased sophistication and growth of the international logistics service providers are a part of improvements in logistics services elsewhere. They and other developments such as transport deregulation and commercialization have enabled significant reductions in logistics costs.[8] To remain competitive, the economies of East Asia need comparable improvements in their domestic and international logistics capabilities. Changes in the global business environment are placing new pressures on logistics costs and on the capabilities of firms to function effectively in a supply chain management environment. These changes elsewhere heighten the need for improved logistics in East Asia. Investments and policies should be considered not as isolated initiatives but as integrated programs to reduce costs and encourage innovative changes in logistics.

One means by which logistics service is being improved elsewhere and in East Asia is through the expanding role of specialized logistics service companies. The extent to which these companies or their equivalent are able to develop in the economies of East Asia is an important question. The opportunities for their growth are influenced greatly by public policies and corporate attitudes. These tend to vary nationally and are examined for the three countries considered later.

The recent and expected expansion of the economies means that added infrastructure in transport and communications is needed. Investments are needed to expand capacity and to improve the speed and reliability of logistics services. Although these investments are location specific, they are considered in the next section rather than in the country case studies.

PORT AND AIRPORT INFRASTRUCTURE

The past and current needs for additional infrastructure in East Asia are huge. As such, they are frequently the subject of large and specialized studies, most of which are specific to particular areas. Some have been carried out under the auspices of the World Bank (see, for example, World Bank 2001a, 2001b, 2001c, 2002).

Good infrastructure is a necessary, but not a sufficient, condition for good logistics. As such, a general review of infrastructure is appropriate. The focus here is on investments in the hub facilities of ports and airports. Although communications infrastructure is critical to effective logistics,

8. Quantitative measures of logistics costs are sorely lacking. However, a valuable statistical series on logistics costs for the United States shows the improvement that occurred in logistics performance during the 1980s, a period of major regulatory reform. Logistics costs dropped from more than 16 percent of GDP in 1981 to about 10 percent of GDP after 1992 (Delaney and Wilson 2001).

information technology is not examined here. In many East Asian economies, the geography is such that highways are the most significant investments. To the extent that congestion is a common problem around ports and airports, urban road or rail improvements are needed. In Bangkok, for example, off-dock container sites have been built, and the ability to move containers between the port and these facilities by rail is a great improvement. In China the expansion of rail capacity, improvement in lines and yards to handle intermodal traffic, improvements in information technology capabilities, and modernization of management practices are all needed. However, it is not practical to cover the great diversity of transport and information technology investments required in the region. Treatment of some inland transport issues is included in the section on China.

Development of Port Services

The rapid growth of trade within Asia and between Asia and the rest of the world places tremendous pressure on the ports ("Asian Ports" 2001). A study of container traffic in East Asian ports by Ocean Shipping Consultants forecasts total volume of containers handled to increase from 107 million TEUs (20-foot equivalent units) in 2000 to between 254 million and 306 million TEUs in 2015. Over the period, the greatest rate of growth is forecast for Southeast Asia and the least is forecast for Northeast Asia. A recent study by the United Nations Economic and Social Commission for Asia and the Pacific estimates that the region will need another 427 container berths by 2011. In a presentation to the terminal operations conference in Portugal in June 2001, Mr. Quan Peitao of the Chinese Ministry of Communications noted that the value of Chinese imports and exports is expected to grow from $470 billion in 2000 to $680 billion in 2005 and that the volume of port containers is expected to increase from 40 million TEUs to 60 million TEUs over the same period ("China Plans Development" 2001). In fact China's total trade in 2003 surpassed $850 billion. Matching the availability of container berths with the demand created by growth in the manufacturing sector is a harder task than matching bulk berths for commodities such as coal or oil. The latter are readily linked to the needs of specific projects, although the lumpiness of port investments aggravates the difficulty of forecasting demand. Therefore, the emphasis here is on developments in the container business (Gooley 2001).

There is wide public and political recognition of the critical role played by international trade and, therefore, by ports in the East Asian economies. Building new ports and container terminals is seen as a matter of survival. Consequently, the process of planning and building port facilities faces fewer restrictions and is less time consuming in East Asia than in developed countries.

New ports have sprung up across the region. A decade ago, bustling container ports like Laem Chabang in Thailand, Tanjung Pelepas in Malaysia, Tanjung Priok in Indonesia, and Yantian, Chiwan, and Shekou in China did not exist. At Laem Chabang, construction on five new terminals is scheduled to begin in stages between 2002 and 2008 so that the port can serve Thailand and the Mekong Valley. In Indonesia, Tanjung Priok handled nearly 3.3 million TEUs in 2000, up from 2.1 million TEUs the previous year, and Tanjung Pelepas moved nearly 1 million containers in the first quarter of 2004, compared to 750,000 in the first quarter of 2003. By the end of 2003, Tanjung Pelepas had the physical capability to handle nearly 5 million TEUs. The new ports in China are also investing heavily to boost container-handling capability.

Older ports, too, are quickly adding terminals and berths. Significant additions, for example, are currently under way in Singapore; Subic Bay (Philippines); Kaohsiung (Taiwan, China); Hong Kong (China); Shanghai (China); and Busan (Korea). Perhaps, the most notable expansion is in Shanghai, where investment in the port is part of a multifaceted initiative to establish the community as a major transportation and logistics hub. The port handled 5.6 million TEUs in 2001 and is targeted to move more than 20 million TEUs by 2010 following investment in new capacity amounting to $12 billion. When completed the offshore Yangshan port will have a total of 52 berths.

The investment in and modernization of the ports has been aided greatly by the participation of foreign capital and management. With the exception of Singapore, itself an investor, this has become the norm. Foreign companies are attracted for a number of reasons. Trade is booming, so that more ports have potential for direct rather than feeder services. Foreign participation in port terminals can be introduced conveniently as terminals are often new, relatively discrete operations. They are site specific, although the ramifications of their business are widespread. Finally, a number of international companies have been looking for opportunities to expand in the relatively new global business of container terminal management. The result is that CSX World Terminals, Hutchison Port Holdings of Hong Kong (China), P & O Ports, and the Port of Singapore Authority are major participants in port developments. The container terminal managers have had major beneficial effects in their ports.

The awarding of concessions for terminal management within ports is commonly a difficult matter because of the small number of terminals and uncertainties about the effectiveness of competition among ports. This is an issue in the port of Manila, where the Philippine Ports Authority has a poor record of efficient management and is trying to determine the number of terminal operators that should be allowed. Economics suggest two, although other considerations may lead to a single operator being chosen (Consilium International 2001). The level of local and foreign ownership may also be an issue. In China, after the granting of majority ownership of the Yantian Terminal in Shenzen to Hutchison Port Holdings raised concerns of "foreign" control of ports, local governments have retained majority ownership in subsequent contracts ("Ports in a Storm" 2001). The effect of such joint venture arrangements on performance is not clear. Much depends on the roles of senior executives appointed by the parties as specified in contracts and on the objectives of local government.

The container terminal managers have brought wide experience in terminal management and have introduced new technology-related initiatives. For example, in September 2001, PSA celebrated the successful implementation of the Dalian PortNet System, a customized version of PSA's e-commerce network for the shipping and port community, and opened a building housing PSA's first overseas information technology establishment (PSA Corporation 2001). The container terminal managers have also dealt with concerns about external factors that are important to efficient terminal throughput, for example, inland connections and customs operations.

The Asia-Pacific Economic Cooperation (APEC) Congestion Point Study (APEC 1997) notes that congestion of traffic to and from terminals is a common problem that is difficult to correct. Intermodal connections may not be readily available, or their services may not be of the quality needed. A current study is looking at an expansion of intermodal rail service between Hong Kong (China) and China. It is expected that the planned Shenzhen–Hong Kong (China) western corridor project will double the handling capacity for vehicles moving between the two important regions. The expectation is that container-handling activity, once concentrated in Hong Kong (China), in the future will spread more rapidly outward toward China. With the exception of Singapore and Hong Kong (China), customs procedures need substantial improvement. For Hong Kong (China), the clearance of transshipment cargo with China needs improvement.

No major constraints on the provision of ports are evident in the region, although considerable pressures remain on port capacity because of the rate of growth of trade. Competition between local governments could become an issue affecting the pattern of development. There is a lack of well-located deepwater sites, so development costs will be high. Current issues in port development concern the detailed characteristics of contracts with container terminal managers, the quality of customs services, and the adequacy of land-side connections, including intermodal services in China.

Development of Airport Services

Air cargo is the most rapidly growing freight business. This is not surprising in view of the real decline in airfreight rates (Hummels 1999), the increase in the value to weight-volume ratios of many products, and the higher value of speed in logistics. The demands of airfreight have put additional pressures on the capacity of major airports, although the main demands on capacity are exerted by the passenger airlines. The importance of the passenger business is the first main distinction between infrastructure planning for ports and for airports. The second distinction is that there are fewer new airports than ports. The third distinction is that public policies play a large role in development of the airfreight business.

New airports have added capacity, reduced congestion, introduced better freight-handling conditions and clearance procedures, and made 24-hour airport operation possible. Other airports, such as Singapore's Changi Airport, have made substantial investments to accommodate growth and maintain their status within the international network. Changi Airport has benefited substantially from infrastructure investments and policies that have created operational efficiencies consistent with Singapore's role as a logistics hub. The free trade status of the Airport Logistics Park of Singapore is an excellent example.

In July 1998 Chek Lap Kok opened in Hong Kong (China), becoming the world's busiest cargo airport in 2000—a ranking it retained through 2003 with a throughput of 2.6 million tons. This occurred in spite of concerns about the high level of airport fees. The airport's computer-linked system of customs clearance has been working well. By transmitting cargo data to customs electronically, the majority of freight can be cleared before the plane arrives ("Companies Vie" 2000; Sowinski 1999). The volume of freight moving through Hong Kong (China) is influenced by six main factors. First, the cost of using the airport must be balanced against the second factor, the quality of the services provided. Third, the airport has a high "local" level of business. Fourth, Hong Kong (China) is a good gateway to a large and productive region. Fifth, a significant amount of overseas traffic is routed through Hong Kong (China) by air or by land because of public policies, for example, quotas that apply in China but not in Hong Kong (China) and the preference of many vendors for doing business through Hong Kong (China) rather than a Chinese city to minimize administrative bottlenecks. Finally, the operating rights of airlines (considered later) are vital to understanding the pattern of service provided.

In 1999 the Pudong International Airport was opened in Shanghai, and the volume of air cargo has grown faster than that of any other airport in

Asia (Zhang 2001). The area's conditions are reviewed below as part of the coverage of China.

In Korea the new Incheon International Airport was opened in 2001, relieving congestion and making flight times for air cargo more convenient. As with the other new airports, it is hoped that the facility will become a logistics gateway in Korea.

Unlike shipping, the services of airlines are limited by their operating authorities, which are derived from the international bilateral negotiation regime dating back to the Chicago Convention of 1944. The rights of national airlines to offer services or to fly into or over countries are negotiated between governments, and individual governments allocate the rights so gained to their carriers. In spite of domestic deregulation in a number of developed countries, the international regime has been slow to change. Relaxation of constraints is faster for all cargo services than for passenger services. A consequence of the regime is that the liberalization of trade in air services is often interrelated with the domestic policy regime. It is also more likely that a large country such as China will have difficulty changing its international regime because of the intricacies of domestic airline policy. The United States is leading the world in pushing for more "open skies."

In 1997 Singapore was the first Asian economy to enter into an open skies agreement with the United States, followed by Korea in 1998. Taiwan (China), the Philippines, Brunei, and Singapore have entered into seven freedom rights agreements for cargo (the open right of a carrier from a third economy to provide service between two other economies) to facilitate mini hub operations for FedEx and UPS (Zhang 2001). In an interview, (Chanda 2001) UPS chairman and chief executive officer, James Kelly, commented:

> We are looking now at expanding our intra-Asian service. We have been waiting for the right time to make those kinds of investments . . . Taipei is our hub, but it is not an international hub. Geographically, the Philippines is the right place . . . one of the issues we continue to have in Hong Kong [China] is trying to serve Asia quicker . . . We are trying to get fifth freedom—the right to fly beyond Hong Kong.[9]

China's air policies have resulted in the slow entry of freight services (Einhorn 2001). DHL was the first international air express company to form a joint venture with a Chinese cargo transport agency: Sinotrans, China's largest cargo transport agency. Emery Worldwide recently announced that it has entered into a joint venture with a subsidiary of state-owned China

9. Subsequently, in 2003, UPS was able to launch its own direct service from Hong Kong (China) to its intra-Asia hub in the Philippines.

National Electronics Import and Export Corporation and will subsequently apply for a class-A license. UPS, in contrast, which has served China since 1988 by passing traffic on to Sinotrans through Hong Kong (China), continues its relationship with Sinotrans as part of its new direct services to Beijing and Shanghai, which commenced in 2001. With its entry into China, like other foreign companies, UPS expects to improve services by providing real-time tracking for customers such as apparel and electronics firms (Morton 2001).

LOGISTICS DEVELOPMENTS IN THREE COUNTRIES

The air policies of Singapore and China illustrate the diversity of policies affecting logistics in East Asia. A review of conditions related to logistics in China, Singapore, and Thailand identifies more contrasts and brings to light practices and policies affecting the development of logistics services.

Singapore

Singapore was founded as a trading post in 1819, and its location and deep-water harbor have been keys to its economic opportunities ever since. From its founding as a separate crown colony in 1946 to its independence in 1965, the small city-state has been forced to realize the maximum advantage from these natural assets. Programs to attract industry were spearheaded by the Economic Development Board as early as 1968, and emphasis was placed on technology and computer education and on productivity campaigns to ensure a work force able to sustain a trading and manufacturing economy. These directions remain keys to Singapore's success, and in 1997 the logistics industry contributed 7 percent of Singapore's GDP (Chin and Tongzon 2001). The Economic Development Board continues to recognize the logistics and transport cluster as an important sector for investment and employment. The cluster's ability to integrate the various modes of transport as well as logistics and supply chain management services enhances Singapore's attraction for knowledge-intensive and value-added activities.

The Port of Singapore is one of the most efficient in the world, and its investments have benefited shippers and brought significant backward benefits to other sectors of the economy. The port has one of the higher-income multipliers in the economy because it imports few resources (Toh, Phang, and Khan 1995). Several factors have enabled the port and airport to support the success of Singapore as a hub (Chin and Tongzon 2001).

First, Singapore's location on main shipping routes with good centrality in Southeast Asia and Australasia makes it ideally suited to play the role of a logistics hub. An example is the role that the Schenker Singapore Parts Centre plays for Volkswagen in taking care of replenishment, inventory control, and invoicing of distributors and subsidiaries in the region from Korea to New Zealand and India. Second, Singapore's ability to capitalize on its location has meant that it handles a high volume of traffic, resulting in high levels of connectivity. Success brings advantages to hub locations. Connectivity is derived from the number of service links and the frequency of services, both aided by high volumes of traffic. Third, Singapore has been successful in its logistics activities through efforts to ensure speed and reliability of handling.

The high level of services provided in Singapore has been achieved as the result of deliberate programs. Adequate infrastructure has been built to take advantage of the deepwater harbor. In 2001, 15.5 million TEUs were handled with 50 berths. Singapore's traffic includes a lot of reloads, which is consistent with its hub function (80 percent of containers are train-shipped) and lessens the pressure on inland distribution compared with gateway ports such as Hong Kong (China). However, important to the efficient functioning of terminals are soft infrastructure (such as 24-hour operation), a skilled labor force, and supporting organizations (such as customs). It is reported that electronic goods can be delivered to users within six hours of landing at the airport. Customs' services in Singapore are considered "ahead of the game" (Reyes 2001). Excellent information systems are essential to achieving such levels of performance. Singapore's great attention to information technology has supported a well-paid and productive work force and enabled PSA to be a leader in port management. The port's PortNet information system is used in a number of independent ports around the world. For example, in 2001 Seattle announced that it was investing in the system. PortNet is an outgrowth of Singapore's TradeNet, which was built in 1987 under the direction of the Trade Development Board. Singapore is widely recognized as a world-class logistics environment, warranting its characterization as the "intelligent island" (Biederman 2000).

A wide range of government policies has played a role in the success of logistics in Singapore. Although state monopolies are numerous, the government has also created competitive markets in some services, such as ship bunkering. The Economic Development Board and other state agencies have created conditions favorable to the attraction of industries. A good example is the development of Jurong Island, initially for electronics and now for chemicals ("Atofina Profile" 2001; "Singapore, Economic Development Board" 2001). Banyan Logistics Park offers chemical firms on Jurong Island readily accessible, specialized logistics services.

The efficiency of the port and airport services is essential to the flourishing logistics services that have been attracted to Singapore, often located in free trade zones. The attraction of Singapore as a logistics hub has evolved over many years. In 1987 the London Metal Exchange designated Singapore as the first delivery point outside London. Major companies such as Bayer, Black & Decker, Caterpillar, and IBM selected Singapore as a distribution center. Virtually every major logistics service company has a significant presence in Singapore, which is generally the Asian head office. These offices frequently employ teams responsible for developing and managing logistics and supply chain solutions for customers. Air express courier companies such as DHL, FedEx, and UPS have major facilities in Singapore.

However, the success of Singapore has also brought rising costs, which have caused certain types of manufacturing, such as the hard disk drive industry, to migrate to other locations. Shipping lines, notably Evergreen and Maersk SeaLand, are now shifting to the adjacent new port of Tanjung Pelepas in Malaysia. It is probable that the lines are seeking not only to achieve lower charges but also to exert greater influence over operating practices. However, from an economic perspective, the new port may be viewed as an extension of industrialization in and around Singapore and as a reflection of the success of its logistics strategy.

Thailand

Thailand is a country of some 60 million people in which agriculture contributed 10 percent of GDP in 1997. Never colonized, the country offers cheap labor coupled with a relatively stable economic and political climate. This mix has attracted foreign companies that have invested billions of dollars in Thai operations. In accounting for the decision of General Electric and Shin-Etsu Chemical to establish Asia's largest silicone manufacturing unit in Thailand, the senior managing director of Shin-Etsu Chemical said, "With its political and social stability, transportation access, and the availability of natural gas, we are confident with our investment decision in Thailand" (Wood 2001).

However, Thailand has not been as attractive as Singapore for various reasons, of which infrastructure shortcomings play an important role (Jongsuwanrak, Prasad, and Babbar 2001). The present government recognizes the importance of improving infrastructure for the country's overall economy. It has privatized telecommunications services and sought help from private local and foreign companies to cope with the road shortage through build-operate-transfer arrangements for highways in and around Bangkok (Tam 1999).

The later and slower path to a more competitive and productive economy in Thailand than in Singapore affects all industries, including logistics services. The contrast is indicated by some features of the privatization of state-owned enterprises. As early as 1960, the Royal Thai government expressed an intention to privatize poorly operated and inefficient state-owned enterprises (Dempsey 2000). However, implementation in the early years proceeded at a snail's pace. It was not until 1986 that more significant attempts to privatize began because of the increasing inability of state enterprises to self-finance their own expansion. These efforts resulted in the sale or lease of a few state-owned enterprises to the private sector, but most of the larger state enterprises saw little change. This was partly because workers resisted change but more likely because the need for reform seemed less pressing, as Thailand had the fastest-growing GDP in the world between 1985 and 1995. The economic recession of 1997 led to renewed calls for action. In the summer of 1998, the government unveiled new plans for privatization as part of an overall strategy to recover from its recent financial downturn. The plan grouped state enterprises to be privatized into five categories: four infrastructure sectors—energy, telecommunications, transportation, and tap water—and a fifth category for 31 "miscellaneous" enterprises. Despite the political roadblocks mounted by consumers, labor, and managers of state-owned enterprises, the privatization of state-owned enterprises in Thailand is proceeding, but more slowly than once anticipated.

Another aspect of Thailand's policy regime has been its commitment to free trade mechanisms in Asia and its active profile in activities of the World Trade Organization (WTO). However, in spite of some liberalizing initiatives, Thailand's economy is not as open as that of either Singapore or Hong Kong (China). In 1972, perhaps under duress rather than from self-interest in freer trade in services, Thailand entered into an agreement with the United States that enabled U.S. companies to set up wholly owned businesses in Thailand. But the right was not extended to companies from other countries, a fact that is reflected in the structure of the countries' logistics service sector. A number of U.S. logistics service companies took advantage of the opportunity and have contributed to the logistics requirements of the expanding industries in the country. However, Thailand held back the open and aggressive growth of non-U.S. logistics companies that have been leaders in the global industry. This was to the detriment of the logistics service sector in Thailand and contrary to the interests of manufacturers. Yet it was favorable to the concentration of logistics in Singapore and may have contributed to the contrast in Thailand between the sophistication of logistics in the foreign-dominated automobile industry and the lack of sophistication in many other manufacturing sectors. This is an example of the close relationship between broad industrial policy, as well as practices applicable explicitly to logistics services.

Industrial growth in Thailand, like much of East Asia, has been mainly in export-led industries with high levels of sales volatility. The most notable specialization is in the auto industry, which has led Thailand to be referred to as the "Detroit of Asia" (see chapters 4 and 5). This concentration arose partly because Thailand had positive features for the industry and partly because Thailand was not trying to protect a domestic car. The more open policy attracted foreign companies from Japan. More recently, in 1996, General Motors opened a manufacturing plant on "the strength of the domestic vehicle market, proven infrastructure, and well-established supplier base" (General Motors 1996). In 2000 a new assembly center costing $640 million proceeded, modeled on a state-of-the-art General Motors plant in Germany (General Motors 2000). Ford also has production facilities in Thailand. The development of a complex of suppliers and a qualified labor force has also enabled local firms to develop significant export businesses. For example, Aapico has grown from a low-tech maker of custom-designed jigs for the local market into a producer of export-quality parts in both the high-volume and low-volume sectors (Deyo and Doner 1995; "Fast Lane to Success" 2002).

The plants of the major automobile companies in Thailand operate with fully integrated supply chains supported by excellent information technology and software systems and door-to-door control of transportation arrangements (although local suppliers still lag in this regard). As with the hard disk drive industry, efficient logistics (brought by the manufacturers) are essential to effective integration of the supply chain. The automobile industry appears to be the only industry with this level of integration in Thailand. All other export trades operate on free-on-board shipment terms. There is also a wide gap in the level of information technology capability between Thailand, as for most of Asia, and Singapore. A news release on the Bangkok Freight Logistics Workshop held in March of 2001 is reported on the Website of the Department of Transport and Regional Services, Canberra, Australia. The news release notes,

> At the moment through-chain practices in much of SE [Southeast] Asia are underdeveloped, placing serious constraints on supply chain efficiency. Countries in the region also stand to gain significant domestic savings from improved chain integration, strategic planning, and utilisation of infrastructure. This in turn empowers internationally competitive export chains and an ability to participate effectively in the global economy, underpinning economic growth.

Australian organizations have been working to improve the logistics of agricultural products exported to Southeast Asia. Once perishable products arrive, they move under the control of local distributors to various types of retail outlets. Processes are not well integrated or controlled. Consequently, problems of quality arise. Efforts to achieve better-integrated procedures have produced few results, and customs and other entry procedures are slow. Shifts in the structure of the retail industry now taking place may eventually be the source of fuller integration in the supply chain.

Thailand also faces well-known problems of congestion in Bangkok, its main industrial city and its access point. Congestion affects many costs, including logistics. One approach is to invest in infrastructure. Another is to encourage decentralization of new plants away from Bangkok. Not only do new locations avoid the congestion, but they also are closer to large pools of workers in the countryside (Doner and Hershberg 1999). Industrial parks, many of which have much available space, are being developed to encourage decentralization (Mertens 2000). Foreign investors in cars and electronics have located on industrial estates outside Bangkok in areas where new clusters of related economic activities are developing. Seagate Technologies has constructed a new plant in the northeast city of Korat.

Unlike Singapore, Thailand has not had a strategy for logistics services. Policies applicable to the economy in general have applied. Thailand has lacked the cumulative benefits of initiatives that support one another such as an open economy, investments in infrastructure and information technology, education, land development, and trade policies. Some policies, such as restrictions on non-U.S.-owned enterprises, have particularly limited the growth of logistics services.

China

The challenge of dealing adequately with logistics in China is commensurate with the size of the country. Because the country is so large, it is necessary to consider some aspects of the internal transportation system. The attributes of the rail system limit the development of intermodal services and western development, so they are the focus here.

Development of the rail system and intermodal transport. The operation of the rail system under the Ministry of Railways is a huge undertaking and is associated with different problems than those affecting the ports. The large size of the business, its established practices, and its need for major investments in infrastructure make improvements a massive task. In 2000 the Ministry of Railways invested close to half of its revenue in construction (Xigui 2000). The turnaround in railway performance includes an increase in traffic in 2000 of about 5 percent in tons moved and ton-kilometers overall, including a 17 percent increase in international traffic with ports. Initiatives have also been undertaken to accelerate reform of the railway transport management system and to place more emphasis on the quality of freight service. Infrastructure limitations are greatest for services that need speed and on-time delivery. For example, added track capacity is needed to allow fast trains to proceed without frequent delays. New and appropriately designed yards are needed to service these trains quickly. Primary attention is being given here to issues associated with the development of intermodal transport. Unfortunately, in spite of the substantial investments made and improvements achieved, much remains to be done.

Intermodal rail transport with international containers is a relatively recent phenomenon in China in spite of many reviews, policies, and regulations by various state agencies, both singly and collectively. Between 1982 and 1999, ministries and commissions under the State Council issued 80 major regulations on container transport. The first intermodal train service took place in 1994, from Zhengzhou to Hong Kong (China) in collaboration with OOCL. The development of services has continued to be linked with the involvement of shipping lines. For example, in 1996 APL participated in its first rail-based intermodal service in China, with service between Harbin and Dalian. In 2001 there were some 350 container stations that could handle international containers and more than 20 railway container routes. However, the use of these routes in the overall marine container trade is low. In 2000 railroads carried only 330,000 TEUs in international trade, only 3 percent of the international traffic moving through the ports (Dekker 2001). More than 85 percent of port container traffic is handled by trucks, often to and from consolidation facilities. Intermodal operations are hindered generally because enterprises focus on their own activities, including the use of their own forms of waybills and bills of lading. The result is redundant documentation, errors, and delays.

There are a number of other reasons for the low utilization of rail services. First, service is unreliable, a particularly costly attribute for the movement of manufactured goods in international supply chains. Container trains do not have sufficient priority in the scheduling of traffic, and many are delayed during times of congestion. However, the problem is also part of the generally poor rail service. Unfortunately, no data are available regarding on-time performance.[10] However, not only is the arrival record of trains per-

10. In 1997 for agricultural products the rail transit time to Shanghai was 21 days from Chengdu (1,462 miles) and 16 days from Wuhan (960 miles). The Ministry of Railways is now indicating that freight from Wuhan to Shanghai will be available within 14 days.

ceived as poor relative to the cost and service options available by barge and truck, but ancillary services are not consistent with today's logistics management requirements. For example, there is no notification of the arrival of goods at rail yards so the consignee or agent must check frequently. The shipper is "definitely a supplicant even today" (Ralph Huenemann, University of Victoria, personal communication, 2001). There is neither automated car-tracking capability nor integration of information technology, even between the provincial railways. In part, the shortfall in information capability is the result of an overemphasis on investments in rail container infrastructure and insufficient attention to software development and process restructuring. Only a tiny proportion of the investment in container transport goes to the information system. Consequently, information systems are poor and differ among the various agencies. Even if the railways had the will to allow shippers to trace their traffic, it would not be practical to do so.

Inland conditions are generally poor for intermodal operations. The number of terminals is few, their quality is low, loading and unloading facilities are inadequate, and the management skills and organization of technical personnel are poor. These problems inhibit intermodal development in inland areas, with a negative effect on the national strategy of developing the west. Boxes are often used in one direction only, because few imports move inland, especially from the newer ports away from Hong Kong (China). Consequently, boxes for loading tend to be in short supply inland, causing delays in loading. Domestic freight is not supposed to be moved in international containers. Even if it were, there would be little opportunity for balanced loads.

In spite of the service conditions, rates are high, which makes competition from other modes effective. Competition from trucking is significant for a number of reasons. Much of the container traffic is associated with the coastal areas for which the length of haul favors trucking. Shipment quantities are often smaller than container loads. The absence of inland consolidation facilities (in part, because of licensing constraints on forwarders) results in the dominance of truck movements. More than half of road transport is private, and entry is easy, unlike the monopolistic rail market. The barge services on the Pearl and Yangtze rivers are well used, with services offered by state-owned big or mid-size enterprises and some private businesses. The market sets truck rates; the Ministry of Communications sets guidelines on barge rates. Recently, rail rates often have been raised to more closely recapture capital invested and to shed excess demand (Huenemann 2001). Traffic is moved in the limited number of solid container trains at a considerable premium. The Ministry of Railways is responsible for rail rates and generally charges set mileage rates without regard to competitive conditions along the route.

Other public policies also contribute to the dominance of road over rail transport of containers. Highway improvements have been substantial as a result of annual investments of $25 billion under the Ninth Five-Year Plan. Many of the new roads are toll roads in the coastal area. Moreover, strong hierarchical structures in state and local governments often result in decisions favoring narrow industrial or local interests rather than the broader interests that are so important to intermodal systems.

The differences in rates and service conditions among the different modes of transport mean that the greatest challenges to enhancing logistics services are present in inland areas, where improved rail service is needed most. Opening China more to the presence of international logistics companies will help to support consolidation and distribution facilities and to develop information technologies within which improved rail service can operate. However, considerable improvement is needed in a range of rail infrastructure and in processes and priorities.

Constraints on the entry of firms into business. The licensing of firms is designed to control entry into a business. It may be used to control the number of firms in a business and the attributes of those firms. It may also, deliberately or otherwise, cause firms to focus narrowly on a certain type of business consistent with the narrow permission of the license. This may affect the vertical or horizontal structure of an industry. Both characteristics are present in China. The particular aspect of interest here is the use of licensing to control the presence of foreign companies.

China is not alone in controlling the participation of foreign interests in transport and forwarding, but China's restrictions are very severe compared with those of other countries.[11] Nevertheless, China has come a long way in a decade. It was only in 1992 that foreign shipping companies were able to establish their own offices in China (APL and SeaLand were leaders). Now many have done so. However, it will be some time before the benefits of integrated logistics services are available widely in China. The results of interviews undertaken in the research for this chapter indicate that China's licensing system is a major factor. The regulations seriously limit the functions of logistics businesses. They limit the extent of operations of firms that have a presence, which is a serious disadvantage in a country the size of China. Large networks of services are desirable because of the large size of the country and the number of large urban areas.

11. Many countries have restrictions that get less international attention than those of China. For example, the United States requires all domestic shipping firms to be majority owned by American citizens.

The regulations do not just hamper foreign companies from participating in businesses; they also slow substantially the change in current logistics practices in China. While the accession of China to the WTO requires China to open up its logistics industry over three to four years, it is in the interest of China to do it much sooner. Only in 2005 will wholly owned foreign enterprises be allowed throughout China in forwarding and logistics services. They are allowed in free trade zones, such as Waigaoqiao in Shanghai, under local approval. Improved logistics would bring benefits not only to China's international trade but also to its domestic trade, in which the same attitudes, methods, and technologies are applicable.

The regulations promulgated in 1996 by the Ministry of Foreign Trade and Economic Cooperation (renamed the Ministry of Commerce in March 2003) required foreign-funded enterprises seeking to serve as an agent for international cargo transport (including firms, such as OOCL, located in Hong Kong [China]) to do so in a joint venture in which the majority interest is held by an established Chinese company—in effect a state-owned enterprise—with related expertise. The process of obtaining a license is long, complex, and costly. Without the class-A license, it is not possible for a foreign firm to issue a bill of lading or an invoice or to collect payment. The system essentially forces freight forwarders into a minority stake in a joint venture or a fee-for-service relationship with Chinese firms that have a class-A license. Foreign forwarders say that the charges for this service are "excessive," because of the limited competition among the licensees. Danzas was the only forwarder to have entered into a joint venture as of 2001. Other firms operate on the basis of representative offices that legally are for liaison and research activities; they are not allowed to engage in sales or operations. Under the licensing system, the office of a shipping line may not serve as a representative office; separate licenses and operations are required for freight forwarding.

The system has a number of important ramifications. Forwarding and logistics service companies in China have not invested in systems in the way that they would if ownership were allowed. They do not believe that they can make optimal use of capital investments in China when they have only a minority interest in the business. Most operate their business in Hong Kong (China) and contract for consolidation and distribution services in Chinese ports, airports, and inland cities where the volume of traffic warrants. They have representative offices and agents in China. In practice, the law turns a blind eye to a number of practices that make day-to-day business easier than the law would require. However, there is no doubt that the expansion of logistics capabilities in China has been seriously delayed by the absence of foreign firms and the demonstration effect that they would bring.

The expansion of trade and the accession of China to the WTO are encouraging firms to position themselves more aggressively for fuller entry into the Chinese market.[12] This may account for recent initiatives by both APL Logistics and Maersk Logistics. APL Logistics has entered into a memorandum of understanding with the Eastern China Railway Express and the Shenyang Transportation Group in order to extend its supply chain network in inland China and expand the network's information technology capabilities and supply chain tools. The new relationships are seen as a means to allow one-stop shopping domestically and internationally (APL Logistics 2001.) APL Logistics has also been appointed to the Modern Logistics Advisory Board of the State Economic and Trade Commission ("APL Logistics Appointed" 2001). The APL Logistics Asia–Middle East president, Rick Moradian, has said, "We welcome the opportunity to contribute at the policy and governmental relations level of China's logistics industry. Our customers are looking to us to ensure that they receive superior end-to-end supply chain solutions at the global, national, and local levels."

Maersk Logistics has also been expanding its operations in China. It recently opened a national distribution center in the Jiuting economic development zone of Shanghai. The facility is viewed as a key piece in Maersk's emerging China network ("Maersk Opens" 2001). As reported above, the airfreight and courier services also have been affected by licensing laws and the general regime of bilateral airline rights.

The need for foreign companies to expand their logistics capacities is evident in the announcement last year by Chen Shanru, director of the Guangdong Economic and Trade Commission, that the province is searching for new foreign cooperation to increase its logistics capabilities. As a result, the Shanghai-based Sinar Mas Logistics is a new subsidiary of the Sinar Mas Group, an Asian business group with $15 billion in assets. Also, a new organization, the China International Freight Association, was formed in 2000 to represent the growing industry. However, China has a long way to go to translate the wide recognition now given to the word "logistics" into effective practice. Policies and attitudes are an important part of bringing about change.

Soft infrastructure. Important limitations still affect the soft infrastructure of China. In spite of significant progress, much improvement is needed

12. The shifts in policies and the actions of firms since accession to the WTO are not covered in this chapter.

in some institutional practices and work force skills. Unfortunately, doing business in China faces many obstacles.

Customs procedures. Customs procedures are unduly burdensome to international trade. The experience of Chiaphua, a maker of domestic appliances in Hong Kong (China), illustrates the concerns of firms. The firm felt that it kept running up against the great wall of Chinese customs regulations when shipping to China. Regulations require up-front declarations of all materials that companies like Chiaphua plan to import for six months ahead and details of how much actually enters. Before exports are cleared, the company must balance this equation: imported materials must equal exported materials contained in each shipment. Discrepancies mean fines or bans. In the end, Chiaphua designed its own software to keep track of the required information (Reyes 2001). In general, customs and related documentation procedures are still cumbersome and can give rise to significant problems, especially if goods move from one district to another. While the regulations are supposedly national, provincial jurisdictions impose local interpretations. Many goods move from inland points in bond to avoid the need for a representative of the issuer of documentation in the region of origin to be present in the export location. Even with a movement in bond, there is the risk that personnel in different customs districts will insist on reviewing all of the documents, resulting in delays. Finally, imports are sometimes delayed when goods are not moved in bond as arranged from the first point of landing because that office has not achieved its overall revenue quota. Although the delay is not legal, the revenue quota gives an unfortunate incentive to holding cargo in the hope that duties will be paid locally rather than inland.

One means by which buyers seek to ease customs procedures and to avoid some of the difficulties of crossing multiple local jurisdictions is to buy through vendors based in Hong Kong (China). In this way, the issuance of documents maintains the historic institutional dominance of Hong Kong (China) and the concentrated flow of trade. It contributes to the continuance of Hong Kong (China) as the transshipment center for China.

Local governments make the development of national logistics systems difficult by creating barriers to trade (among others described in chapter 6). The nascent trucking-logistics industry faces irritating local barriers to developing long-haul routes. Some provinces and municipalities make it so onerous for outside trucking firms to secure licenses that shipments must be reloaded onto the next jurisdiction's trucks. Thus the problems of vast geography and poor infrastructure are compounded by policies to promote local self-interest. Toby Marion, managing director of LCP Asia, a logistics consultancy firm, says, "Every province is like a different country. Everyone wants you to come to their [sic] door for approvals" (Tanzer 2001).

Local governments imposed tariff and nontariff barriers to keep out one another's products, although these are being dismantled. Such strategies are linked to bottom-up revenue sharing that requires localities to submit only a portion of their revenues to the upper levels and then allows them to retain all, or most, of the remainder (Doner and Hershberg 1999). Shanghai, for instance, imposed huge license fees on competing Citroen cars from Hubei Province to protect the locally made Volkswagen Santana and its own stake in the Volkswagen joint venture. Faced with budget constraints, local administrators look for new sources of economic growth. In view of the state's involvement in the economy, it is probable that local governments wear two hats: one as administrator and one as entrepreneur.

There is a wide perception among those doing business with China that logistics are the flavor of the month but that people do not understand them. This is not surprising given the length of time that managers and writers in North America have struggled with their real meaning! In China it is often seen as simply more trucking, warehousing, or intermodal services. Managers are seen as focusing on operations or as being driven by production within their own business. This denies them the opportunity to realize tradeoffs and integration. Warehouses are being built, but too often the local administrator is designing them as storage rather than as throughput facilities. People's lack of experience with information technology goes along with the problem of getting people to use it. One company facing these challenges is Hunt Corporation, based in Statesville, North Carolina, and a distributor of office supplies and graphics products. Hunt uses faxes and telephone calls to process transactions with some of its suppliers in Asia, a process that slows the movement of goods through the company's supply chain (Songini 2001). Getting companies to switch to electronic messaging is slow for smaller companies. As in Thailand, major international companies can have more success in imposing logistics standards and methods. This is particularly true in the more competitive coastal cities, especially in the new environment in Shanghai.

Further views of buyers on logistics in China. The difficulty of managing logistics in China is reflected in the almost universal practice among foreign firms of buying on free-on-board terms, with this interpreted in the trade as goods being delivered to the shipping line, air carrier, or consolidator. Some buyers who use the terms of ex-works in other trades to ensure the level and cost of their choice find doing so impractical in China. Forwarders or other logistics companies with whom they regularly do business do not have their own services inland. Furthermore, local problems arise frequently, and local firms are best equipped to handle them. As a consequence, logistics in China remains rather a black box, with the best that firms can do being to monitor the progress of purchase orders, generally with less difficulty than used to be the case.

Buyers comment that they have particular difficulty getting practices consistent with the logistics requirements of new manufacturers away from the coastal areas. "Forcing" manufacturing inland through the use of quota allocations is not likely to engender more market-responsive attitudes. Buyers attribute the lack of responsiveness variously to the absence of an awareness of competitive market requirements and the lack of appreciation of logistics. Examples are an unwillingness to reduce run sizes of a product and excessively long order cycles. Of course, both may be caused by high costs of change. Long cycle times could well be the result of poor inbound logistics to plants inland in China.

Credibility is given to the difficulty of dealing with China by the consistency of this view and the contrasting positive comments expressed about Hong Kong (China) and also about Shanghai. The commitment of the government of Shanghai to be successful as a financial, trade, and logistics center is reflected in the community at large. In its Five-Year Plan for 2001–05, Shanghai expects to create a world-class logistics hub with the Waigaoqiao free trade zone as the core. Shanghai has invested and is investing substantially in three platforms: transport, information, and policy. The APEC Shanghai Model Port Project will introduce a computer-based customs clearance system that will have a direct impact on the movement of goods into and out of Shanghai. A 24-hour customs clearance center is planned for the airport. Other plans include a better-educated work force capable of meeting international business standards quickly (McDaniels and Zhao 2001). The strategy in Shanghai is comparable to the programs of Singapore and Hong Kong, China.

CONCLUSIONS

The conclusions of this chapter are presented in three parts. First, it is appropriate to highlight the importance of global trends in the competitiveness of industries that have implications for logistics services and the public policies that affect logistics. Second, it is essential to review the importance of a multifaceted approach—the five-pillar approach—in public policy to shift the logistics capability of a country or region. Finally, the particular conditions in Singapore, Thailand, and China are reviewed. The contrast among the countries demonstrates the importance of multifaceted but focused public policy initiatives.

The Characteristics of Logistics and Implications for Public Policies

Economic trends are bringing new forces to bear on national economies and are transforming structural relationships among businesses. This is true within an advanced economy such as the United States (Berger International 1999) and in the developing economies of East Asia. The effects are evident in most industries. Examples are the hard disk drive, computer, apparel, and automobile industries, which illustrate several features of the competition among regions to manufacture and deliver goods competitively.

Competition among businesses is more global and dynamic than formerly. Firms face competition as members of supply chains in which the integration achieved through logistics services has strategic importance. The integration in supply chains results in new and closer relationships among manufacturing, assembly, and logistics operations. In particular, the amount and reliability of time required have become more important. The heightened strategic value of information and logistics technologies is reflected in the growth of specialized logistics service companies that are playing a growing role in the design and management of logistics activities. The performance of these companies and of logistics services generally is recognized in theory and practice as important to economic growth.

As a consequence of these conditions, understanding the competitiveness of a firm in a location requires considering how well attuned all processes are to supply chain requirements. Competitiveness cannot be viewed as a function just of production or of logistics. Competitiveness must also be focused on the particular needs of customers, whoever and wherever they may be.

The needs and expectations of customers are complex and vital to understanding sourcing decisions. It is important to understand the policies of individual firms in a supply chain. For example, some buyers leave the search for suppliers in China in the hands of vendors based in Hong Kong (China), while others are involved more directly with suppliers. Different policies among buyers concerning risk management affect decisions on the number of vendors used in total and in particular countries or regions, the duration of relationships with vendors, and the availability of alternate transportation services. Some firms have a stronger preference than others for controlling the supply chain from factory of origin rather than just from port of export. Firms give different levels of responsibility to third-party logistics providers.

These conditions have various implications for policies to encourage industrial development. First, logistics are vital but only one important component of an economic development strategy. Second, efficient logistics are about more than physical assets. Safe skills, processes, and institutional matters affecting logistics are important. Therefore, a multifaceted strategy is needed for the development of efficient logistics services.

Case for the Five-Pillar Approach to Logistics

The five-pillar approach is consistent with evidence of conditions in East Asia. It is consistent with the practices of successful gateways such as Singapore, Hong Kong (China), and the Netherlands. It is consistent with the expectations revealed in interviews with shippers and service providers. And, last but not least, it is intuitively logical. The five pillars are the provision of efficient infrastructure and services in transportation; the provision of efficient infrastructure and services in communications; the passage of appropriate public policies and institutional practices; the availability of effective private sector logistical organizations; and the existence of a work force knowledgeable of logistics skills and concepts. The cumulative benefits of improvements concurrently in each of these dimensions are more than additive. Change and innovation in logistics and in the affected manufacturing and marketing activities will enable multiplicative gains elsewhere. However, program initiatives need to focus on the individual elements, as in the following discussion.

Transport infrastructure and services. The presence of transportation able to provide efficient services is an essential condition for efficient logistics. The pattern of private investment in ports shows that, where actual or latent demand is evident, facilities and services at hubs can be provided with the support of private capital. The amounts and performance of the investments are influenced by policies affecting them, for example, the conditions of concession agreements. Policies concerning concessions are important. The nature and effects of contracts in Chinese ports are one example. The concession in Manila is another.

However, there is a danger of placing too much emphasis on the provision of infrastructure or services without regard to the logistics requirements of shippers. For example, investment in physical infrastructure by the Ministry of Railways in China, although urgently needed, may be too focused to the detriment of investments in information technology. Similarly, shippers suggest that investments in warehouses by public enterprises in China are often of traditional design and ill suited to the need for modern logistics. Also, excessive focus on a particular mode or technology can give rise to problems at their interface with other modes. Interfaces between modes of transport are common sources of problems, evident in congestion about ports and delays in intermodal transport.

Particular capacity and service problems exist in rail service. They appear to be harder to deal with than problems in other modes, a phenomenon evident at various times in many other countries as well. The difficulties of handling many commodities over a common track and the size and inertia of railway organizations are major problems. The problems lie, in part, in the need for capital investment, but they also lie in the need to establish the right structure of incentives for management and employees to improve productivity and be driven by customer service.

The provision of adequate transport infrastructure and services is the outcome of direct investment and of various policy measures. It is important to recognize the role of policies as well as investment in the development and provision of services. The most striking examples in airfreight and freight forwarding are discussed below.

Communications infrastructure and services. The provision of efficient infrastructure and services in communications is an increasingly important element in the design and management of supply chains. Its value goes well beyond logistics. Communications affect both the grand design of supply chains and the specific location of facilities. They affect the grand design of supply chains because the transmission of information can affect the amount and location of production and of inventories in supply chains. They also affect the choice of site because buildings have to be wired (Kanwerayotin 2001). A frequent example of the role of information in logistics performance is its effect on the potential for customs to provide fast and reliable clearance of goods. Information systems are at the heart of efficient logistics services that provide visibility for information about sales and purchase orders. Such information is essential to the long-run design and operation of lean manufacturing and retailing systems and to the short-run adjustment of those systems to unexpected events.

Public policies and institutional practices. Appropriate public policies are vital to the development of conditions conducive to good supply chain structures and practices. There are several reasons for this. First, infrastructure investment needs to be attended to. In some cases, this requires public investment, for example, in roads. In other cases, it requires appropriate regulatory policies affecting ownership and competitive conditions in an industry, as in telecommunications. Second, regulatory policies affecting the entry and structure of firms in the transport and logistics industries are crucial. Costly restrictions are associated with the structure of freight (and passenger) airline routes resulting from bilateral agreements concerning air traffic rights. Fortunately, no equivalent regime exists in shipping. The bilateral air regime is a major public policy impediment to the efficient provision of transport and the development of logistics in the region. The restrictions on the effective participation of logistics service companies in China are dealt with more fully later. Third, the policies of some public agencies directly affect logistics performance. This is most obvious in customs, which should be seen as a logistics service with the purpose of facilitating efficient and safe trade. Efficient customs procedures are a competitive advantage, as is evident in the case of Singapore and the Netherlands. The use of electronic clearance of cargo is a case where the whole concept of customs processes can be redesigned to the benefit of better logistics and safer trade (Heaver 1992). Fourth, public policies are important in guiding resources broadly, for example, to education to ensure a knowledgeable and skilled work force. Finally, public policy is one source of leadership and vision that facilitates development of momentum toward a goal whether the progress involves much public presence, as in Singapore, or little public presence, as in Hong Kong (China). It may be tangible, as in the form of industrial estates and free trade zones, or intangible, as in the promotion of gateway concepts and capabilities. It is easier to achieve such leadership in a small and more homogeneous area than in a large state. However, fragmentation of jurisdictions without appropriate leadership or rules may lead to counterproductive rivalry, a danger in China.

Private sector logistics service providers. An important contributor to the improvement of international logistics performance over the last 15 years has been the growth of the international logistics service industry. The growth of the industry has coincided with the expansion of outsourcing and the evolution of distributed manufacturing. Most prominent in the industry are the integrated global players that have grown from two main roots: the freight-forwarding industry (for example, Schenker) and the transportation industry (for example, Maersk Logistics and UPS Logistics). However, the industry is very dynamic, with national providers going international and with new, substantial national firms developing. The industry is also diversified in the range of services provided. Some firms specialize in a service such as the development of software for supply chain management or the operation of ocean shipping (for example, CP Ships).

The integrated logistics service firms are a vital part of managing the processes in logistics. Their presence affects the attractiveness and efficiency of locations for industry. The severe restrictions placed on forwarding in China are an impediment to the improvement of logistics services in that country. Not only do the companies have direct effects through the services they provide, they also have indirect effects through their example and their development of a knowledgeable work force. Knowledgeable workers end up with a variety of firms, and some will start their own businesses.

A work force with logistics skills and concepts. A common observation is that logistics in Asia are the "flavor of the month" but that the requirements for better logistics performance are poorly understood. This is particularly true of China, but not of Singapore. Too often, too much reliance is placed on the development of facilities, such as warehouses and truck depots, and of large infrastructure projects, such as ports, in which local governments can invest. Too little attention is given to integrating logistics activities within the context of supply chain management. A lack of appreciation is evident particularly among managers located away from the coastal areas and with limited exposure to the needs of international trade. Often, these managers focus on their own operations and use traditional approaches that do not reflect today's needs. Respondents give examples of difficulties in getting reliable information about production schedules and delivery times, lack of flexibility in scheduling, and inability to get faster services. In general, it is too difficult to improve relationships across businesses, which are needed in closer supply chain relationships. The situation is fostered by the separation of businesses under China's business licensing system.

More generally, it is easier to do business with firms in Singapore or through Hong Kong (China) than directly with firms in China for various reasons. These include language difficulties, limited exposure to international trade, and limited use of information technology.

The patterns of industrial development in East Asia exemplify the importance of a skilled work force. The development of pools of skilled labor is not just for employees working in a particular type of job or firm. Pools of labor have the effect of developing a heightened awareness of the needs and conditions in an industry of value to suppliers and customers. As manufacturing becomes structured in more complex supply chains, understanding a particular business activity from the perspective of another becomes more important. The accumulation of skills in a work force has contributed to the growth of clusters in many industries, including electronics and the automobile industry in East Asia. Labor force skills and attitudes, and institutional cultures consistent with them, are appropriately viewed as a part of the soft infrastructure of a community that significantly affects its logistics performance.

The evolution of a community with these skills and attitudes is hard to achieve. As such, communities with these attributes enjoy a competitive advantage for some time. The conditions cannot be created quickly by investments in physical assets. They are created by a sustained, multifaceted approach over time.

Benefits of the five-pillar approach to logistics. For communities to have the capacity and reliable performance expected of logistics systems today requires excellence in each of the pillars identified. The presence of all pillars is necessary to realize the potential of location, resources, and human assets in economic development. The range of attributes is wider, and their level of sophistication is greater than in prior decades. This is evident in the level of information technology required, the participation of new logistics firms, and the need for a supply chain perspective rather than an individual-enterprise perspective.

These conditions make it harder for new centers to develop excellent logistics. They have to put more attributes in place than formerly, and they often need to do so without the benefits of economies of scale, which contribute to the efficient use of infrastructure, service frequency, and an experienced work force.

Logistics Challenges Specific to Singapore, Thailand, and China

The particular requirements for and economic benefits of improvements in logistics are location specific. They reflect the range of physical, social, political, and economic conditions in a region and country. However, the core ingredients in the multifaceted approach need to be present. They may be delivered in various ways consistent with the time and place.

Examination of conditions in three Asian countries provides an opportunity to explore the conditions required for modern logistics services and the particular challenges in these countries. The challenges of the three countries differ greatly because of differences in their geography and history. However, examination of the conditions in the countries provides useful insights into logistics needs generally.

Singapore. Singapore is widely recognized as having excellent logistics services. It has benefited from deliberate programs to build on the state's inherent geographic advantages to achieve a leading role in logistics services. Its success is a pointer for all countries in the required attributes of infrastructure and services. How they are achieved can be expected to vary from one country to another.

The success of Singapore, with its limited land and population base, is causing costs to increase, leading to spillover of activities elsewhere. Spillover to other places is not necessarily bad for Singapore, as long as it is able to provide the knowledge-based activities required of a central hub. This process has been taking place in manufacturing. It is only now evident in transportation and logistics. Building on its reputation as an "Intelligent Island," Singapore is seeking to maintain a high productivity for key services rather than to offset the community's high costs. Singapore is expected to retain its position as a major regional logistics center.

Thailand. The combination of political stability, low costs, and a reasonably open economy enabled Thailand to enjoy very rapid growth in the decade after 1985. The rate of growth based on the existing attributes of the country may have slowed the implementation of programs that would have better positioned the country to realize the benefits of further economic growth. The absence of a multifaceted program to advance capabilities has resulted in the limitations evident in logistics activities today.

Thailand needs to advance its logistics capabilities in a number of ways. It needs to invest in infrastructure and improve the knowledge base of the work force in information technology and logistics. Advances in the application of logistics can be accelerated by opening the country more fully to international logistics service companies. Achieving greater efficiency in logistics also requires changes in public services—for example, in customs and in the economy generally—and in public policies—for example, by promoting competition. Finally, although the size of the country sets the scale of infrastructure developments lower than in China, investments to make specific infrastructure more accessible are necessary and must address the congestion associated with the growth of Bangkok.

China. The challenges facing China in logistics development are much greater than those facing Singapore and Thailand by virtue of the country's size and restrictions on doing business. Significant initiatives are needed in each of the five pillars.

The size of the country and economy affects not only the scale of investment in infrastructure but also its allocation. In particular, China needs greater intermodal rail capability. The improvement of intermodal services calls not only for making continued investment but also for dealing with important policy matters. These include the level of priority given to international container traffic; the use of international containers in domestic trade; the management structure of the railway to achieve better integration among the operating units; an incentive structure to encourage greater productivity and better customer service; and a market-sensitive pricing system for all modes of transport.

The rail services and the economy in general need to emphasize information technologies as a means to change the design of systems and to speed up processes. Investment in information technology needs to be supported by programs to accelerate its use. Like intermodal rail, the benefits of information technology in overcoming distance can be greatest in central and western China.

Accession to the WTO is the symbol of new economic policies in China. It requires or is consistent with a number of policies to facilitate trade through improved logistics. First, extending the freedoms of air cargo and courier services would be very beneficial. Second, relaxing the entry conditions for foreign-owned freight forwarders and other logistics service companies would increase the capacity and improve the quality of logistics services. The Chinese economy is a major market opportunity for such firms, but public policy severely limits their participation and effectiveness. Third, fewer restrictions on the lines of business of carriers and logistics service providers, regardless of the owner, are desirable to enable greater integration across logistics services. Finally, improvements in government services that are a part of logistics—for example, customs—are needed. A demonstration project is the current initiative in Shanghai.

China's policy environment is complicated by the powers of the regional and local governments. It is desirable to reduce internal barriers to trade and certainly to resist their escalation if the full benefits of better logistics services and freer trade are to be realized.[13]

The transition of thinking involved in moving from a planned economy to an open market economy is profound. It is no wonder that buyers find producers less responsive to market needs than they desire. The implication is that significant education requirements exist so that there is full

13. In Canada internal barriers to trade persisted among the provinces well after the North American Free Trade Agreement went into effect with Mexico and the United States. However, they did not get worse and have now been largely eliminated.

and common understanding about business concepts, including supply chain management and logistics. A potential, largely unexplored, source of change in the thinking about logistics and in the logistics industry in China is the changing structure of retail industry ("A Hyper Market" 2001). This will have ramifications for the structure of supply chains.

It is appropriate to conclude with some observations about improved logistics and regional economic development in the context of central and western China. Unlike Thailand, the distances from the coastal to the inland regions are associated with high enough transfer costs to create obstacles to development.

The reduction of logistics costs through improved transport services, especially intermodal services, and better information technology services should be of greatest benefit to locations with long average lengths of haul. The benefits would come in the form of lower costs, less time, and more reliability. However, there is an important qualifier to this: the volume of traffic by rail significantly affects the frequency and costs of service. The same issue affects the strategic benefits associated with critical mass to get economies of scale in production and logistics. Within Central China, large cities provide a scale in which the development of new national and international businesses is latent. Improved logistics would likely trigger an effective growth in volume. The potential to reduce the time to move goods by rail from a city such as Wuhan is substantial when the transit time for general intermodal service from Wuhan to Shanghai is "within 14 days." Further west, however, where economic activities are fewer and communities smaller, the potential for reducing the costs of distance is less clear. Improvements in processes at intermediate hubs are particularly important to the traffic of these communities. Any transition will be gradual. Particular attention should be given to the drive of first movers to meet the varied needs of customers so that interest in sourcing in the west is gained, not lost.

REFERENCES

The word *processed* describes informally reproduced works that may not be commonly available through libraries.

Abernathy, Frederic H., John T. Dunlop, Jannice H. Hammond, and David Weil. 2001. "Globalization in the Apparel and Textile Industries: What Is New and What Is Not." Harvard University, Center for Textile and Apparel Research. Cambridge, MA. Processed.

APEC (Asia-Pacific Economic Cooperation). 1997. "APEC Transportation Working Group Congestion Points Study (Sea and Air). Asia-Pacific Economic Cooperation Group, Final, Draft, and Technical Reports February 1995 to February 1997." Processed.

APL Logistics. 2001. "Agreements Designed to Extend Supply Chain Reach and Enhance Inland China Capabilities." *PRNewswire*, June 28. Processed.

"APL Logistics Appointed to China's SETC Board." 2001. *American Shipper*, November 6. Available at ShippersNewsWire@AmericanShipper.com. Processed.

Armstrong and Associates. 2001. "3PL/Contract Logistics Market." Available at http://www.3plogistics.com/logmkt.htm. Processed.

"Asian Ports to Triple Box Volumes by 2015." 2001. *American Shipper*, June 19. Available at ShippersNewsWire@AmericanShipper.com. Processed.

"Atofina Profile: Competitive Advantage in Singapore." 2001. *Chemical Week*, May 23, pp. 20–22. Processed.

Bangkok Freight Logistics Workshop. 2001. News release, March 21–23. Available at www.dotrs.gov.au/logistics. Processed.

Berger International. 1999. "Economic Trends and Multimodal Transportation Requirements." Report 421. Transportation Research Board, National Cooperative Highway Research Program, Washington D.C. Processed.

Biederman, David. 2000. " 'Intelligent Island' Links the Globe." *Traffic World*, December 4, pp. 28–31.

Chanda, Nayan. 2001. "Outside the Box." *Far East Economic Review*, February 15, p. 35.

Chin, T. H., and Jose L. Tongzon. 2001. "Transportation Infrastructure Management for Attracting Global and Regional Distribution Centres in Singapore." KOTI-NEAEF (Northeast Asia Economic Forum) Conference on Developing Regional Logistics Centers in Northeast Asia, Honolulu, Hawaii, August. Processed.

"China Plans Development of Coastal Hubs, Waterways." 2001. *American Shipper*, June 21. Available at ShippersNewsWire@AmericanShipper.com. Processed.

"Companies Vie for HK Logistics Center." 2000. *Air Transport World, ATW's Airport Equipment and Technology* 3(2, Summer):28–29.

Consilium International. 2001. "Port Privatization: Single Terminal Operator versus Two Terminal Operator. Study of Manila North Harbour." Calgary, Montreal, and Toronto. Processed.

Cottrill, Ken. 2001. "Speed Merchants." *Traffic World*, August 13, pp. 24–25.

Danzas. 2001. "Danzas Joint Venture with the Philippines: Royal Cargo Corporation." News release, October 2. Processed.

Dekker, Neil. 2001. "Carving a Niche." *Containerisation International* (November):91–93.

Delaney, Robert V., and Rosalyn Wilson. 2001. "Twelfth Annual State of Logistics Report." Available at Cassinfo.com/bob.html. Processed.

Dempsey, John R. 2000. "Thailand's Privatization of State-Owned Enterprises during the Economic Downturn." *Law and Policy in International Business* 31(2):373–402.

Deyo, Frederic C., and Richard F. Doner. 1995. "Networks and Technology Development: A Case Study of a Thai Autoparts Manufacturer." *Journal of Asian Business* 11(1):125–39.

"Distribution Hub for French Industrial Group LEGRAND." 2001. *Schenker Singapore News*. Available at http://www.schenker.com.sg/news. Processed.

Doner, Richard F., and Eric Hershberg. 1999. "Flexible Production and Political Decentralization in the Developing World: Elective Affinities in the Pursuit of Competitiveness." *Studies in Comparative International Development* 34(1):45–82.

Einhorn, Bruce. 2001. "Hong Kong Can't Keep Its Skies Closed Forever." *Business Week*, international on-line ed., August 6.

"Fast Lane to Success." 2002. *Far Eastern Economic Review*, September 12.

Feng, Cheng-Min, Chia-Juch Chang, and Kai-Chieh Chia. 2001. "A Survey of Supply Chain Adjustment for Taiwanese Information Technology Firms." National Chiao Tung University, Taiwan (China). Processed.

Forrester, Jay W. 1958. "Industrial Dynamics: A Major Breakthrough for Decision Makers." *Harvard Business Review* 36(4):37–66.

General Motors. 1996. "General Motors Selects Thailand for Manufacturing Plant." News release, May 31. Processed.

———. 2000. "General Motors Inaugurates Thailand Assembly Center." News release, August 3. Processed.

Gooley, Toby B. 2001. "Asian Ports Fight to Keep Ahead of the Game." *Logistics Management and Distribution* (August):55–61.

"Guangdong Seeks Foreign Logistics Partners." 2001. *SML News Focus*. Available at Avwww.sml.com.cn. Processed.

Heaver, Trevor D. 1992. "The Role of Customs Administration in the Structure and Efficiency of International Logistics: An International Comparison." *International Journal of Logistics Management* 3(1):63–72.

———. 2002. "Supply Chain and Logistics Management: Implications for Liner Shipping." In Costas Grammenos, ed., *Maritime Economics and Business*, pp. 375–96. London: Lloyds of London Press.

HLB Decision Economics. 2001. "Relationship between E-business, Advanced Transportation Logistics, and Canadian Industrial Economic Performance." Report T8080-00-1335, submitted to the Canada Transportation Act Review Panel, Ottawa, Ontario. Processed.

HLB Decision Economics, ICF Consulting, and Louis Berger Group. 2002. "Benefit-Cost Analysis of Highway Improvements in Relation to Freight Transportation: Microeconomic Framework." White paper. Final report submitted to the Federal Highway Administration. Processed.

Hong Kong Port and Maritime Board. 2001. "Study to Strengthen Hong Kong's Role as the Preferred International and Regional Transportation and Logistics Hub." Executive summary. Hong Kong. Processed.

Huenemann, Ralph W. 2001. "Are China's Recent Transport Statistics Plausible?" *China Review* 12(4):368–72.

Hummels, David. 1999. "Have International Transportation Costs Declined?" Unpublished mss. University of Chicago. Chicago, IL. Processed.

———. 2001. "Time as a Trade Barrier." Unpublished mss. Purdue University, Department of Economics. West Lafayette, IN. Processed.

"A Hyper Market." 2001. *The Economist*, April 7, pp. 68–69.

ICF Consulting and HLD Decision Economics. 2002. "Economic Effects of Transportation: The Freight Story." A report for the Federal Highway Administration. Processed.

Johnson, Eric. 2001. "Learning from Toys: Lessons in Managing Supply Chain Risk from the Toy Industry." *California Management Review* 43(3):106–24.

Jongsuwanrak, Wanida, Samer Prasad, and Sunil Babbar. 2001. "Inventory Systems of Foreign Companies in a Newly Industrialized Country." *Multinational Business Review* 9(2, Fall):47–56.

Kanwerayotin, Supapohn. 2001. "Buildings Feel the Heat of Hi-Tech." *Far East Economic Review*, June 21, p. 64.

Kittiprapas, Sauwalak, and Philip McCann. 1999. "Regional Development in Thailand: Some Observations from the Thai Automotive Industry." *ASEAN Economic Bulletin* 16(2):190–207.

Lakshmanan T. R., and W. P. Anderson. 2002. "Transportation Infrastructure, Freight Services Sector, and Economic Growth: A Synopsis." Federal Highway Administration. Available at http://ops.fhwa.dot.gov/freight/econben/Summary%20report%20 Lakshmanan.doc. Processed.

Levy, David. 1995. "International Sourcing and Supply Chain Stability." *Journal of International Business Studies* 26(2):343–60.

Lewis, David. 1991. *Primer on Transportation Productivity and Economic Development.* Report 342. Washington D.C.: Transportation Research Board, National Cooperative Highway Research Program.

"Maersk Opens National Distribution Center in Shanghai." 2001. *American Shipper*, June 19. Available at ShippersNewsWire@AmericanShipper.com. Processed.

McCann, Philip. 1993. "The Logistics-Cost Location-Production Problem." *Journal of Regional Science* 33(4):503–16.

———. 1998. *The Economics of Industrial Location: A Logistics-Costs Approach*. Berlin: Springer.

McDaniels, Iain, and Sophie Zhao. 2001. "Shanghai Snapshot." *China Business Review* (September–October):42–45.

McKendrick, David G., Richard F. Doner, and Stephan Haggard. 2000. *From Silicon Valley to Singapore.* Palo Alto, Calif.: Stanford University Press.

Mentzer, John T., and others. 2001. "Defining Supply Chain Management." *Journal of Business Logistics* 22(2):1–24.

Mertens, Brian. 2000. "SMEs Find Good Deals in Thai Estates." *Asian Business* 36(9):23–24.

Mohring, Herbert, and Harold F. Williamson Jr. 1969. "Scale and Industrial Reorganization Economies of Transport Improvements." *Journal of Transport Economics and Policy* 3(3):251–71.

Morton, Roger. 2001. "Services and Software Turning to the East." *Transportation and Distribution* (April):35–38.

Nemoto, Toshinori, and Hironao Kawashima, eds. 2000. *Logistics Integration in the Asia-Pacific Region.* Report prepared by the OECD TRILOG Asia-Pacific Task Force. Tokyo: Institute of Highway Economics. Processed.

Panalpina. 2001. "Supply Chain Management for IBM Asia." *Panalpina News Release* 22(October).

Parker, John. 2000. "Samsung Signs UPS." *Traffic World* 264(10, December 4):20.

Porter, Michael E. 1985. *Competitive Advantage: Creating and Sustaining Superior Performance.* New York: Free Press.

"Ports in a Storm." 2001. *Economist*, April 14, p. 57.

PSA Corporation. 2001. "PSA Steps up Technology Presence in China." *PSA News*, September 26. Processed.

QAD. 1998. "Ford Thailand Shifts Asian Operations into High Gear with QAD's ERP Solution." News release, September 24. Available at www.qad.com. Processed.

Reyes, Renato. 2001. "Business-to-Bottleneck." *Far East Economic Review*, July 12, p. 36.

"Schneider Logistics, Hellmann Worldwide Create Alliance." 2001. *American Shipper*, October 23. Available at ShippersNewsWire@AmericanShipper.com. Processed.

SembCorp Logistics. 2000. "Proposed Acquisition of 20 Percent Interest in the Enlarged Issued Share Capital of Kuehne & Nagel International AG (KNI)." Announcements, November 27. Available at http://www.Semblog.com/home_frame.htm. Processed.

"Singapore, Economic Development Board." 2001. *Chemical Week*, September 26, pp. 118–19.

Songini, Marc L. 2001. "Global Supply Chains Rife with Challenges." *Computerworld*, March 12, pp. 1, 16.

Sowinski, Lara. 1999. "The U.S.-Asia Trade Lane: Moving Goods by Sea and Air." *World Trade* 12(10):54–56.

Tam, C. M. 1999. "Build-Operate-Transfer Model for Infrastructure Developments in Asia: Reasons for Successes and Failures." *International Journal of Project Management* 17 (6, December):377–82.

Tanzer, Andrew. 2001. "Chinese Walls." *Forbes*, November 12, pp. 74–75.

Toh, Rex S., Sock-Yong Phang, and Habibullah Khan. 1995. "The Port Multipliers in Singapore: Impact on Income, Output, and Employment." *Journal of Asian Business* 11(1):1–9.

Weisbrod, Glen, and Frederick Treyz. 1998. "Productivity and Accessibility: Bridging Project-Specific and Macroeconomic Analyses of Transportation Investments." *Journal of Transportation and Statistics* 1(2):65–79.

Weisbrod, Glen, Donal Vary, and George Treyz. 2001. *The Economic Implications of Congestion.* Report 463. Washington, D.C.: Transportation Research Board, National Cooperative Highway Research Program.

Wong, Joon San. 2001. "IT Bridge System Boon to Asia-Focused Merchants." *South China Morning Post*, special supplement, May 18.

Wood, Andrew. 2001. "GE, Shin-Etsu Pick Thailand for Siloxanes Project." *Chemical Week*, February 14, p. 21.

World Bank. 2001a. "National Railway Project, China." Project appraisal document. World Bank Report 23317. Washington, D.C., December 20. Processed.

———. 2001b. "Road Maintenance Project, Laos." Project appraisal document. World Bank Report 21773. Washington, D.C., January 31. Processed.

———. 2001c. "Third Inland Waterway Project, China." Project appraisal document. World Bank Report 21909. Washington, D.C., May, 24. Processed.

———. 2002. "Third Xinjiang Highway Project." Project appraisal document. World Bank Report 24645. Washington, D.C., August 12. Processed.

Xigui, Ren. 2000. "Development of Chinese Railways in the 'Ninth Five-Year Plan' and Working Considerations for the 'Tenth Five-Year Plan' Period." Ministry of Railways, December 15. Available at http://www.china.org.cn/e-news/news12-15.htm. Processed.

Zhang, Anming. 2001. "Identifying Barriers to Trade in Air Services: The Case of China." KOTI-NEAEF (Northeast Asia Economic Forum) conference on developing regional logistics centers in Northeast Asia, Honolulu, Hawaii, August. Processed.

CHAPTER 8

TECHNOLOGY AND INNOVATION IN DEVELOPING EAST ASIA

Hal Hill

This chapter examines the processes of technological change and innovation in selected East Asian developing economies whose importance is increasing for the reasons set forth in the earlier chapters. It aims to survey and synthesize a very large literature on a diverse set of economies and institutional contexts. The topic is necessarily broad and complex and impinges on practically all major facets of economic analysis. Perhaps more than any major area of public policy, technology and innovation require an understanding of the interaction between long-term economic development, the internationalization of an economy, government intervention, and the microeconomics of firm-level innovation. These constitute the major themes studied in this chapter.

There are also practical reasons to study the topic. In the more advanced East Asian economies, at least 1 percent of GDP is being spent on research and development. In the Republic of Korea the figure is close to 3 percent. Government supplies most of these funds, either directly or indirectly, and a substantial proportion entails direct government implementation. Poorer countries in the region have aspirations to support funding on such a scale within a decade or so. In addition, the effective reach of related technology and innovation programs is greater still.

Among the questions to be addressed, with reference to the East Asian experience,[1] are the following: How do we define technology, innovation,

For helpful comments on earlier drafts, the author wishes to thank Shahid Yusuf, Lewis Branscomb, Mark Dodgson, and participants at the conference on East Asia's Future Economy, Harvard University, October 1–2, 2001.

1. The primary focus in this chapter is on the four Asian newly industrialized economies—Hong Kong (China), Republic of Korea, Singapore, and Taiwan (China)—and the ASEAN-4—Indonesia, Malaysia, Philippines, and Thailand, members of the Association of South East Asian Nations (ASEAN). These correspond to the original "miracle" economies, in addition to the Philippines.

and research and development (R&D)? What sorts of comparative quantitative proxies are available to measure these concepts? What are the mechanisms and magnitudes of cross-border flows, including especially the role of multinational corporations? What has been the approach to technology and innovation in East Asia, and is there an East Asian "model"? How does technology and innovation policy interact with a broader array of economic policy and institutional settings? Finally, is there an analytical case for government intervention in technology markets, and, if there is, what sorts of interventions are likely to be most effective?

There is no universally accepted definition of technology and innovation. The most common approaches define technology as "a collection of physical processes that transform inputs into outputs and knowledge and skills that structure the activities involved in carrying out these transformations" (Kim 1997, p. 4). Technological progress may be regarded as a better way of doing things or as a means of producing more from less, by employing new technologies and generating new products and processes. At every point, one encounters blurred definitional boundaries among technology, innovation, and science policies. Dodgson (2000, pp. 230–31), for example, defines science policy to be primarily investments in universities and laboratories, whereas technology policy focuses on the development of generic technologies, such as information technology and biotechnology. Innovation policy, in this schema, involves new products, processes, and services. Kim and Nelson (2000a) point out that the distinction between formal innovation and creative imitation is often a fine one.

There is a tendency to equate technology and innovation with path-breaking scientific discovery and research at the frontiers of knowledge. In reality, much R&D involves not the "R" of fundamental research, but rather the "D" of acquiring, adapting, and modifying frontier technology. Rosenberg (1994, p. 13), for example, cites estimates for the United States that just 8 percent of R&D expenditure is devoted to basic research, compared with 25 and 67 percent for applied research and development activities, respectively. For economies well within the frontiers of technology, the proportion devoted to basic or applied research is obviously much lower. In these cases, development and diffusion are much more important than basic research, and public policy should be directed to these areas.

A key strand in virtually every major study of technology is that the process of acquiring and mastering technical know-how is evolutionary, gradual, and long term. Governments can expedite this process in ways discussed below, but there are no simple shortcuts. This may seem so obvious as to hardly require emphasis. But since these tenets are at odds with much official thinking and popular discussion in East Asia, and with many ambitious high-tech projects, they deserve restatement. Moreover, all detailed studies of technological change emphasize that it is a slow, laborious, and uncertain process (Lall 2000a). Mistakes frequently occur. Even seemingly simple cases of reverse engineering are in reality complex and time consuming. Economists' notions of infant industries may involve more protracted periods of endeavor than is commonly supposed. These two features—uncertainty and long gestation periods—immediately underline the links between innovation and a country's political and institutional structures. That is, potential innovators need a nurturing environment, in which, at the very least, property rights, financial institutions, and the political system are supportive.

The range of policies that impinge on technology and innovation is much broader than is commonly supposed. Indeed, a government's officially enunciated technology policy may well be a relatively unimportant part of a country's broader technological effort. Economies may invest significant resources in technology, but if the economic policy environment is distorted, the payoff may be minimal. The former Soviet Union is an obvious example of this proposition. By contrast, small, nimble, highly open economies like Hong Kong (China) may invest very little in formal R&D programs but still achieve rapid growth based on high-quality service delivery.

A central argument in this chapter is that there is no single East Asian model of technology and innovation (Pack 2000). East Asian economies vary significantly in their levels of development, resource endowments, historically embedded openness, and institutional quality. For lower-income countries, like Indonesia and the Philippines, the principal task of government is to provide an enabling commercial environment and to supply crucial, growth-enhancing public goods. Thus macroeconomic stability, public education and health, protection of property rights, openness to the world economy, a functioning financial sector, and efficient physical infrastructure are all essential ingredients.

Programs to foster innovation become more important as real wages begin to increase, and there is a loss of comparative advantage in simple, labor-intensive activities. This period approximately corresponds to the newly industrialized economies in the early 1980s, Malaysia a decade later, and Thailand immediately prior to the economic crisis. There are, moreover, broader historical and institutional differences within East Asia. At comparable levels of development, for example, Japan and Korea invested more heavily in public education and were less open economies than is the case in much of Southeast Asia.

This chapter is organized as follows. The first section introduces the analytical and policy framework. The second surveys some widely used indicators, with reference to the eight economies. The third provides brief profiles of these economies. The fourth section addresses a range of policy and analytical issues, providing also some examples from a few economies. These include international technology flows and FDI; industry policy; enclaves and electronics; small- and medium-size enterprises and agriculture; and the role of public research institutes.

I largely eschew discussion of the recent crisis, as it is not particularly germane to the subject matter. But there are, of course, connections. Since 1997 the governments of the four crisis-affected countries have been preoccupied with recovery and reform, pushing technology and innovation issues off the immediate agenda. And, focusing on the opposite direction of causality, technology policies may have been marginal contributors to the crisis.[2] But in both cases the links are rather tenuous and, as recovery proceeds, these longer-term developmental issues will again become important.

ELEMENTS OF AN ANALYTICAL AND POLICY FRAMEWORK

A number of factors shape the process of technological change and innovation. As the East Asian experience illustrates (see, for example, Kim and Nelson 2000b; Nelson 1993), they differ in their relative importance both over time and across countries:

- *Openness.* More than 90 percent of the world's R&D is undertaken in OECD (Organisation for Economic Co-operation and Development) economies, and thus openness to global R&D capacity is critical for borrowers and latecomers. There are, of course, myriad mechanisms of transmission. FDI and participation in global production networks are among the major channels of technology flows. Technology is also embodied in the trade of goods, as it is in the trade of nonfactor services. Labor migration and associated flows of human capital are increasingly important.[3]
- *Human capital.* This is a critical factor at all levels of development, in creating the ability to absorb, assimilate, and diffuse imported technology

2. One can, for example, readily think of connections, involving either too much or too little attention to technology issues. In Korea there was arguably too much intervention under the broad umbrella of technology and innovation. To the extent that it involved directives to the finance sector and highly geared borrowings by the *chaebols*, it increased vulnerability to the crisis (see, for example, Smith 1998). Conversely, it was commonly recognized that Thailand underinvested on the supply side, especially in postprimary education. Thus, with real wages rising rapidly in the 1990s, the country began to encounter major competitiveness problems, only partly eased by its open labor market (see Warr 2000).

3. For a detailed East Asian study, see Athukorala and Manning (1999).

and to innovate on the basis of these imports. At early stages of development, mass literacy and universal primary education deserve priority, while upgrading strategies at middle incomes require more attention to higher education, to quality, and to the composition of educational output among disciplines.

- *Infrastructure and institutions.* This encompasses a diverse set of factors. An efficient business operation able to connect to the global economy requires high-quality telecommunications and physical infrastructure. There needs to be a commercial climate that is conducive to long-term planning horizons and investment decisions. Property rights and an adequately functioning legal system are also important. Financial institutions, domestic and foreign, need to be present. Macroeconomic stability is crucial.
- *Commercial environment.* Competition is vital to innovation, as a spur to raise productivity. Monopolies invariably prefer a quiet life and should therefore be a cause for concern, unless they are Schumpeterian in nature or operate in an open international environment. Whether industrial structure more generally (for example, the level of foreign or state ownership, the size distribution of firms) matters is a more complex question with no clear answers.
- *R&D institutions.* These constitute the formal component of most governments' technology and innovation policy. Important questions here concern the scale of these institutions, their funding, and their interaction with enterprises and the education system. For latecomer economies, their role focuses almost entirely on the modification and diffusion of essentially off-the-shelf technologies.

All five elements are important ingredients in the overall schema. Success in two or three may be insufficient, especially if other facets of the policy framework are not supportive. For example, economies may invest substantially in R&D, but these investments may be rendered ineffective by poor macroeconomic management or barriers to international commerce (for example, China and India before reform). Governments may invest in expensive high-tech projects while starving the public education system of resources (for example, India again and Indonesia, to some extent). To focus just on technology policy, narrowly defined, therefore misses this broader dimension.

What is the rationale for government intervention to develop an economy's technological capability? The principal one has to do with the public good characteristics of the capability (see, for example, Lall 2000a; Pack and Westphal 1986). Left to the market, there is likely to be underinvestment in activities for which private agents are unable to appropriate adequately the returns from their investments. Such investments tend to enter the public domain quickly, a process facilitated in part by classic free-rider problems. The returns are diluted, especially in developing countries, by a weak legal system that is unable to protect intellectual property rights. High mobility of workers between firms, especially for workers whose skills are in short supply, exacerbates the problem, since firms are reluctant to invest in R&D programs if there is a high probability that key staff embodying these investments will quit before returns can be appropriated.

These problems are all the more serious in an economic and political environment characterized by great uncertainty. In the best of environments, investments in R&D are highly uncertain and generally slow yielding. Where private agents heavily discount future earnings, whether owing to high real interest rates, lack of credible macroeconomic management, or political uncertainty, the incentives to invest are weaker still. Thus, while reported rates of return on industrial R&D are high in the few studies to investigate the subject in developing countries (see, for example, Evenson and Westphal 1995), these factors inhibit such investments and probably explain their limited scale. The inability of firms to appropriate fully the returns on these investments results in social rates exceeding private rates and hence provides a justification for government intervention.

Other forms of market failure provide a further rationale for government intervention (see, for example, Stiglitz 1996). Where markets are underdeveloped—incomplete or missing—price signals do not function adequately. This may arise because of poor information flows, limited entrepreneurial capacity, inadequate property rights, or physical infrastructure bottlenecks. Institutions (such as producer cooperatives or industry associations) that foster the cooperative behavior sometimes needed to achieve improved outcomes may not exist. Coordination functions, especially in dealing with international markets, may be poorly developed. The cooperative and coordination functions do not necessarily need to be supplied by government, but public policy can hasten their efficient evolution.

The new growth theories, with their emphasis on the determinants of long-term growth rates, have revived interest in technological change and technology policy. "The suspicion that [growth differentials among countries] may have something to do with technology has been around for a long time," as Fagerberg (1994, p. 1147) observes. However, the new theories have yet to develop powerful policy relevance. Empirical refinement is pushing in this direction, examining, for example, the growth-stimulating effects of investments in R&D and schooling and attempting to measure the R&D spillovers from international trade and investment.

QUANTITATIVE INDICATORS

Here I present a set of factors that, consistent with the analytical framework adumbrated above, indicate the technological and innovation capabilities of the eight economies covered in this chapter.[4]

There is no widely used summary measure of technological competence. All technology indicators have serious conceptual and empirical limitations, which explains why a basket of such measures is typically employed in making comparisons over time and across countries. The usual approaches include both input and output measures. The former focus on resources devoted to the development of technological capacity, such as expenditure on R&D and investment in human capital. The latter concentrate on outcomes of this capacity and include such indicators as patents and the proportion of output or exports originating from technology-intensive activities.

Many of these indicators are empirically slippery. For every variable presented here, two or more alternatives could feasibly be employed. Many others could be added, to capture elements that are germane to the process of technological change and innovation.[5] The purpose here is to provide a rough sketch, and a recent snapshot, for each economy. Nevertheless, a reasonably robust and plausible picture emerges. Extensive refinement of the measures would arguably not fundamentally change the basic picture.

Tables 8.1 to 8.4 present indicators grouped under four major headings: R&D, openness, human capital, and institutions and infrastructure. I consider each in turn.

As expected, the R&D indicators in table 8.1 show the Asian newly industrialized economies to be mostly well ahead of the ASEAN-4. The former's formal R&D effort is generally at median OECD levels or more, with Korea among the highest in the world. Taiwan (China)—and now Singapore—invest proportionately more in R&D than several middle-income OECD economies. The major outlier among the newly industrialized economies is Hong Kong (China), which is perhaps the world's prime example of a rich technology-borrower economy, in which strategic location

4. The database is deficient in numerous respects. Korea, Singapore, and Taiwan (China) now produce annual statistical series on technology and innovation and, as an OECD member, Korea is included in major comparative series for developed economies. However, the R&D database for the ASEAN-4 countries is very weak, and it is difficult to obtain internationally comparable series for East Asian economies as a whole. Moreover, it is virtually impossible to quantify the overall magnitude of international technology flows and the major constituent elements (see Soesastro 1998).

5. There is much discussion of these indicators and their meaning in the literature. The work of Sanjaya Lall (for example, Lall 1998a, 2000a) is particularly important. See also UNIDO (2002) for a recent and comprehensive set of comparative indicators.

Table 8.1 Indicators of R&D

Economy	Ratio of R&D to GDP	Patents (thousands)	High-tech exports	Tech index
Asian newly industrialized economies				
Hong Kong, China	0.3	2.4	21	0.6
Korea, Rep. of	2.8	130	27	0.8
Singapore	1.1	37.7	59	2.0
Taiwan, China	1.9	—	—	0.9
ASEAN-4				
Indonesia	0.1	4.5	10	–0.7
Malaysia	0.2	6.4	54	1.1
Philippines	0.2	3.6	71	0.5
Thailand	0.1	5.4	31	–0.1

—Not available.

Note: For this and the following tables, data refer to the latest year available, as cited in the relevant publication. In most cases it is late 1990s. In some cases, more recent country data are readily available, but to preserve the comparisons, the original data are used. Ratio of R&D to GDP refers to research and development as a percentage of GDP. Patents refer to the number of foreign and domestic patents filed in 1997. High-tech exports refer to the percentage of manufactured exports that were high technology in 1998. Tech index refers to the technology index developed in World Economic Forum (2000); it ranges from 2.02 (United States) to –1.84 (Bolivia).

Source: World Bank (2000); World Economic Forum (2000).

Table 8.2 Indicators of Openness

Economy	Ratio of exports to GDP	Average tariff (percent)	Import fees	Ratio of foreign direct investment to GDP
Asian newly industrialized economies				
Hong Kong, China	132	0	1.1	60.8
Korea, Rep. of	42	7.9	1.9	5.1
Singapore	> 200	0.3	1.0	90.0
Taiwan, China	—	3.2	1.5	7.5
ASEAN-4				
Indonesia	54	9.5	2.2	46.8
Malaysia	124	8.1	1.5	50.4
Philippines	56	10.2	2.3	12.8
Thailand	57	5.9	1.8	15.2

—Not available.

Note: Import fees attempt to measure all import barriers and range from a score of 1 (Luxembourg and Singapore) to 3.4 (India), as reported in World Economic Forum (2000). Foreign direct investment as a percentage of GDP measures the stock of inward foreign direct investment (as reported in UNCTAD (2001) as a percentage of GDP).

Source: World Bank (2000); World Economic Forum (2000).

Table 8.3 Human Capital Indicators

Economy	Years of school	Enrollment		R&D employment	Ranking
		Secondary	Tertiary		
Asian newly industrialized economies					
Hong Kong, China	9.5	81	22	—	21
Korea, Rep. of	10.5	102	60	2,193	14
Singapore	8.1	90	39	2,318	1
Taiwan, China	8.5	94	67	—	3
ASEAN-4					
Indonesia	4.7	56	11	182	39
Malaysia	7.9	59	12	93	22
Philippines	7.6	77	29	157	49
Thailand	6.1	56	22	103	38

—Not available.

Note: Years of school refer to average years of schooling by population 25 years of age and older, as reported in Barro and Lee (1996). Enrollment—enrollment ratios at secondary and tertiary levels—refer to numbers enrolled as a percentage of the relevant age group. Numbers can be misleading when large numbers of students are enrolled abroad. R&D employment refers to scientists and engineers in R&D per million population. Ranking refers to assessments of math and basic science education, ranging from 1 (Singapore) to 50 (South Africa), as reported in World Economic Forum (2000).

Source: World Bank (2000); World Economic Forum (2000).

Table 8.4 Institutional and Infrastructure Indicators

Economy	Property rights	Infrastructure	Stock market	Information technology hosts
Asian newly industrialized economies				
Hong Kong, China	6.4	6.4	377	163
Korea, Rep. of	4.9	4.3	78	60
Singapore	6.3	6.7	208	452
Taiwan, China	5.6	4.4	—	—
ASEAN-4				
Indonesia	3.7	3.3	54	1
Malaysia	5.2	5.3	188	25.4
Philippines	4.2	2.3	62	1.6
Thailand	4.7	3.8	48	6.5

—Not available.

Note: Property rights refer to the legal protection of property rights, scored from 6.8 (Luxembourg) to 2.6 (Russian Federation), as reported in World Economic Forum (2000). Infrastructure refers to overall quality of infrastructure, scored from 6.7 (Singapore) to 1.4 (Bolivia), as reported in World Economic Forum (2000). Stock market refers to stock market capitalization at the end of 1999 as a percentage of 1999 GNP. Information technology hosts refer to Internet hosts per 10,000 population, as of January 2000.

Source: World Bank (2000); World Economic Forum (2000).

and the provision of extremely high-quality governance, infrastructure, and service activities are the main competitive advantages. The ASEAN-4 countries invest very little in R&D, although Malaysia is beginning to do so.

Patent activity is more vigorous in the newly industrialized economies than the ASEAN-4, and these economies rate more highly according to a broad estimate of technology capabilities.[6] Within the newly industrialized economies, according to this index, Korea's superiority disappears (and it falls below even Malaysia), raising the question of whether it is investing its R&D effort productively. The composition of exports also presents an ambiguous picture. I return to these issues in the next section.

According to the various indicators of openness, the distinction between the newly industrialized economies and the ASEAN-4 becomes blurred (table 8.2). Singapore most of all, together with Malaysia and Hong Kong (China), stand out as the economies most open to trade and investment. This partly reflects their size—small economies by definition trade more than larger ones—but it also indicates open FDI regimes and, in the case of the city-states, very low formal trade barriers.

Among the other economies, trade barriers are generally low, certainly in comparative developing-country context. Barriers are somewhat higher in the ASEAN-4, although the difference between them and the other two newly industrialized economies is not great. Historically, of course, the patterns varied much more, with Indonesia, Korea, and the Philippines displaying high levels of protection. One crucial distinction, not captured in these data, is that almost all economies in the sample have placed exporters on something close to a free-trade footing.

The major difference occurs with respect to FDI regimes. The ASEAN-4 mostly, and the city-states always, have been more open to foreign direct investment than Korea and Japan. These data have to be treated with caution for several reasons,[7] but the general conclusion still holds.

---

6. Patent data in the table refer to patents filed in each country, by both resident and nonresident entities. In all cases, nonresident entities constitute the major source and, in all but Korea, are more than 90 percent of the total. There is likely to be an element of "double counting" present in these data, to the extent that multinational corporations file for the same patent in more than one economy. Nevertheless, these data convey a picture broadly indicative of patent activity in each economy. Alternative indicators (for example, patents filed in the United States by economy of origin) provide a similar ranking.

7. For example, the data reflect past investment decisions and regimes; Korea, notably, is now much more open to FDI than in the past. Some economies are large recipients of FDI in part because of the structure of their economies (for example, Indonesia has a large FDI–intensive natural resource sector). Moreover, low levels of FDI may indicate either restrictive regimes or unattractive investment climates (for example, the latter is the major factor in the low figure for the Philippines).

There are many other channels for the international diffusion of technology, and reasonably accurate quantitative indicators are available for some of these. Examples include skilled labor migration, non–FDI licensing arrangements, and imports of capital equipment. I discuss some of these in later sections.

Human capital indicators present a mixed picture (table 8.3). Most East Asian economies score well on basic education achievements, certainly in comparative developing-country context. The Asian newly industrialized economies are already at OECD levels, with Korea in the lead. Very high levels of secondary enrollments are evident among the newly industrialized economies and the Philippines (Booth 1999). At tertiary levels, the newly industrialized economies also generally stand out. The figures for Malaysia and Singapore are greatly understated, because about one-third of their tertiary enrollments are abroad. These quantitative indicators reveal little about quality. International rankings (admittedly confined to math and science examinations) do confirm general impressions that quality is much higher among the newly industrialized economies. The Philippines slips markedly according to this criterion.

Finally, a range of institutional and infrastructure indicators is presented in table 8.4. This is a very large area, and the few variables presented here can provide only a glimpse of the bigger picture. In general, the newly industrialized economies perform quite well and are ahead of the ASEAN-4, although Malaysia, in some respects, belongs more to the former than the latter. Thus, for example, the three former British colonies score best on property rights and legal institutions. Indonesia ranks poorly. The physical infrastructure of the newly industrialized economies and Malaysia is generally very good, with Singapore regularly topping international comparisons.

One indicator of financial depth (stock market capitalization) shows the city-states and Malaysia to be the most advanced. This is, of course, a partial indicator, since one needs to know more about other financial institutions and prudential-regulatory standards. With regard to information technology capacity, the city-states again emerge as the strongest.

Concluding this discussion, I draw attention to two widely used indicators that can be quite misleading. The first is export composition and "revealed comparative advantage" based on factor intensity classifications. These have been widely used in the literature, from Lary (1968) onward. However, owing to the rapidly expanding "slicing" of international production activities and the dominance of the electronics industry in East Asian exports, these indicators have become practically irrelevant. Even finely disaggregated export statistics cannot capture the fact that the electronics industry encompasses the full gamut of factor intensities, ranging from highly

R&D-intensive activities to technologically simple packaging and assembly operations. This proposition is illustrated in table 8.1. The country that exhibits the highest proportion of high-tech exports is the Philippines, which has one of the weakest technological capacities in East Asia. As electronics are becoming ever more important in a country's exports, this indicator is essentially a measure of the reliance on electronics exports.[8]

The second is the use of total factor productivity (TFP) growth as an indicator of technological progress. The limitations of TFP growth are well known. TFP growth moves closely with the business cycle. As component inputs, and particularly their quality, are measured more accurately, recorded TFP growth becomes smaller. In fact, the Krugman-inspired "perspiration versus inspiration" debate about TFP reveals very little about trends in technological capabilities in East Asia (Chen 1997). The TFP numbers mask the microeconomics of technical progress, for example, the mastery of products and processes that, in many cases, did not exist in these countries a decade or more ago (see Nelson and Pack 1999 for a detailed discussion of this issue). For example, the proposition that there was little technological progress or innovation at the enterprise level in Singapore during the 1960s and 1970s because measured TFP growth was negligible contradicts all available evidence pointing to that country's evolution toward technologically advanced, export-oriented industrialization.

PROFILES OF EAST ASIAN ECONOMIES

This section presents several economy profiles, complementing the statistical indicators of the previous section and serving as a prelude to the discussion of a range of topical issues in the following section. I focus mainly on Indonesia, Korea, Malaysia, Singapore, and Taiwan (China), a sample that captures the diversity of East Asian development levels and policy experiences. They are among the better documented cases in the region.

Korea

Korea attracts attention for several reasons: its extraordinary pre-crisis development achievements; its serious economic crisis in 1997–98 (the worst

8. It is surprising how frequently and loosely this misleading indicator is still used. The more knowledgeable analysts of export composition data (for example, Lall 2000b) are, of course, careful to point out its limitations.

among the Asian newly industrialized economies); its controversial and highly interventionist industrial policy regime; and its major commitment to R&D (R&D expenditure rose from 0.3 percent of GDP in 1971 to 2.8 percent in the mid-1990s) and to education at comparatively modest levels of income per capita. Its record is also by far the best documented in developing East Asia.[9]

Kim (1997) emphasizes several key ingredients in the Korean experience that are readily transferable. These include export orientation, as a means of ensuring that assisted firms are able to quickly meet some sort of market test; a strong commitment to education; and reasonably sound macroeconomic management.

More controversial has been Korea's strategy of promoting the *chaebol* (Amsden 1989). Much of the national R&D effort was concentrated within these conglomerates (for example, Samsung, Hyundai, LG) rather than government-funded institutes, and their links to universities and small- and medium-scale enterprises have been weak. Even before the crisis it was obvious that these *chaebols* were both an "asset and burden" (Kim 1997, p. 196). While effective in pooling resources and marshaling inputs, there were large costs in terms of political corruption and neglect of small- and medium-scale enterprises. The restrictive FDI regime hampered the access of Korean firms to international know-how. Such a strategy would be even more costly in economies that have not invested so heavily in scientific education. Moreover, it is not obvious that, in the current international commercial policy environment, such a restrictive regime would even be possible in the latecomer industrializers.

More generally, as Kim emphasizes, Korea's political and historical development was unique and therefore difficult to replicate. The harsh Japanese colonial rule, the Korean civil war, and the ever-present threat from the North bequeathed a regime with an unparalleled commitment to development, an intrusive, authoritarian political system, and a government bureaucracy that actively cajoled firms and individuals to meet highly ambitious targets.

Moreover, the crisis highlighted significant weaknesses in the Korean system. It appears that Korea is not obtaining the full benefits from its very large investments in R&D and education because of weaknesses in the incentives regime and its institutions. These include inadequate conditions for the generation and exploitation of knowledge and information (for example, the

9. See, for example, Dahlman and Andersson (2000), Kim (1997, 2000), Lee (2000), and references cited therein, together with much comparative material (especially Matthews and Cho 2000; Wong and Ng 2001).

protection of intellectual property rights). There are also insufficient competition, flexibility, and diversity (for example, among the *chaebols* and in the finance and education sectors). Moreover, there is a misallocation of investments, including duplication in public R&D investments and probable overinvestment in public education of limited quality. Finally, a consistent theme in the literature on Korea is the need to become more international, in its education system, R&D institutions, and global commercial networks.

The Korean model suggests that successive governments were remarkably successful in lifting the country from very low levels of income to middle-income status, but that the government's role now needs to be redefined. That is, the policy framework is highly effective at "catch-up," but less useful as the country approaches the technological frontiers. As Kim (2000, p. 336) observes, in the past decade the government shifted "from an effective orchestrator to a rigid regulator." Markets need to be able to function more effectively, and the economy needs to be less government led and *chaebol* dominated. The recent literature (see especially Dahlman and Andersson 2000) makes a number of recommendations. These include a more modern legal system; more competition in product markets; better supervision and transparency in financial markets; more flexible labor markets; removal of the still significant informal barriers to FDI; greater autonomy for educational institutions; stronger linkages between educational institutions and research agencies; and liberalization and improved regulation of telecommunications.

Malaysia

Malaysia has grown rapidly for most of its post-independence history. It possesses relatively strong institutions. It is very open to both international trade and FDI (apart from, recently, the special case of short-term capital) and has been a vigorous exponent of export processing zones. Combined with good infrastructure, political stability, a liberal investment regime, and proximity to Singapore, this has resulted in a huge electronics industry. Its base of human capital is good, but not outstanding. Its business environment is open, although the government's ethnic restructuring objectives do impose some restrictions on firms' commercial freedom. In consequence, it has consistently lost high-level non-*Bumiputera* (ethnic Malay) human capital. The country had notable early strengths in agricultural extension services (particularly for cash crops), which for many years underpinned its status as the world's largest producer of natural rubber and palm oil.

More recently, the government has embarked on a number of ambitious projects.[10] The current 20-year plan seeks to lift the country to OECD status, particularly in knowledge-based activities, including computer infrastructure, "infostructure," education and training, and R&D and technology. Just prior to the Asian financial crisis, the government announced its plans to develop a Multimedia Super Corridor, which included the aspiration to develop a "Silicon Valley of Southeast Asia." The government is taking the initiative in this project, through the provision of subsidized telecommunications infrastructure, fiscal incentives, university-based incubators, and related government investments. The timing of the Multimedia Super Corridor's establishment—just prior to the crisis—was unfortunate. It is still at an early stage of development, and it may be premature to offer any evaluation (however, see Chapter 3).

Malaysian industry comprises several distinct segments. One is electronics, which has been extremely successful in terms of growth and exports, but now must make the transition upstream, integrate more with the rest of the economy, and develop its base of human resources. I return to these issues below. Another segment is the automotive industry, whose record is the subject of considerable controversy. Here it seems that Malaysia has paid a very high price by attempting "back-to-front" industrialization—that is, by embarking on a prolonged period of protection for the assembly industry rather than following the more successful route of, for example, Taiwan (China) and Thailand. Thailand lowered protection a decade ago, has performed particularly well in the components industry, and is now emerging as Southeast Asia's leading automotive nation. A third segment is the agro-processing industries, where Malaysia has been able to develop on the basis of its traditional production strengths and a high-quality, publicly funded (or at least mandated) R&D base.

There is considerable regional diversity in the Malaysian industrial sector. The state of Penang, in particular, has been the leading performer, as the home of the earliest export-oriented investments and the location where the strongest base of small- and medium-scale enterprise subcontracting has developed. Its commercial and business history, together with effective political leadership and public institutions, are generally regarded as the key to its success.[11] As a state within a federal political structure, its government does not, of course, have access to the full array of policy

10. For a description, see Economic Planning Unit (2000, ch. 5). Jomo and Felker (1999), and Jomo, Felker, and Rasiah (1999) provide a comprehensive analysis of the country's technological capabilities and innovation system. See Lall (1995) for an earlier analysis.

11. See Rasiah (2001) and Toh (2002) for interesting accounts of Penang's success and challenges with export-oriented industrialization.

instruments, and it remains to be seen how it will adapt to the challenge of upgrading.

In many respects, Malaysian approaches have mirrored those of Singapore, albeit with a lag. That is, it has sought to build on its very good physical infrastructure and openness to foreign trade and FDI. Initially, FDI in electronics and other labor-intensive industries was seen as a means of providing employment for a labor-surplus economy; the country has long been home to some electronics majors, including Intel and Motorola. More recently, the government has pursued a strategy of technological leverage vis-à-vis MNCs by developing linkages through supply networks, joint venture training projects, and interfirm worker mobility. While generally successful, its public policy capacities have never matched those of the city-states, clouded by a less secure legal environment, corruption, and the politics of affirmative action. Thus there is a general consensus that, notwithstanding its rapid growth, Malaysia could have done more by way of leveraging the presence of MNCs to its own advantage (Westphal forthcoming).

Singapore

Singapore has notable strengths in the area of technology and innovation policy. It is an extremely open economy (except for some services, mostly now being opened up too), its bureaucracy is one of the best in the world, its physical infrastructure is likewise absolutely world class, and its education system is good. And, as it contemplates upgrading, it has larger fiscal reserves, relative to the size of the economy, than any country in the world.

Its technology and innovation policy has evolved rapidly as it has shifted quickly from using to attempting to create these technologies.[12] It began to lose comparative advantage in labor-intensive activities in the late 1970s, hastened by the government's aggressive high-wage policy. Singapore embarked on its first serious R&D effort in the late 1980s. Until then, it had lagged the OECD countries as well as Taiwan (China) and Korea. An "applied R&D expansion" phase lasted for the decade commencing in 1989, which saw the beginnings of innovation and local capacity and more government funding. A big push for higher education, high-tech industry, and basic R&D got under way in the late 1990s.

R&D expenditure has increased rapidly, about sevenfold in 1987–99, reaching 1.84 percent of GDP in 1999. Most of it is in the private sector but

12. See Wong (2001, 2003) for a useful analytical history, on which we draw. NSTB (1997) provides a succinct official perspective, while NSTB (2000) and earlier publications in this series provide a statistical summary.

features much state inducement. During the 1990s, funding for public research institutes and universities grew quickly. There was also some administrative reorganization, with the National Science and Technology Board (NSTB) focusing on basic R&D, and the Economic Development Board directed to applied and commercial areas. With its high salaries and open labor market, Singapore institutions are able to recruit on the international labor market. In 1999, for example, 17 percent of science and technology personnel were foreigners; many more were migrants who had become permanent residents and citizens.

Singapore's ownership structure is unique, with its large foreign- and government-owned segments, and its technology trajectory has been shaped by this fact. MNCs have always been a dominant feature of the modern Singapore economy, generating about three-quarters of manufacturing value added. The government has actively sought to induce Singapore-based MNCs to upgrade and innovate, to use the country as a regional headquarters, to undertake product and process innovation, and to build and fund joint venture technology institutes to diffuse the technology of MNCs.

In certain designated industries, it has aggressively attracted foreign companies. This applies to electronics in general (see Matthews and Cho 2000). A particular case of success is the hard disk drive industry, in which Singapore spearheaded its relocation to Southeast Asia (see McKendrick, Doner, and Haggard 2000 for a detailed case study). The initial attraction for the industry was the region's cheap labor and good infrastructure, in addition to which Singapore was motivated by a desire to overcome the economy's historically high rates of unemployment. Public policy played an active role not only in attracting the industry but also in nurturing and developing it beyond its initial unskilled labor-intensive phase, when it might otherwise have migrated to other low-wage locations. A combination of push and pull factors was at work. Agglomeration and momentum factors were both important, as was Singapore's growing "country reputation" in attracting other MNCs. The Singapore government as host played a key role in developing complementary support facilities and in having officials who understood the needs of the industry and were able to make quick decisions. Singapore's high-quality institutions made a difference.[13]

13. The Singapore Economic Development Board is one of the most important government institutions in East Asia in the realm of technology and innovation policy. For a positive, quasi-official assessment of its first 25 years, see Low and others (1993). The one perhaps contentious aspect of Singapore's record is the liberal use of fiscal incentives, which in other institutional settings would almost certainly be susceptible to abuse.

Schemes were introduced to train employees and develop supplier industries. The government employed fiscal incentives to induce MNCs to use the country as a regional headquarters. Grants from the NSTB were offered to MNCs to undertake R&D. A Skills Development Fund levied on firms provided funds for training. Perhaps more than any government in the world, officials are attentive to the needs of foreign companies and think creatively about the means of leveraging their presence. Few countries in the world can match Singapore in the time taken between investment approval and the commencement of production operations.

There is some controversy about the size and role of government-linked corporations, particularly as these corporations are increasingly active investors abroad. They are large, equivalent to at least one-quarter of GDP and probably a good deal more. The government has employed some of the larger government-linked corporations as agents of innovation, especially the Changi Airport and the Port of Singapore Authority. (Both regularly receive commendation as among the most efficient service providers in the world.) Defense receives a high priority in the budget, and this too has important implications for R&D.

Among future challenges, Wong (2001, 2003) identifies developing links between enterprises and both the universities and the public research institutes. Links between universities and public research institutes are already strengthening. Venture capital funds are emerging but remain rather limited. International R&D and educational links need to be developed. There is the question of what R&D to fund given that Singapore's small size dictates specialization. The government laments the lack of domestic entrepreneurship, while vigorously recruiting the best and the brightest from tertiary institutions. With a large presence of well-paying multinational corporations, it is perhaps not surprising that local privately owned firms are the country's weak link.

Taiwan (China)

Taiwan (China) stands out for its open economy, resilient small- and medium-size enterprises, very high saving rates and international reserves, high educational achievement, good physical infrastructure, and some effective support for innovation. Taiwan (China) also weathered the Asian economic crisis more successfully than Korea. What Chu (2000) terms a "competent, authoritarian government" promoted rapid export-oriented industrialization from the late 1950s, after a brief period of import protection and significant state ownership. Most of the major labor-intensive export industries of the 1960s to 1980s received at most modest levels of protection and in some cases were taxed. Examples include garments, footwear, travel goods, toys, and bicycles (in which Taiwan [China] became the world's largest exporter by the mid-1980s). A crucial feature is that firms were quickly subjected to the market test, through either domestic competition or export performance. Taiwan (China) was consistently less interventionist than Korea and embarked on few massive heavy-industry projects. Instead, the keys to its success appear to have been the extraordinary resilience and flexibility of its small- and medium-scale enterprises, combined with a government-supported R&D effort from the early 1980s and a capacity among small- and medium-scale enterprises to absorb and diffuse new technologies very quickly. Taiwanese (Chinese) firms score very highly in terms of their flexibility, their ability to quickly penetrate, and switch between, international marketing channels, and their ability to rapidly absorb and diffuse technology introduced by foreign direct investment or government research institutes.

An early and major investment in education at all levels was important, including especially strong science and technology (Lin 1998). Increasing numbers of students went abroad beginning in the 1960s, principally to the United States. Many stayed abroad for decades, returning not only with formal education but also with extensive international commercial experience. This is a feature in all the Asian newly industrialized economies, but it is arguably most important of all in Taiwan (China).[14]

In many respects, Taiwan (China) differed significantly from Korea (Smith 2000; Westphal forthcoming). It lacked the large conglomerates within which major R&D initiatives could be funded. Historically, government-supported R&D was not large. Its FDI regime was neither as restrictive as that of Korea nor as open as that of the city-states. From the mid-1980s, government support for science and technology increased sharply. Key centers at universities and public research institutes were funded. There was concern that the small and medium enterprise–dominated industrial structure would be slow to innovate and would experience difficulty absorbing foreign technology directly. Therefore, cooperation between enterprises and research institutes was given high priority. Some successful consortia emerged, combining basic R&D and commercialization. The Industrial Technology Research Institute was a key actor. The Hsinchu Science-Based Industrial Park, established in 1980, likewise played a major role, especially in the development of a

14. Pack (2000) draws attention to this phenomenon. See Yusuf with others (2003) for a review of the evidence.

more sophisticated electronics industry (see Matthews and Cho 2000, p. 257 ff).

Indicators of its R&D output point to significant achievements; for example, its firms have the largest number of U.S. patents among the four newly industrialized economies. However, there is concern that, as in Korea, the government is not moving quickly enough to develop the institutions required to support a modern, technologically progressive economy, including a high-quality legal system and adequate supervision of financial institutions and the stock market.

Indonesia

Indonesia is the poorest country in this East Asian sample and the one most affected by the recent crisis. It is experiencing a major and prolonged economic and political transition, and thus technology and innovation issues are likely to be of secondary importance for at least a decade.

In the 20 years to 1997, its research and technology minister, Dr. B. J. Habibie (who was also briefly president, 1998–99), completely dominated Indonesian technology policy.[15] Indonesia had ambitious plans for technological development over this period. For example, in its Second Long-Term Development Plan (PJP II, 1994–2019), drawn up well before the recent crisis, expenditure on R&D was projected to rise sharply, from about 0.2 percent of GDP to 2 percent. Government expenditure on R&D before the crisis was quite modest, about $400 million, but was dwarfed by the huge investments in the nation's "showcase" aircraft factory, IPTN, which totaled approximately $3 billion. Reflecting in part the background of the key personalities involved, program objectives also emphasized technological self-sufficiency through a four-stage evolution from basic manufacturing capacity to the mastery of leading-edge technologies as embodied in advanced manufacturing.[16]

According to all the standard indicators, Indonesia is very much an industrial and technological latecomer. It is only since the late 1960s that it has been open to international technology markets and has had a government commitment to the development of any sort of scientific base. The

15. See the contributions to Hill and Thee (1998) for detailed analyses of the pre-crisis situation. Okamoto and Sjoholm (2003) provide a recent survey of the issues. The best study of the Habibie mega projects, though now dated, remains McKendrick (1992).

16. Dr Habibie's disdain for labor-intensive industries was well known. He frequently singled out the garments industry, arguing that, in spite of its prominence in early-stage East Asian industrialization, Indonesia should not aspire to be just "tailors for the world."

government is the dominant actor in formal R&D activity, both as a funder and as an "implementer." It provides about 80 percent of the resources and directly carries out more than 60 percent of the activities. International comparisons suggest that the Indonesian government's involvement is unusually large.[17] Other indicators also underline Indonesia's status as latecomer. Owing to its size, it has the largest stock of scientific personnel in ASEAN, but when adjusted for quality of training, hardware support, and active scientific activity, its R&D effort is much closer to that of the Philippines and Thailand.

Over the period 1967–97 the government achieved notable success in macroeconomic management, promoting near-universal education at primary and junior secondary levels and pursuing a reasonably open trade and investment regime. But its microeconomic interventions, including technology and innovation policy, often resulted in gross distortions. Program implementation has always been a major challenge. Programs have often lacked coordination across ministries. Corruption has been a continuing and serious problem. State enterprises have a poor financial record and achieve few of their supposed "developmental objectives" (including those with a technology mission). Finally, important services that only government can provide (such as an incorruptible police force, an efficient and clean judiciary, and a well-regulated financial sector) have not been supplied.

SELECTED ISSUES

This section addresses a range of policy and analytical issues, also providing some country illustrations. These include international technology flows and FDI; industry policy; enclaves and electronics; small- and medium-size enterprises and agriculture; and the role of public research institutes.

International Technology Flows and Foreign Direct Investment

International connections critically shape a country's pattern and rate of technological development. This is a vast topic with many dimensions. First, R&D activity is heavily concentrated in a handful of rich OECD economies (the G-7 account for more than 90 percent of global R&D expenditure), and so a capacity to tap into global technology markets is the first requirement for net technology importers in East Asia. Second, the

17. For example, the comparable figures for Korea are 17 percent (of funds) and 4 percent (of implementation), although government's indirect influence is certainly much higher.

limited documentation of international strategic technology alliances concludes that it is overwhelmingly an intra-Triad phenomenon (that is, within the European Union, Japan, and the United States; see Duysters and Hagedoorn 2000). Third, international technology flows are increasing rapidly and, like trade and investment, almost certainly are growing faster than global output. Fourth, there is a steadily increasing array of suppliers in the international technology marketplace, although, in certain technology-intensive activities, markets are still oligopolistic in structure and information flows are inherently imperfect. Finally, more so than for any other form of major international commercial transaction, technology flows are poorly documented, not only because of conceptual difficulties in defining just what constitutes a "flow of technological services" but also because of empirical limitations. In most cases, there is practically no means of quantifying these flows in any sort of detail in aggregate, much less by source country, channel, or the recipient sector.

An important lesson from the Asian experience is the need to link technology strategy to international commercial policy (Soesastro 1998). Open economies with a good base of human resources may achieve high payoffs from carefully targeted science and technology policies. Taiwan (China), Hong Kong (China), and Singapore illustrate this proposition. They are all nimble, resilient, and open economies with (until recently) below-average R&D expenditures relative to per capita income but with effective innovation strategies. By contrast, India and Russia devote large budgets to R&D activities, but historically in the context of closed economies, which thereby minimized the opportunities for a well-developed domestic research industry to interact with international best practices. This issue has been emphasized by OECD (1997), which proposes that a nation's technological effort should explicitly include acquired (imported) technology together with a national R&D effort. The OECD estimates suggest that the magnitude of such acquired technologies may be at least as large as that of domestically generated R&D activities. In developing countries, with their limited local R&D expenditure, acquired technology would be a larger proportion of the total.

FDI is generally among the most significant components of international technology flows.[18] The literature on foreign direct investment and technology transfer—or spillovers—has proceeded in two main directions, a newer one that is macro, econometric, and inferential and a tra-

18. Some of the material in this subsection draws on Athukorala and Hill (2002). See Chapter 9 on the mechanisms of technology transfer. Urata (2001) provides a good survey of the issues relating to trade and FDI in East Asia.

ditional one that is more micro, qualitative, and firm based. The two approaches are particularly data intensive and so rarely are both employed in one study. The first of these employs a large secondary data set in which foreign and domestic firms are identified separately. It examines trends in productivity (either total or partial) between the two groups and across industries to discern whether the foreign presence affects levels and growth rates among domestic firms (see Blomström and Kokko 1998 for a survey). These studies are generally not concerned with the mechanism of transmission, and they are unable to estimate the relative importance of FDI among other factors explaining productivity growth in domestic firms. However, they do provide presumptive evidence of causation. The results vary across countries and industries, indicating that such spillovers do not occur automatically. It is hypothesized, and sometimes empirically demonstrated, that spillovers will be positively associated with the level of competition (which pushes firms to adopt improved technology) and negatively associated with the productivity gap between foreign and domestic firms (on the assumption that a very large gap renders absorption by domestic firms more difficult).

Few studies have employed this approach in developing East Asia. Fan and Warr (2000) examine spillovers, as measured by TFP growth, to state-owned enterprises and enterprises that are collectively owned (mainly the town and village enterprises) in China. Their results highlight the importance of the domestic policy and absorptive environment in maximizing the gains from FDI. Among state-owned enterprises, the spillovers are found to be negative. They attribute this to these firms' operating environment, including their soft budget constraint, a deteriorating base of human capital, and the absence of incentives to improve productivity. By contrast, in the flexible and dynamic town and village enterprises, where appropriate incentive structures are in place, the spillovers from foreign direct investment are found to be positive. Sjoholm (for example, 1999) finds that competitive pressures are significant in explaining inter-industry variations in Indonesian productivity growth. Moreover, domestic competition is found to be more significant than foreign competition. The presumption underlying much of this research is the so-called "Bhagwati hypothesis" that FDI is likely to be more effective in export-oriented than in import-substituting regimes and sectors.

Ideally, one needs to supplement these studies with some more detailed industry-level work to understand better how and why spillovers work. Hou and Gee (1993) attempt to probe technology transfer mechanisms in Taiwan (China). They find labor mobility from foreign to local firms to be a key variable. They also point to the differences between large, medium, and small firms. Whereas the former generally approach technology transfer through formal mechanisms such as joint ventures (including reverse engineering via foreign direct investment abroad) and licensing agreements, the latter rely more heavily on informal channels such as copying and business liaisons.

There continues to be a debate about the effectiveness of Japanese technology transfer via FDI. The analysis has matured from the somewhat polarized debate in the 1970s, with the general conclusion being that Japanese firms are no better or worse than MNCs from other countries, especially after account is taken of linguistic differences and the industrial location of FDI. For example, in a study of Japanese FDI in the Indonesian auto industry, Sato (1998) finds a substantial transfer of managerial know-how from the Japanese investor to, and beyond, the Indonesian partner, which became deeply familiar with the technology and adapted it to Indonesian conditions.

There is a clear consensus in the East Asian literature on the importance of "leveraging" the presence of MNCs as a means of maximizing the benefits for the domestic economy. Singapore has arguably had the greatest success in this regard, as alluded to earlier. The key feature of its policy regime has been to adjust the policy settings as the economy has shifted quickly from its labor-intensive industrialization phase to one that is highly technology intensive. Its government anticipated the shift out of low-wage activities and developed several programs to upgrade local capacities. In the case of the hard disk drive industry, it set up the Data Storage Institute (see McKendrick, Doner, and Haggard 2000). It also recognized the support of clusters and a base of small, specialist suppliers. Thus, for example, Seagate, the major multinational in the industry, "started from nothing" in 1982, but within a decade a large industry cluster was operating effectively.

This was a successful example of efficient industrial policy. The government introduced a Local Industry Upgrading Programme as a means of tapping into the expertise of MNCs. Technical skills were upgraded continuously through technical, vocational, and tertiary education. As the country began to lose comparative advantage in labor-intensive sectors, the government worked with MNCs to induce them to stay and upgrade. On-the-job training was facilitated by the Skills Development Fund, financed in part by a levy on foreign workers. The Economic Development Board introduced schemes to fund the local R&D activities of MNCs. Two additional elements of the Singaporean experience with MNCs are relevant. First, the technology transfers appear to have been both horizontal and vertical, the former occurring through emulation, demonstration effects, and interfirm worker mobility and the latter via subcontracting and other supplier relationships. Second, there were distinct benefits for Singapore's neighbors from its multinational corporation–intensive, export-oriented strategy. It created a regional (Southeast Asian) reputation in the global economy. There were also notable cross-border spillovers, as Singapore-based firms relocated labor-intensive segments to nearby Johor (Malaysia) and the Riau Islands (Indonesia).

While Singapore's geography and history are unique, other countries can learn from its success. Five features of its record deserve emphasis in this context. First, its economy is completely open, and so firms are immediately subject to some sort of market discipline. Second, as part of the package to attract MNCs, it offers the world's best physical infrastructure and an entirely predictable and business-friendly investment climate. Third, the government has demonstrated an unrivaled capacity to walk away from mistakes. A highly open economy reveals these mistakes quickly, and Singapore's meritocratic government is not hostage to the usual set of vested interests that constrain governments. Fourth, the government has revealed a willingness to open its labor market to an extent unparalleled among modern nation-states. At least 25 percent of its work force is foreign, and a much higher percentage was born overseas. With its high salary structure, it is able to recruit in the international labor market. Finally, Singapore has a seemingly completely incorruptible civil service. Its public sector remuneration is one of the highest in the world, and it is insulated from political pressures. Thus a selective industrial policy is more likely to succeed there than in practically any other country in the world.[19] Enthusiasts of Singapore's success and its lessons for other countries need to be cognizant of these features. The more follower countries depart from them, the less this model is likely to be replicable.

A final general observation on technology and MNCs is that it is now very difficult for latecomer industrializers to achieve high rates of export growth without the participation of MNCs. The earlier literature on this subject, in which Nayyar's (1978) is the dominant study, argues that the involvement of MNCs in export expansion from the newly industrialized economies has been low by international standards. However, there is clear evidence that the strong export performance of developing

19. The only caveat I would attach to this extraordinary record of development is that the resources devoted to the country's R&D effort are not particularly transparent. The presumption is that internal checks and balances ensure that there is no corruption.

economies since the 1970s has been closely associated with the involvement of MNCs. Nayyar estimates the share of MNCs in total manufactured exports from developing economies to be not more than 15 percent circa 1974. By contrast, a similar calculation, based on unpublished estimates prepared by Prema-Chandra Athukorala at the Australian National University, suggests that MNCs accounted for 24 percent of total manufactured exports from developing economies circa 1980. This figure had increased to 36 percent circa 1990 and probably surpassed 50 percent by the turn of the century.

Industry Policy

What role has industrial policy—defined here as a deliberately non-neutral incentives regime—played in East Asia's rapid industrialization? This is one of the most extensively debated issues in development policy.[20] At the risk of oversimplification, two contending paradigms have emerged. According to one, the major contribution of governments has been in getting the fundamentals right: macroeconomic stability, predictable and stable policy regimes, improved physical infrastructure and education, a reasonably adequate system of property rights and legal infrastructure, and increasing openness to international commerce. An alternative paradigm accepts some or all of these prescriptions[21] but argues that it is an insufficient recipe for industrial success. This school rejects an emphasis based primarily on economic liberalism and static comparative advantage. It argues that Japan, Korea, and Taiwan (China) owe their success to selective industrial policies—targeting industries, picking winners, and deliberately getting prices wrong through fiscal incentives, subsidized credit, import protection, and direct investment.

There has been less work on this issue among the ASEAN economies, and it is therefore useful to briefly survey the literature. The standard tools

20. The literature on this subject is very large. Smith (1995) provides a comprehensive survey of the East Asian literature from a largely neoclassical perspective. Influential East Asian case studies among the interventionist school include Amsden (2001) in general, Amsden (1989) and Chang (1994) on Korea, and Wade (1990) on Taiwan (China). Hughes (1988) provides a dated, though still important, airing of competing views. The contributors to Wong and Ng (2001) provide comprehensive analyses of Japan and the four Asian newly industrialized economies, much of it sympathetic to interventionist paradigms. Yusuf (2001, pp. 20–25) provides a succinct and balanced summary of the literature.

21. Indeed, this school arguably takes achievements in these areas too much for granted and downplays the fact that, with limited high-level bureaucratic resources available, a more activist strategy may be implemented at the cost of diminished performance in the core areas.

of industrial policy have been employed in all three economies. Inter-industry variations in protection have been considerable, especially in Indonesia. A large state enterprise sector has been present in several of them. Subsidized credit and interest rate controls have been present, as have fiscal incentives.

In reality, however, promotional measures in ASEAN have been prone to abuse, implementation has been sporadic and often short-lived, and there has been little systematic attempt to prescribe conditionality in the sense of linking incentives to tightly defined performance criteria. It is hardly surprising that studies that have examined the relationship between inter-industry variations in government assistance and subsequent (lagged) performance, according to a variety of measures, have found little evidence of causality. In the case of Indonesia, a survey of selective policy instruments introduced or extended in the 1970s and early 1980s—protection, credit subsidies, state enterprises—detected very little evidence of such a strategy having "worked" according to a range of subsequent performance criteria (Hill 1996). Similarly, Warr's (1995) study of several policy correlates of export performance in Thailand over the period 1970–89 revealed a remarkably consistent picture of negative coefficients for all variables and time periods. In the case of Malaysia, it is very difficult to find any evidence in support of the proposition that the government's promotion of heavy industry over the period 1978–86 produced efficient industrial growth (Athukorala and Menon 1996). The most concerted attempt at explicit industry policy has been applied to the auto industry, which, after more than two decades of high protection, still shows little sign of being internationally competitive.

The economic crisis has weakened still further the case for an interventionist industrial policy. That is, in the wake of Korea's deep economic crisis and Japan's protracted recession, the "Northeast Asian model" is looking a lot less attractive. Attempts to finance uneconomic, highly politicized industrial projects through a shaky financial system, or through foreign borrowings with implicit government guarantees, contributed somewhat to the crisis.

Of course, as argued above, markets do fail. The challenge for policymakers is to intervene effectively in a rapidly globalizing world economy and to overcome market failure without introducing government failure. Fortunately, as industrial latecomers, the lower-income Southeast Asian countries can learn from the successes of others. Elsewhere in this chapter, I look at cases of effective public policy, especially in leveraging the presence of MNCs and the provision of high-quality industrial extension services. Much, of course, depends on political will and bureaucratic capacity.

Enclaves and Electronics

The electronics industry is of particular interest because it dominates much of East Asian manufacturing exports, in several economies constituting more than half of merchandise exports. As Hobday (1995, 2001) points out, in the four Asian newly industrialized economies, virtually all combinations of policy regimes have been evident. These include (a) extensive intervention and the presence of large firms in Korea and Singapore, which have been distinguished, respectively, by closed and open policies toward FDI, and (b) small local firms alongside larger foreign ones in Taiwan (China) and Hong Kong (China), in the context, respectively, of partially liberal and laissez-faire FDI and trade policies. It is important, however, to emphasize the common elements across the four economies: export orientation, strong investments in human capital, generally sound macroeconomic management, and liberal policies toward technology imports (at least for export-oriented firms).

There is a presumption in much of the literature, especially in Malaysia,[22] that the "Southeast Asian model" of an electronics industry led by MNCs is somehow inferior to that of Korea, which aimed to build up domestic technological capabilities with a quite limited presence of multinational corporations. Critics see the former as a case of shallow, enclave-based, foreign-dominated industrialization. The industry is sometimes admonished for the slow pace of backward integration. However, Hobday (2000, 2001) and others have cast doubt on such a notion. First, for the late-comer industrializers, labor-intensive electronics exports initially provided much-needed employment when unemployment levels were high.[23] This growth also established a form of "country reputation" that was crucial in attracting other foreign investors (see Wells 1994 on this point in the East Asian context). Moreover, a good deal of innovation is already occurring within the industry, in response to rising real wages and increased domestic competence. In the case of Malaysia, according to Hobday (2001, p. 20), "Contrary to popular wisdom, the Malaysian (and to a lesser extent the Thai evidence) shows that TNC [transnational corporations]-led development has proved to be a remarkably successful strategy." Finally, arguments supporting the superiority of the Korean approach to electronics overlook its weaknesses, some of which were alluded to above: the marginalization of

22. See Mohd Nazari Ismail (1999, especially pp. 30–34) for an interesting and balanced account of the industry, with much useful material on instances of "technological deepening."

23. For example, Singapore from the late 1960s onward, Malaysia from the late 1970s, and the Philippines from the mid-1990s.

small- and medium-scale enterprises, major weaknesses in design and marketing, and competence in a narrow range of goods.

This is not to argue that this Southeast Asian model is inherently superior. It, too, has weaknesses. The absence of backward linkages is a cause for concern. There has been an underinvestment in technical and vocational education in Malaysia and Thailand. Fiscal incentives for firms within the export-processing zones have arguably been excessively generous.

Small- and Medium-Scale Enterprises

Small- and medium-scale enterprises typically employ most of the industrial labor force in East Asia and generate a significant share of its output. Yet the literature on industrial innovation typically ignores these firms. It is useful, therefore, to refer to some recent research on success stories, mainly from Indonesia. These include Cole's (1998) study of the Bali garment export industry and research on the export-oriented small- and medium-scale furniture manufacturers in the town of Jepara, Northern Central Java.[24]

The Bali garment industry, which grew spectacularly in the 1980s and is based almost exclusively on small firms, was practically an "accidental" case of industrialization. Foreign tourists, mainly surfers wishing to support a recreational lifestyle, saw commercial opportunities in Balinese garments and its indigenous design capacity. They were able to act as marketing intermediaries, connecting local producers with retail outlets abroad, in the process dispensing important information on designs and production techniques. Later, as the island's fame spread, these links developed quickly, and the industry mushroomed from its seasonal, cottage origins to larger production units and some local design capacity. The Jepara furniture industry had its origins further back, but it too began to grow quickly in the 1980s. The industry lacked the connection with tourism, but it did have a good base of local skills together with access to raw materials, and foreigners quickly saw the opportunities for profitable export as deregulation proceeded.

These studies suggest a model of successful and innovative development of small- and medium-scale enterprise in which the following ingredients appear to be important. First, there was some basic industrial competence in a particular field of activity (sewing and woodworking in these cases). Second, there was generally a conducive macroeconomic environment (except at the peak of the 1997–98 crisis), including especially a competitive exchange rate. Third, there was reasonably good physical infrastructure,

24. On the latter, see Berry and Levy (1999); Sandee, Andadari, and Sulandjari (2000); Schiller and Martin-Schiller (1997).

including proximity to adequately functioning import and export facilities. Finally, of importance also were injections of technical, design, and marketing expertise linking small producers to new ideas and major markets.

With the possible exception of the first ingredient, all four elements are directly amenable to public policy. These case studies also have important implications for government policy. Neither resulted from any deliberate government promotional measures. The government did play an important role in providing a supportive macroeconomic environment and a rapidly improving infrastructure. In Bali the local government generally adopted a fairly open policy toward the presence of foreign entrepreneurs, and export procedures were not unduly burdensome most of the time. The June 1994 reform of regulations regarding FDI made it easier for small foreign investors to operate in the country without harassment. These, of course, hardly constitute "contributions" from government, except in the negative sense of avoiding a harshly restrictive regulatory regime.[25]

Agriculture

It is often forgotten that arguably the most important technological revolution to occur in East Asia over the past four decades was the green revolution.[26] Beginning in the 1960s, it resulted in unprecedented increases in food production and transformed this sector in many countries, vastly improving the precarious food supply equations. Although criticized at the time on both environmental and equity grounds, it boosted agricultural productivity and rural development, especially as it was largely scale neutral in its application and impact. This was an excellent illustration of effective international assistance combining with local innovation capacities (Pingali 2001).

Agricultural sectors are now very much smaller, but they still employ at least one-third of the work force in the poorer East Asian economies studied here. There is no reason in principle why, with modification, such a model cannot continue to energize agriculture. National agricultural research institutes are now much less significant actors. They also need to be reformed (Byerlee and Alex 1998): there need to be more linkages with the private sector and with universities, more competition for grants, a more demand-driven outlook, and a critical review of staffing and functions. As public sec-

25. As Cole (1998, p. 277) puts it, "Beyond these points, the role the government played seems more positive in its absence than in its actions."

26. Evenson and Gollin (2001) provide a good retrospective.

tor support, both domestic and international, has dried up, attention has focused on ways of attracting more private funding. While in principle desirable, it is important not to lose sight of the public good dimensions of international support for agricultural research, that is, its nonrival and nonexcludable nature (see Herdt 1999). Shifting to private sector funding on a large scale runs the risk of jeopardizing the key elements of the past record. Of course, much agricultural research is still conducted in OECD-country institutes, but their work focuses on an economic and ecological setting that is generally very different from that of low-income tropical agriculture. The work of international centers with a specific developing-country focus deserves much more support.

A second strand in the literature on agricultural innovation and diffusion focuses on the cash crop sector. Here there have not been the major breakthroughs or large-scale international support that were evident in the food crop sector, but rather a more incremental and gradual process. Experience shares much in common with the industrial small- and medium-size enterprise sector (Barlow 2000, 2001). In the case of the natural rubber industry, Malaysia established its global supremacy in part owing to its excellent research institutes and extension networks. As Malaysia began to lose comparative advantage in rubber, Thailand, a relatively minor producer as recently as 1960, became the world's largest producer by 1990. Unlike its two neighbors, the Thai industry is almost entirely smallholder. Several ingredients underpinned the Thai success. Its agricultural extension staff were effective, and there was excellent officer-farmer interaction. Physical infrastructure, especially roads, was developed, and this in turn led to the rapid spread of motorized transport. Central government nurseries were established initially to distribute improved seed varieties, but these spread rapidly to many small private operations. There was also a need to overcome the moral hazard problem of overtapping of rubber trees—which leads to their premature aging—and the assumption that the government would come to the rescue. This was achieved through effective local community organization. These initiatives were generally not high cost and were funded by a combination of a small levy on rubber exports plus development assistance and central government funding.

Indonesia's record in rubber was less successful (Barlow 2000). Government-owned estates dominated the industry. These were bureaucratic, politicized, and subsidized entities, with low levels of efficiency and little incentive to innovate. Moreover, the various schemes to promote smallholders prior to the 1980s were largely unsuccessful. By contrast, Indonesia's cocoa industry has been a success story (Takamasa and Nishio 1997). Comprising mainly smallholders in Sulawesi, output has grown rapidly since

1980. The key factors appear to have been the provision of reasonably good physical infrastructure (especially roads), a competitive exchange rate, and the absence of government distortions in marketing.

The lesson for technology and innovation policy emerging from these agricultural studies is that the key role of government is to spur innovation and to make markets work. Incomplete markets are clearly a major barrier to innovation. An effective public policy package, suitably modified for the local ecological and institutional context, therefore appears to include the following: (a) good-quality physical infrastructure, which introduces more competition among traders and lowers transport and information costs; (b) efficient financial intermediation, particularly small-scale and informal credit,[27] that *inter alia* rests on secure and transparent property rights; (c) particularly in the early stages, rural research institutes, which can assist small farmers to absorb, assimilate, and adapt new technologies, and effective agricultural extension staff; and (d) more controversially, modest and strictly time-bound subsidies, which may be useful to induce risk-averse farmers to adopt new technologies and commercial inputs.

Public Research Institutes

Public research institutes are critical actors in the early stages of technology and innovation. Typically, most commercial entities are too small to sustain a major R&D effort. A culture of R&D has generally not yet taken root in developing East Asian economies, and thus these institutes can act as catalysts. Moreover, the institutional supports necessary to maintain a modern R&D effort—from protection of intellectual property rights to a venture capital market—are generally underdeveloped.

East Asian governments have experimented with a wide variety of institutional arrangements in support of public research institutes. The general record has been one of limited success, since the institutes are typically poorly funded and their links to universities and firms are weak.

A widely cited success story of clever and nimble intervention is Taiwan's (China's) Industrial Technology Research Institute (ITRI).[28] Founded in 1973, the institute expanded rapidly in the 1980s; by 1994 it had 6,000 employees, 560 of whom possessed doctorates. ITRI is now the largest industry-oriented research institution in Taiwan (China) and has working relations with some 20,000 companies. It has received significant public sector

27. For an informative Indonesian case study on small-scale credit, see Patten, Rosengard, and Johnston (2001).

28. This paragraph draws on Lin (1998). See also Chu, Chen, and Chen (2001).

funding, absorbing about 25 percent of the government's non-defense technology projects between 1983 and 1994. The Hsinchu Science-Based Industrial Park, established in 1980, has been an effective means of disseminating new technologies. In both cases, Taiwan's (China's) industrial structure was such that government funding of these initiatives was a necessary prerequisite for upgrading. Possessing few very large conglomerates, the Korean strategy of promoting R&D within large firms was not an option.

Central to their success, according to Lin (1998), were six factors: a strong national base of human resources on which to draw, access to international technology markets, a base of competent domestic R&D, an emphasis on diffusion and commercialization of domestic and international know-how, strong ties to the private sector, and supportive intellectual property rights. The technology diffused very quickly to small- and medium-scale enterprises, through the movement of personnel between firms and research institutes and direct observation and learning. Joint ventures with foreign firms adjacent to these scientific facilities hastened the process. As Westphal (forthcoming) notes, ITRI entirely funded Taiwan's (China's) first DRAM (dynamic random access memory) production facility, which in turn underpinned much of the subsequent success in electronics (Matthews and Cho 2000).

Perhaps more typical of the broader developing-country context are the much less effective R&D institutions in Indonesia (see Lall 1998b; Thee 1998) and the other lower-income Southeast Asian countries. Two interrelated weaknesses are generally present. First, their base of funding is generally inadequate. This results in salaries that are too low to attract high-caliber staff, facilities that do not meet best practice requirements, and (especially in those agencies working in remote rural communities) capacity that is too limited to engage in meaningful outreach for the targeted client groups. The second problem relates to the objectives and functioning of these institutes. Their mission statements and philosophies are typically supply rather than demand driven. Their ties with the private sector are generally weak: there is little staff mobility, and even their location may not be based on the needs of their client groups. Frequently, the institutes have a welfare rather than an efficiency orientation.

The rest of East Asia is typically located somewhere between Taiwan (China) and Indonesia in terms of the capacity and effectiveness of its public research institutes. Malaysia has a mixed record (Rasiah 1999). As noted, it was historically the most successful exponent of high-quality agricultural extension services for tropical cash crops, and there have been some notable success stories in the development of Penang's electronics industry. But elsewhere, the linkages between public research institutes and domestic firms have been weaker, in part perhaps because of the schism between a Malay-dominated public sector and an ethnic Chinese–dominated sector of small- and medium-size enterprises. In Korea most R&D has been undertaken within the *chaebols* rather than public research institutes, and links between the two, and with universities, have been rather weak. Very little of the country's spectacular advances in electronics, for example, have originated from public research institutes (Matthews and Cho 2000). Since the crisis, a change in policy direction is evident. Similarly, these links have not been well established in Singapore (Wong 2001). This, too, is changing, and in any case the public sector (both the so-called government-linked corporations and government departments) has been a particularly active innovator, with rapid diffusion to private firms.

SUMMING UP

This chapter has focused on a number of issues. One is that technology and innovation are gradual processes of evolution, especially for developing East Asian economies, which with a few exceptions are "followers" and net technology importers. Especially among the poorer economies, the challenge is to import, absorb, assimilate, and diffuse technology rather than to undertake new R&D at the frontiers.

Second, these are extraordinarily diverse economies in their history, level of development, international orientation, and institutional capacity. As a result, there is no unique path of technological development. It is clear, though, that the role of the government shifts over time. Among the late industrializers in Southeast Asia, the principal role of government is to provide competent macroeconomic management, an open economy, good basic public services (education and health especially), and infrastructure. Explicit technology and innovation policies are not central to this mission, although extension services and other agencies that diffuse technology are important.

Third, as economies approach middle-income status, and there is a loss of comparative advantage in labor-intensive activities, governments need to play a role in the upgrading process by strengthening institutions, building supply-side capabilities, and encouraging the emergence of national innovation systems. The major challenge here is to decide where markets work or do not work and to intervene cautiously. Much depends in this context on administrative capacity.

Finally, the region presents a mixed record of notable strengths and persistent weaknesses, and it is useful to summarize some of the key challenges arising from the experiences of five economies examined in this chapter.

Korea stands out for its major commitment to building domestic R&D capacity, its high level of educational achievement, and its ability to contemplate daunting development challenges. Nevertheless, it has paid a high price for restricting firms' access to international know-how, and in consequence its firms have difficulty developing international networks. Its industrial policy has been adventurous and sometimes costly. It lacks the flexibility of a dynamic base of small- and medium-scale enterprises and a capacity to incubate start-ups. Linkages between *chaebols* and both government institutes and universities are weak, while universities' research capacities are limited. Finally, corporate governance (including minority shareholder confidence in the stock market) is poor by OECD standards.

Singapore scores highly for its extremely open economy, its high-quality public administration, its major recent commitment to higher education and research, its aggressive innovative practices in the public sector, its insulated public policy processes, and its capacity to leverage the presence of multinational corporations to its own advantage. But domestic entrepreneurship and creativity are still lacking, reflected in part by the fact that the government has avowedly sought to attract the top talent to its ranks, with the world's best-paid civil service. Moreover, an overbearing government may stifle the very creative spirit it seeks to ignite. Links between public research institutions and universities, on the one hand, and firms, on the other, are still rather weak. Finally, although fiscal incentives are in all likelihood awarded effectively and contribute to technological upgrading, there is little transparency in the process.

Indonesia's impressive development achievements during the first quarter-century of the Suharto era were marred in the case of technology policy by the massive investments in the Habibie mega projects, particularly the aircraft factory, while basic education and agricultural and industrial extension services were starved of resources. In the post-Suharto era, the challenge is to overcome the financial and fiscal crisis, maintain past investments in physical and social infrastructure, and reestablish a business-friendly environment in the context of highly fluid and uncertain political structures. Explicit technology and innovation programs will inevitably receive a low priority during this period of transition.

Taiwan (China) has achieved unparalleled development success over more than four decades and has survived the recent Asian crisis more successfully than any other newly industrialized economy. Its model of industrial and technology development is unique, with its emphasis on a highly flexible, resilient, and internationally oriented small- and medium-scale enterprise sector. Although not as open to FDI as the city-states and Malaysia, its international connections are well developed, aided by the return in recent years of much high-level human capital. It was able to develop a strong R&D base through public funding in the 1980s, which diffused quickly to small- and medium-scale enterprises. Apart from the general challenge of integration with Greater China, its major challenge appears to be weak corporate and political governance.

Malaysia too has been a striking success story for most of its independent history but now has to manage the challenge of upgrading its R&D base and capacity for innovation. Arguably, its formal industry and technology policy, especially its heavy industry push, has contributed little to its rapid growth. The quality of higher education may well have deteriorated since the 1970s. The elite generally attend universities abroad, and there continues to be an exodus of highly educated Malaysians who are not *bumiputeras*. Finally, it could have done more to leverage the presence of MNCs and maximize potential spin-offs.

REFERENCES

The word *processed* describes informally reproduced works that may not be commonly available through libraries.

Amsden, Alice. 1989. *Asia's Next Giant: South Korea and Late Industrialization.* New York: Oxford University Press.

———. 2001. *The Rise of the Rest: Challenges to the West from the Late-Industrializing Economies.* New York: Oxford University Press.

Athukorala, Prema-Chandra, and Hall Hill. 2002. "Host Country Impact of FDI in East Asia." In Bijit Bora, ed., *Foreign Direct Investment: Research Issues.* London: Routledge.

Athukorala, Prema-Chandra, and Chris Manning. 1999. *Structural Change and International Migration in East Asia: Adjusting to Labour Scarcity.* Melbourne: Oxford University Press.

Athukorala, Prema-Chandra, and Jayant Menon. 1996. "Foreign Direct Investment and Industrialisation in Malaysia: Exports, Employment, and Spillovers." *Asian Economic Journal* 10(1):29–44.

Barlow, Colin. 2000. "Institutions of Change in Rural Development: Mediating Markets for New Crop Technologies in Sumatra." In Colin Barlow, ed., *Institutions and Economic Change in Southeast Asia.* Cheltenham, U.K.: Edward Elgar.

———. 2001. "The Role of Institutions in Planting Improved Smallholder Rubber." Unpublished ms., Australian National University, Canberra. Processed.

Barro, Robert J., and Jong-Hwa Lee. 1996. "International Measures of Schooling Years and Schooling Quality." *American Economic Review* 86(2):218–23.

Berry, R. Albert, and Brian Levy. 1999. "Technical, Marketing, and Financial Support for Indonesia's Small and Medium Industrial Exporters." In Brian Levy, R. Albert Berry, and Jeffrey B. Nugent, eds., *Fulfilling the Export Potential of Small and Medium Firms.* Dordrecht, the Netherlands: Kluwer Academic Publishers.

Blomström, Magnus, and Ari Kokko. 1998. "Multinational Corporations and Spillovers." *Journal of Economic Surveys* 12(3):247–77.

Booth, Anne. 1999. "Education and Economic Development in Southeast Asia: Myths and Realities." *ASEAN Economic Bulletin* 16(3):290–306.

Byerlee, Derek, and Gary E. Alex. 1998. *Strengthening National Agricultural Research Systems: Policy Issues and Good Practice.* Washington, D.C.: World Bank.

Chang, H. J. 1994. *The Political Economy of Industrial Policy.* London: Macmillan.

Chen, E. K. Y. 1997. "The Total Factor Productivity Debate: Determinants of Economic Growth in East Asia." *Asian-Pacific Economic Literature* 11(1):18–38.

Chu, Y.-P. 2000. "Markets Grew and Matured as a Result of State Actions: The Taiwan Experience." Unpublished mss., Academia Sinica, Taipei. Processed.

Chu, Y.-P, T.-Y. Chen, and B.-L. Chen. 2001. "Rethinking the Development Paradigm: Lessons from Taiwan—The Optimal Degree of State Intervention." In Poh Kam Wong and Chee Yuen Ng, eds., *Industrial Policy, Innovation, and Economic Growth: The Experience of Japan and the Asian NIEs.* Singapore: Singapore University Press.

Cole, William. 1998. "Bali's Garment Export Boom." In Hall Hill and Kian Wie Thee, eds., *Indonesia's Technological Challenge.* Singapore: Institute of Southeast Asian Studies.

Dahlman, Carl J., and Thomas Andersson. 2000. *Korea and the Knowledge-Based Economy: Making the Transition.* Washington D.C.: World Bank.

Dodgson, Mark. 2000. "Policies for Science, Technology, and Innovation in Asian Newly Industrializing Economies." In Linsu Kim and Richard R. Nelson, eds., *Technology, Learning, and Innovation: Experiences of Newly Industrializing Countries.* Cambridge, U.K.: Cambridge University Press.

Duysters, Geert, and John Hagedoorn. 2000. "International Technological Collaboration: Implications for Newly Industrializing Economies." In Linsu Kim and Richard R. Nelson, eds., *Technology, Learning, and Innovation: Experiences of Newly Industrializing Countries.* Cambridge, U.K.: Cambridge University Press.

Economic Planning Unit. 2000. "Developing Malaysia into a Knowledge-Based Economy." In *The Third Outline Perspective Plan, 2001–2020,* ch. 5. Kuala Lumpur.

Evenson, Robert E., and Douglas Gollin. 2001. "The Green Revolution at the End of the Twentieth Century." Paper prepared for the Technical Advisory Committee, Consultative Group on International Agricultural Research (CGIAR). Processed.

Evenson, Robert E., and Larry E. Westphal. 1995. "Technological Change and Technology Strategy." In Jere Behrman and T. N. Srinivasan, eds., *Handbook of Development Economics,* vol. 3. Amsterdam: Elsevier.

Fagerberg, Jan. 1994. "Technology and International Differences in Growth Rates." *Journal of Economic Literature* 32(3):1147–75.

Fan, Xiaoqin, and Peter G. Warr. 2000. "Foreign Investment, Spillover Effects, and the Technology Gap: Evidence from China." Working Paper in Trade and Development 00/06. Australian National University, Division of Economics, Research School of Pacific and Asian Studies. Processed.

Herdt, R. W. 1999. "Enclosing the Global Plant Genetic Commons." Paper presented at a conference at the Institute for International Studies, Stanford University, Palo Alto, Calif. Processed.

Hill, Hal. 1996. "Indonesia's Industrial Policy and Performance: 'Orthodoxy' Vindicated." *Economic Development and Cultural Change* 45(1):147–74.

Hill, Hal, and Kian Wie Thee, eds. 1998. *Indonesia's Technological Challenge.* Singapore: Institute of Southeast Asian Studies.

Hobday, Mike. 1995. *Innovation in East Asia: The Challenge to Japan.* Cheltenham, U.K.: Edward Elgar.

———. 2000. "East versus Southeast Asian Innovation Systems: Comparing OME- and TNC-led Growth in Electronics." In Linsu Kim and Richard R. Nelson, eds., *Technology, Learning, and Innovation: Experiences of Newly Industrializing Countries.* Cambridge, U.K.: Cambridge University Press.

———. 2001. "The Electronics Industries of Pacific Asia." *Asia-Pacific Economic Literature* 15(1):13–29.

Hou, Chi-Ming, and San Gee. 1993. "National Systems Supporting Technical Advance in Industry: The Case of Taiwan." In Richard R. Nelson, ed., *National Innovation Systems: A Comparative Analysis.* New York: Oxford University Press.

Hughes, Helen, ed. 1988. *Achieving Industrialization in East Asia.* Cambridge, U.K.: Cambridge University Press.

Ismail, Mohd Nazari. 1999. "Foreign Firms and National Technology Upgrading." In K. S. Jomo, Greg Felker, and Rajah Rasiah, eds., *Industrial Technology Development in Malaysia: Industry and Firm Studies.* London: Routledge.

Jomo, K. S., and Greg Felker, eds. 1999. *Technology, Competitiveness, and the State: Malaysia's Industrial Technology Policies.* London: Routledge.

Jomo, K. S., Greg Felker, and Rajah Rasiah, eds. 1999. *Industrial Technology Development in Malaysia: Industry and Firm Studies.* London: Routledge.

Kim, Linsu. 1997. *Imitation to Innovation: The Dynamics of Korea's Technological Learning.* Boston: Harvard Business School Press.

———. 2000. "Korea's National Innovation System in Transition." In Linsu Kim and Richard R. Nelson, eds., *Technology, Learning, and Innovation: Experiences of Newly Industrializing Countries.* Cambridge, U.K.: Cambridge University Press.

Kim, Linsu, and Richard R. Nelson. 2000a. "Introduction." In Linsu Kim and Richard R. Nelson, eds., *Technology, Learning, and Innovation: Experiences of Newly Industrializing Countries.* Cambridge, U.K.: Cambridge University Press.

Kim, Linsu, and Richard R. Nelson, eds. 2000b. *Technology, Learning, and Innovation: Experiences of Newly Industrializing Countries.* Cambridge, U.K.: Cambridge University Press.

Lall, Sanjaya. 1995. "Malaysia: Industrial Success and the Role of the Government." *Journal of International Development* 7(5):759–73.

———. 1998a. "Technological Capabilities in Emerging Asia." *Oxford Development Studies* 26(2):213–44.

———. 1998b. "Technology Policies in Indonesia." In Hal Hill and Kian Wie Thee, eds., *Indonesia's Technological Challenge.* Singapore: Institute of Southeast Asian Studies.

———. 2000a. "Technological Change and Industrialization in the Asian Newly Industrializing Economies: Achievements and Challenges." In Linsu Kim and Richard R. Nelson, eds., *Technology, Learning, and Innovation: Experiences of Newly Industrializing Countries.* Cambridge, U.K.: Cambridge University Press.

———. 2000b. "The Technological Structure and Performance of Developing Country Manufactured Exports, 1985–98." *Oxford Development Studies* 28(3):337–63.

Lary, Hal B. 1968. *Imports of Manufactures from Developing Countries.* New York: Columbia University Press.

Lee, Won-Young. 2000. "The Role of Science and Technology Policy in Korea's Industrial Development." In Linsu Kim and Richard R. Nelson, eds., *Technology, Learning, and*

Innovation: Experiences of Newly Industrializing Countries. Cambridge, U.K.: Cambridge University Press.

Lin, O. C. C. 1998. "Science and Technology Policy and Its Influence on Economic Development in Taiwan." In Henry S. Rowen, ed., *Behind East Asian Growth: The Political and Social Foundations of Prosperity.* London: Routledge.

Low, Linda, Helen Hughes, Toh Mun Heng, Soon, Tech, Wong, and Tan Kong Yam. 1993. *Challenge and Response: Thirty Years of the Economic Development Board.* Singapore: Times Academic Press.

Matthews, John A., and D-S Cho. 2000. *Tiger Technology: The Creation of a Semiconductor Industry in East Asia.* Cambridge, U.K.: Cambridge University Press.

McKendrick, David. 1992. "Obstacles to "Catch-up": The Case of the Indonesian Aircraft Industry." *Bulletin of Indonesian Economic Studies* 28(1):39–66.

McKendrick, David G., Richard F. Doner, and Stephan Haggard. 2000. *From Silicon Valley to Singapore: Location and Competitive Advantage in the Hard Disk Drive Industry.* Palo Alto, Calif.: Stanford University Press.

Nayyar, Deepak. 1978. "Transnational Corporations and Manufactured Exports from Poor Countries." *Economic Journal* 88(March):59–84.

Nelson, Richard R., ed. 1993. *National Innovation Systems: A Comparative Analysis.* New York: Oxford University Press.

Nelson, Richard R., and Howard Pack. 1999. "The Asian Miracle and Modern Growth Theory." *Economic Journal* 109(July):416–36.

NSTB (National Science and Technology Board). 1997. *National Science and Technology Plan.* Singapore.

———. 2000. *National Survey of R&D in Singapore, 1999.* Singapore.

OECD (Organisation for Economic Co-operation and Development). 1997. *Technology and Industrial Performance.* Paris.

Okamoto, Yumiko, and Fredrik Sjoholm. 2003. "Technology Development in Indonesia." In Sanjaya Lall and Shujiro Urata, eds., *Competitiveness, FDI, and Technological Activity in East Asia.* Cheltenham, U.K.: Edward Elgar.

Pack, Howard. 2000. "Research and Development in the Industrial Development Process." In Linsu Kim and Richard R. Nelson, eds., *Technology, Learning, and Innovation: Experiences of Newly Industrializing Countries.* Cambridge, U.K.: Cambridge University Press.

Pack, Howard, and Larry E. Westphal. 1986. "Industrial Strategy and Technological Change: Theory versus Reality." *Journal of Development Economics* 22(1):87–128.

Patten, R. H., J. K. Rosengard, and D. E. Johnston. 2001. "Microfinance Success amidst Macroeconomic Failure: The Experience of Bank Rakyat Indonesia during the East Asian Crisis." *World Development* 29(6):1057–69.

Pingali, Prabhu L. 2001. "Milestones in Impact Assessment Research in the CGIAR, 1970–1999." Unpublished ms. Available at http://www.cimmyt.org/Research/economics/map/impact studies/milestones/pdfs/milestones_pingali.pdf. Processed.

Rasiah, Rajah. 1999. "Malaysia's National Innovation System." In K. S. Jomo and Greg Felker, eds., *Technology, Competitiveness, and the State: Malaysia's Industrial Technology Policies.* London: Routledge.

———. 2001. "Government-Business Coordination and Small Business Performance in the Machine Tools Sector in Malaysia." Unpublished ms. United Nations University, Institute for New Technology (UNU-INTECH), Maastricht, the Netherlands. Processed.

Rosenberg, Nathan. 1994. *Exploring the Black Box: Technology, Economics, and History*. Cambridge, U.K.: Cambridge University Press.

Sandee, Henry, Ross Kities Andadari, and Sri Sulandjari. 2000. "Small Firm Development during Good Times and Bad: The Jepara Furniture Industry." In Chris Manning and Peter van Diermen, eds., *Indonesia in Transition: Social Aspects of Reformasi and Crisis*. Singapore: Institute of Southeast Asian Studies.

Sato, Yukio. 1998. "The Transfer of Japanese Management Technology to Indonesia." In Hal Hill and Kian Wie Thee, eds., *Indonesia's Technological Challenge*. Singapore: Institute of Southeast Asian Studies.

Schiller, Jim, and Barbara Martin-Schiller. 1997. "Market, Culture, and State in the Emergence of an Indonesian Export Furniture Industry." *Journal of Asian Business* 13(1):1–23.

Sjoholm, Fredrik. 1999. "Productivity Growth in Indonesia: The Role of Regional Characteristics and Direct Foreign Investment." *Economic Development and Cultural Change* 47(3, April):559–84.

Smith, Heather. 1995. "Industry Policy in East Asia." *Asian-Pacific Economic Literature* 9(1):17–39.

———. 1998. "Korea." In Ross H. McLeod and Ross Garnaut, eds., *East Asia in Crisis: From Being a Miracle to Needing One?* London: Routledge.

———. 2000. *Industry Policy in Taiwan and Korea in the 1980s: Winning with the Market*. Cheltenham: Edward Elgar.

Soesastro, Hadi. 1998. "Emerging Patterns of Technology Flows in the Asia-Pacific Region: The Relevance to Indonesia." In Hal Hill and Kian Wie Thee, eds., *Indonesia's Technological Challenge*, pp. 303–25. Singapore: Institute of Southeast Asian Studies.

Stiglitz, Joseph. 1996. "Some Lessons from the East Asian Miracle." *World Bank Research Observer* 11(2):151–77.

Takamasa, Akiyama, and Akihiko Nishio. 1997. "Sulawesi's Cocoa Boom: Lessons of Smallholder Dynamism and a Hands-off Policy." *Bulletin of Indonesian Economic Studies* 33(2):97–121.

Thee, Kian Wie. 1998. "The Determinants of Indonesia's Industrial Technology." In Hal Hill and Kian Wie Thee, eds., *Indonesia's Technological Challenge*. Singapore: Institute of Southeast Asian Studies.

Toh, Kin Woon. 2002. "The Political Economy of Industrialization in Penang." Paper presented at a conference on Malaysian industrialization, Australian National University, Canberra, March. Processed.

UNCTAD (United Nations Conference on Trade and Development). 2001. *World Investment Report*. Geneva.

UNIDO (United Nations Industrial Development Organization). 2002. *Industrial Development Report 2002/2003: Competing through Innovation and Learning*. Vienna.

Urata, Shujiro. 2001. "Emergence of an FDI-Trade Nexus and Economic Growth in East Asia." In Joseph E. Stiglitz and Shahid Yusuf, eds., *Rethinking the East Asian Miracle*. New York: Oxford University Press.

Wade, Robert. 1990. *Governing the Market: Economic Theory and the Role of Government in East Asian Industrialization*. Princeton, N.J.: Princeton University Press.

Warr, Peter. 1995. "Myths about Dragons." *Agenda* 1(2):215–28.

———. 2000. "What Happened to Thailand?" *World Economy* 22(5):631–50.

Wells, Louis T. 1994. "Mobile Exporters: New Foreign Investors in East Asia." In Kenneth A. Froot, ed., *Foreign Direct Investment.* Chicago, Ill.: University of Chicago Press.

Westphal, Larry E. Forthcoming. "Technology Strategies for Economic Development in a Fast-Changing Global Economy." *Economics of Innovation and New Technology.* Special issue on the political economy of technology policy in developing countries.

Wong, Poh Kam. 2001. "From Leveraging Multinational Corporations to Fostering Technopreneurship: The Changing Role of S&T Policy in Singapore." In Linda Low and Douglas M. Johnston, eds., *Singapore Inc: Public Policy Options in the Third Millennium.* Singapore: Asia Pacific Press.

———. 2003. "From Using to Creating Technology: The Evolution of Singapore's National Innovation System and the Changing Role of Public Policy." In Sanjaya Lall and Shujiro Urata, eds., *Competitiveness, FDI, and Technological Activity in East Asia.* Cheltenham: Edward Elgar.

Wong, Poh Kam, and Chee Yuen Ng, eds. 2001. *Industrial Policy, Innovation, and Economic Growth: The Experience of Japan and the Asian NIEs.* Singapore: Singapore University Press.

World Bank. 2000. *World Development Report 2000/2001.* Washington, D.C.

World Economic Forum. 2000. *Global Competitiveness Report 2000.* Geneva.

Yusuf, Shahid. 2001. "The East Asian Miracle at the Millennium." In Joseph E. Stiglitz and Shahid Yusuf, eds., *Rethinking the East Asian Miracle.* New York: Oxford University Press.

Yusuf, Shahid, with M. Anjum Altaf, Barry Eichengreen, Sudarshan Gooptu, Kaoru Nabeshima, Charles Kenny, Dwight H. Perkins, and Marc Shotten. 2003. *Innovative East Asia: The Future of Growth.* New York: Oxford University Press.

CHAPTER 9

TECHNOLOGY TRANSFER IN EAST ASIA: A SURVEY

Kaoru Nabeshima

Faced with globalization and the emergence of China as a world manufacturing platform, East Asian firms need to move to the next stage of their development, building on the solid foundations that these economies have accumulated over the years and focusing on growth based on productivity and, hence, innovation. Among the factors that will contribute to this stage, the ability to rapidly assimilate technology from abroad will be uppermost, as has been the case in the recent past. A better understanding of technology transfer in the current environment is imperative.

This chapter surveys the literature on the various channels of technology transfer, focusing especially on the current and likely future significance for East Asia of the import of capital goods, reverse engineering, inward foreign direct investment (FDI), licensing, the role of the production for original equipment manufacturers (OEM), labor mobility, and the contribution of research and development (R&D). This chapter draws on firm-level case studies and empirical data mainly from East Asian economies to illustrate how companies are primarily responsible for adopting, assimilating, and absorbing technology and developing new technology as they respond to the challenges highlighted in earlier chapters.

The general finding, which reinforces the message of previous chapters, is that the lack of human capital development continues to be one of the principal factors limiting technology transfer because technology does not flow freely or automatically, "but requires" recipients to possess some prior knowledge and to exert great effort in mastering new technology. The structure of the chapter is as follows. The first section selectively reviews the empirical literature on various channels of technology transfer. The second summarizes the main findings, focusing on the hurdles faced by firms in developing economies. Then the chapter suggests possible future policy directions, and a fourth section concludes.

SELECTIVE REVIEW OF EMPIRICAL RESEARCH

This section reviews the literature on the various avenues of technology transfer. It suggests that some avenues offer better scope for learning than others, and in turn require differing levels of local capacity for effective technology transfer.

Capital Goods

The import of capital goods was the primary conduit for technology transfer to East Asian economies; its importance is still large.[1] Studies by Coe and Helpman (1995) and Coe, Helpman, and Hoffmaister (1997) investigate international R&D spillovers using data from developed countries in the Organisation for Economic Co-operation and Development (OECD) and from developing countries and find that domestic and foreign R&D contribute to total factor productivity growth in OECD and developing countries.[2] However, the contribution through foreign R&D depends on the size of a country. On the one hand, for larger countries, such as the United States, domestic R&D is more important than R&D conducted by their trading partners. On the other hand, smaller countries rely more on foreign R&D efforts than their own. Similar results are obtained by Frantzen (2000) and van Pottelsberghe de la Potterie and Lichtenberg (2001). Furthermore, Coe, Helpman, and Hoffmaister (1997) find that, compared with other countries, East Asian economies benefited the most from international R&D efforts.

Similarly, Eaton and Kortum (1999) find that, through their R&D efforts, Japan and the United States together contributed at least two-thirds

1. In this section I concentrate on capital good imports. However, by allowing the import of intermediate inputs, firms can greatly enhance the technological content of their products. With import competition, domestic firms can improve their technological capabilities, contributing substantially to the productivity growth of a country (Lawrence and Weinstein 2001). For example, when Sharp wanted to develop pocket calculators, the only semiconductor producer who agreed to supply integrated circuits was Rockwell International, a U.S. firm. Japanese producers were unwilling to supply Sharp because of the risk involved in high-volume production of new technology. However, after seeing Sharp and Rockwell International succeed, the import of integrated circuits into Japan increased. This surge in imported integrated circuits prompted domestic producers and the Japanese government to improve and develop the necessary technologies (Kimura 1997).

2. Generally, the link between international trade and technology diffusion is measured through the R&D efforts of the exporting country and how they affect the total factor productivity growth of the importing country. See also Madden and Savage (2000), who focus on the role of the trade in information and telecommunication equipment as an important source of technology diffusion.

of the growth in each of the G-5 countries covered in the sample (including their own).[3] The United States is the only country that derives most of its growth from its own R&D efforts, confirming earlier findings that the larger countries tend to depend mainly on their own R&D effort.[4]

However, in this framework, an increase in R&D efforts by exporting countries will automatically lead to an increase in total factor productivity growth in importing countries, even if the share of imported capital goods to total imports remains the same. This growth in total factor productivity is achieved through the increasing sophistication of technology embedded in such imports. This line of research assumes that the transfer of technology is automatic, embedded in capital goods and other intermediate inputs.[5] Thus the characteristics of exporting countries determine what technologies are transferred to the importing countries, without much consideration of the characteristics and capabilities of the importing countries. This framework—and the model specifications based on it—may be adequate for samples consisting mainly of OECD countries, but it is less illuminating when the sample includes developing countries with widely varying characteristics and absorption capabilities.

Paying more attention to the characteristics of developing countries, Navaretti and Soloaga (2001) consider a different specification based on the notion that technology transfer depends on the characteristics of importing countries rather than on those of exporting countries. That is, the sophistication of technology embedded in capital goods imports depends on certain characteristics of importing countries rather than being directly proportional to the exporting countries' stock of R&D.[6] They find that Central and Eastern European countries import simpler machinery (low level of embedded technology) than the United States does, primarily because their stock of human capital is smaller.[7]

3. Countries studied are France, Germany, Japan, the United Kingdom, and the United States.

4. See also Bayoumi, Coe, and Helpman (1999), who integrate the effects of R&D spending, R&D spillover, and trade on total factor productivity growth in an endogenous growth model simulation.

5. Distance can be one of the factors limiting the diffusion of international knowledge. See Baptista (2000) and Keller (2002). Furthermore, Keller (1998) finds that randomly matching trading partners explains more spillover effects than the actual data. This result casts doubt on the validity of the methodologies employed in these research efforts.

6. They use unit value as the measure of machinery sophistication. Eaton and Kortum (2001) use trade-based measures of equipment prices and relate them to the difference in productivity among countries.

7. They investigate the effect of technology embedded in capital goods (mainly machinery) exported from the European Union to Central and Eastern European countries.

These results indicate that (a) the technological sophistication embedded in imported machinery has positive effects on total factor productivity growth, as we would expect, and that (b) developing countries tend to import machinery with less-advanced embedded technology.[8] The implication of these findings is that developing countries tend to experience lower growth than industrial countries that import machinery with more advanced embedded technologies. This, in turn, leads to a widening income gap between developed and developing countries unless developing countries start to import more advanced machinery. Navaretti and Soloaga (2001) speculate that the choice of developing countries results from the difference in human capital, an observation that is partially supported by Caselli and Coleman (2001), who find that educational level matters, at least in the use of personal computers, as measured by imports. In addition, service engineers and highly trained technicians are needed to maintain sophisticated, advanced foreign machinery. The lack of these skilled workers results in an inefficient use of foreign capital, leading to less benefit from the imported machinery. Thus the availability of skills tends to determine the direction of technology adoption. With a larger supply of skilled labor, the technologies adopted are more skill biased. This may explain the correlation between imported capital goods (which are predominantly produced in developed countries) and human capital development (Acemoglu 2001). The implication is that, as countries develop their human capital, the adoption of capital goods made in developed countries is much easier since the endowment of skills is less dissimilar.

Reverse Engineering

Another channel of technology transfer employed extensively by East Asian economies is reverse engineering. Most of the empirical evidence on this front comes from detailed case studies of firms in Japan and the Republic of Korea that were successful in reverse engineering. The strategy that Japan adopted in the past is a prime example of this method. Japan often limited the importation of components to first-generation plants, forcing subsequent plants to use domestically produced equipment for construction. In order to be able to do so, Japanese firms, especially capital goods manufacturers, needed to acquire a solid understanding of the technology used in these prototype plants. Taiwan (China) and Korea both used similar strategies to im-

8. They estimate two regressions: one based on the share of imported machinery over investment and the other on the technological sophistication of the machinery. Their results indicate that both of these variables lead to an increase in total factor productivity growth, with the latter having more robust results.

prove their understanding of foreign technologies (von Tunzelmann 1998). In the area of factory-automation technology, for instance, Korean firms imported the first batch of factory-automation equipment to improve efficiency. Once they assimilated and familiarized themselves with the foreign technology, Korean firms developed their own equipment to meet future needs through internal R&D efforts (Kim 1998).[9]

There are abundant examples of firms acquiring technological capabilities through reverse engineering. For instance, LG Electronics, the first consumer electronics firm in Korea, entered into the production of transistor radios in 1958 by reverse engineering (Kang 2001). Samsung gained a foothold in the semiconductor industry also through successful reverse engineering.[10] The company achieved this by establishing a semiconductor R&D laboratory in 1982. In the early stages, Samsung successfully reverse engineered dynamic random access memory (DRAM) chips up to 256Kb. Gradually, it was able to produce 1Mb DRAM on its own, albeit more slowly than U.S. and Japanese competitors. From 4Mb DRAM on, Samsung's capability as an innovator was close to or even surpassed that of its American and Japanese competitors. In short, Samsung was gradually able to move from an imitator to an innovator in the semiconductor industry. Samsung has subsequently used this experience to climb the ladder to more complex technology. All this could not have been possible without in-house R&D efforts. However, although there are many ways to build up a base of knowledge, one factor that contributed significantly to Samsung's success was the reverse migration of high-caliber scientists and engineers of Korean origin from the United States (Kim 1997). A more detailed discussion of labor mobility comes later.

As illustrated by the case of Samsung, when successful, reverse engineering enables firms to learn and assimilate imported technologies much more fully than by simply relying on the imported components or turnkey plants. But within East Asia, Japan, Korea, and Taiwan (China) were the only economies truly successful in reverse engineering.[11] However, even

9. Korea's efforts in reverse engineering were not limited to capital goods. For example, most of the televisions assembled in Korea entered the market through reverse engineering (Kim 1995). These reverse engineering efforts were accomplished by "poaching" experienced workers from other firms.

10. Once it had succeeded in reverse engineering the microwave oven (with help from General Electric) in the late 1980s, Samsung became the world's largest producer of microwave ovens (Hobday 1995a).

11. Although at the beginning of industrialization (around the time of Meiji Restoration), Japan faced many difficulties in reverse engineering. The Japanese attempted to reverse engineer the steam engine but were unsuccessful (Hayashi 1990). In the modern era, Japan continued to lag behind in aviation and aerospace technologies. Only in 1994 did Japan succeed in launching rockets entirely made domestically (NASDA 1997).

though these economies were successful, not all of their reverse engineering efforts yielded the desired results. Samsung was able to reverse engineer microwave ovens to some extent but then encountered technical difficulties that the firm could not solve. These difficulties were solved only after Samsung became an original equipment manufacturer for General Electric, which provided technical assistance. Additionally, catching up to Japan in the capital goods industry has proven difficult for Korean firms (Hobday 1995b). This highlights some of the limitations of and difficulties associated with reverse engineering.

In light of these experiences, to what extent does reverse engineering lead to the development of a country's own innovative capacity? Has Samsung, through its reverse engineering experience in DRAMs, established a foothold in chip design? The answer is yes, but reverse engineering was a first step. It was the subsequent investment in R&D that helped Samsung to build the capability to innovate.

Inward Foreign Direct Investment

Another major channel by which firms can acquire technology is through FDI.[12] The evidence stemming from this area is also presented by Hill in chapter 8 of this volume. The entry of MNCs provides several potential channels of technology transfer, such as training of local employees, demonstration effects, imitation by local firms, and forward and backward linkages to local firms. In turn, the mode of FDI—whether "greenfield" or joint venture—affects the effectiveness of these channels in host economies. First, what are the consequences of FDI on productivity in the host country? Xu (2000) examines the effects of MNCs on productivity growth and

12. Of course, firms can also acquire technology through outward investment, especially through the acquisition of foreign firms. Two modes of entry are available for MNCs to access technology available in a host country. One is to establish "greenfield" investment. The other is to acquire an existing firm that possesses the technology of interest to quickly gain the expertise of that firm, especially when technology is evolving rapidly (Ahuja and Katila 2001), and this provides an easy way to tap into country-specific, firm-embodied technological advantages (Shan and Song 1997). In the context of East Asia, only firms in Japan and possibly in Korea are capable of acquiring firms in industrial countries. Although Perusahaan Otomobil Nasional (Proton) bought Lotus Group International in order to gain access to design and engineering capabilities, it remains to be seen whether Proton can successfully integrate such capabilities in its own engines ("Can Proton Deliver" 2003). Therefore, I do not focus on this issue in this chapter. However, to summarize the findings in the literature, acquired affiliates have significantly higher R&D intensities than affiliates established as greenfield investments (Belderbos 2001), and the more patents a firm has, the more likely that this firm will be targeted for foreign equity investment (Shan and Song 1997).

finds that most industrial countries reap productivity gains from technology transferred by U.S. MNCs,[13] whereas developing countries do not enjoy such benefits.[14] In explanation, Xu suggests that recipient countries need to attain some minimum threshold of human capital for knowledge transfer to be effective. In the absence of adequate human capital, the benefit accrued to learners will be much smaller, if any.

Braconier, Ekholm, and Midelfart-Knarvik (2000) investigate whether inward FDI can convey R&D spillovers, as measured by labor productivity. The benefit of inward FDI is that it introduces new technology to the host country and lowers the diffusion cost of knowledge.[15] However, the speed of diffusion depends on the degree to which MNCs interact with firms in the host country. If a MNC has fear of leakages by local firms, then the diffusion likely will be quite low and vice versa. Diffusion is also directly proportional to labor mobility between MNCs and host-country firms. Overall, the study finds that there is little evidence of R&D spillover, as measured by total factor productivity and labor productivity, transmitted by FDI. Rather, most of the gain in productivity is a result of R&D efforts by local firms and is a function of the labor-capital ratio.[16]

In the context of FDI, Japanese MNCs are important players in East Asia, especially in some ASEAN (Association of South East Asian Nations) countries (for example, Indonesia, Malaysia, and the Philippines). In these countries, Japanese MNCs are responsible for 30 percent of the employment in the electronics and transport equipment industries (Belderbos, Capannelli, and Fukao 2000). Because of their

13. This study differs from previous studies in the measurement of MNCs activities. A MNC's productivity effect can take two different paths: technology transfer and competition intensified by the MNC's presence. Xu differentiates these two paths by using royalties and license fees as a share of subsidiaries' value added as a proxy for the subsidiaries' involvement in technology transfer.

14. The effect of MNCs on domestic productivity growth is ambiguous, especially for developing countries. For instance, Haddad and Harrison (1993) do not find any evidence of this link. Aitken, Hanson, and Harrison (1997) find only small evidence of such an effect on export behavior in Mexico, and Aitken and Harrison (1999) find only a small overall effect in Venezuela. In contrast, Blomström and Sjoholm (1999) find substantial spillover from FDI in Indonesia. However, such spillover is limited to non-exporter domestic firms, suggesting that the spillover stems mainly from competitive pressure arising from the presence of the MNCs rather than from technology transfer.

15. It is generally held that knowledge diffusion is local in scope, with increasing cost as distance increases (see Baptista 2000; Keller 2002).

16. For an overview and an extensive review of the literature on the link between FDI and technology transfer, see Saggi (2002).

prominence, one would expect Japanese MNCs to have a large influence on the technological capabilities of local industries. However, evidence shows that Japanese MNCs do not transfer much technology or knowledge to local firms, because of the lack of backward linkages to local firms in host countries.[17] That is, even though Japanese MNCs may source locally, these suppliers tend to be subsidiaries of other Japanese MNCs (Belderbos, Capannelli, and Fukao 2000), a point also made by Takayasu and Mori in chapter 5. Therefore, local content requirements may be valuable to the host country, since such arrangements can encourage the transfer of know-how to local supplier industries (Belderbos, Capannelli, and Fukao 2000).[18] Using plant-level data for Indonesia, Blalock and Gertler (2002) test whether technology transfer from FDI is through horizontal spillover or through backward linkages. The horizontal spillover is defined as an increase in productivity of firms engaging in the same industry, and empirical results so far have been mixed. Confirming findings in other studies, Blalock and Gertler find no evidence of horizontal spillover, but they do find significant effects stemming from vertical linkages. Thus the presence of MNCs may not help domestic firms that are competiting directly with them, but it has a positive effect on other domestic firms that are supplying parts to these MNCs. Interviews with both MNCs and local firms reveal that the source of increase in productivity comes mainly from technology transfer. This is quite consistent with the incentives of MNCs. They have every incentive to assist these suppliers to deliver quality parts in a timely manner. At the same time, they would try to minimize the leakage of their technological advantages to local firms that are competing against them. Given this background, the supply chain linkage seems to be a more natural conduit of technology transfer than horizontal FDI (Blalock and Gertler 2002).

17. A study on the determinants of backward linkages by Japanese MNCs reveals that this linkage is quite limited. Research also finds that the mode of investment has a large bearing on the degree to which Japanese MNCs procure from local suppliers. Greenfield investment is associated with a low degree of local procurement, whereas joint ventures and acquired affiliates tend to use more local suppliers (Belderbos, Capannelli, and Fukao 2001). For a theoretical treatment of the effect of FDI on industry development, see Markusen and Venables (1999).

18. However, unless there is a compelling reason to locate in that country, constraining MNCs to source locally can turn away the establishments of their affiliates in the first place. Additionally, given the WTO rule against local content requirements, implementation of such requirements will be difficult in the future.

Local linkages not only serve as conduits for technology transfer; they also can improve the technological sophistication of local firms. However, the development of such linkages is not automatic. Comparing Penang and Klang Valley in Malaysia, Narayanan (1999) finds that linkages between indigenous firms serving MNCs are much greater in Penang than in Klang Valley. He attributes this difference in the degree of linkages to the difference in the characteristics of firms in these two locations. Klang Valley is dominated by Japanese subsidiaries that tend not to source locally. Additionally, Penang hosts more component manufacturers that encounter rapid technological change, inducing a more rapid pace of technology transfer. In Klang Valley, many MNCs are in the consumer electronics industry, where technology changes less rapidly. As a result, fewer supporting firms have emerged in Klang Valley, whereas Penang has seen a large increase in the number of supporting firms (Narayanan 1999). Accordingly, the extent of technology transfer is much larger in Penang. Clearly, Malaysia would be well advised to improve on these linkages to foster such transfer (Hobday 1996). In reality, improving supply linkages with MNCs is time consuming and depends on the capabilities of local firms. Through interviews with U.S. and Japanese MNCs, Blalock and Gertler (2002) find that firms spend considerable time and resources in finding suitable local firms. Some MNCs do not source locally, citing the difficulties in finding firms that can produce inputs with the required quality and can ensure timely delivery (see Heaver in chapter 7 of this volume). Hence, policymakers can try to reduce the search costs of local suppliers for MNCs and, at the same time, can assist local firms in attaining the required capabilities through training and other means.

Singapore's move up the technological ladder was aided by FDI. Unlike Korea, which intended to develop its local industry, the Singaporean government actively invited MNCs through various incentives and infrastructure provisions—in addition to supplying an abundance of educated workers.[19] Initially, MNCs were drawn to Singapore by the availability of low-cost labor that could be used to assemble products. Gradually, MNCs upgraded the tasks conducted by Singaporean subsidiaries to include more complex operations, engineering support, and personnel training (Hobday 1994). Singaporean subsidiaries of MNCs began producing more sophisticated products, employing complex manufacturing processes with forward and backward linkages to local indus-

19. Other reasons cited by MNCs in their decision to locate in Singapore include the growth of the Asian region as a whole and efficient transport systems (both air and water) that enable them to export to other countries (Hobday 1994, 1995a).

tries and thus facilitating further technology transfer into the country (Hobday 1994).

The discussion so far has focused on the role played by the greenfield investment of MNCs. However, many economies in East Asia, especially northeastern ones—China, Japan, and Korea—have used joint ventures as the means of acquiring technologies. This preference is based on the view that technology transfer through MNCs is much more effective through joint ventures because it forces MNCs to interact more closely with local firms. The Japanese Ministry of International Trade and Industry (MITI) helped local semiconductor firms by requiring Texas Instruments to form a joint venture with Japanese firms, to relinquish fundamental patents to other Japanese firms, and to restrict output to 10 percent of domestic demand (Chiang 2000). These requirements enabled Japanese firms to obtain key technologies from Texas Instruments and to establish their own semiconductor industry.

In Korea, Samsung and Goldstar relied on the technology supplied by Matsushita and Sanyo, respectively, through joint ventures to produce transistor radios during the early 1960s. Samsung was able to send employees to factories owned by NEC and Sanyo in Japan for training in the production of radios and televisions. Similarly, firms from Hong Kong (China) and Taiwan (China) entered into joint ventures with firms such as Sanyo to produce air conditioning equipment and audio electronics. MNCs trained numerous local engineers and technicians, in the process transferring valuable foreign technologies and management skills to firms such as Samsung and Tatung of Taiwan, China (Hobday 1995a). However, there are exceptions, as in the case of Daewoo, where a joint venture with General Motors did not lead to significant technology transfer (Lee and Lim 2001).

The Chinese government has used joint ventures as one of the main vehicles for technology transfer (Andreosso-O'Callaghan and Qian 1999) in anticipation that they will generate more local linkages compared with wholly owned subsidiaries.[20] However, larger (and especially Japanese) firms often do not generate the degree of backward or forward linkages that the Chinese government seeks.[21] Lan (1996) studies technology transfer by Japanese firms in Dalian (China) and finds that most joint ventures are engaged in

20. For a list of studies on joint ventures in China, see Ying (1996, table 1).

21. Technology transfers from Japan to China often were granted with strict restrictions on exports and supplies of raw materials and parts. In addition, technology transferred by the Japanese tended to be less advanced than what the Chinese government sought. Japanese firms deliberately attempted to offer less-advanced technologies in order to protect their domestic and world markets (Tang 1998). However, these transfers also reflected the relatively inexpensive labor force in China; the component parts tended to be more labor intensive than technology intensive, which was more appropriate for China at this stage of development.

assembly or the production of final goods where the value added locally is low. Furthermore, the technology brought by foreign firms in Dalian seems to be less advanced than what is available elsewhere, reflecting the stage of Chinese development at that time. In general, technology transfer is limited to the introduction of new production hardware. In addition, training is rarely provided, because the technology employed tends to be fairly simple and does not necessitate the training of workers.[22]

Even though attracting FDI can lead to the transfer of technologies, the range of technology transferred can be fairly narrow. Essentially, MNCs will not diffuse their technologies to firms in the host country unless they have incentives to do so. Otherwise, they would rather exploit their technical advantage to the fullest (van Pottelsberghe de la Potterie and Lichtenberg 2001). For instance, Narayanan and Wah (2000) find that the presence of MNCs in Malaysia greatly increased the production capacity and technical quality of Malaysian products. Yet they have some doubts about the long-term future of Malaysian industries, given that local R&D efforts are lacking and FDI has not enhanced the R&D capabilities of Malaysian firms.[23] Kim (1995) shares this sentiment. FDI typically transfers technology related to production but rarely transfers technology related to engineering and innovation. This point is explored by Wah and Narayanan's (1997) in their survey of Malaysian manufacturing firms. This study identifies three stages of technology transfer: adoption, rooting, and diffusion. Adoption involves access to new technology (adoption and installation of new machinery and equipment). Rooting has three levels (in increasing order of difficulty): operation, maintenance and repair, and R&D. In the final stage, the acquired knowledge and technology diffuse to other firms and industries. Wah and Narayanan find that at the rooting stage most indigenous Malaysian workers can perform operations without any help from expatriate personnel. To a lesser degree, maintenance and repair can also be performed without assistance. However, the researchers find that only a handful of Malaysians are capable of performing R&D activities, reflecting a lack of transfer of R&D skills from MNCs to Malaysians.[24]

22. According to Shi (2001), most of the technology transferred by MNCs is older vintage, in an effort to meet the technology transfer requirements imposed by the Chinese government.

23. There is a question about the extent to which FDI enhances the host country's R&D capabilities. Since R&D is skill-intensive, most R&D efforts are concentrated at home. Even though the host county may wish to improve its R&D capabilities, it does not appear that FDI is a meaningful way of achieving this goal.

24. Most of the innovative activities conducted by Malaysian subsidiaries are improvements to products and processes and not basic research. In some cases, these improvements are adopted in plants outside Malaysia (Hobday 1996).

This is partly due to the nature of MNCs operations in Malaysia: FDI in Malaysia tends to be concentrated in production facilities for reexporting, and MNCs typically centralize their R&D operations at home. The internationalization of R&D efforts is a fairly recent phenomenon, and most MNCs heavily concentrate their R&D facilities in highly industrial countries.

It is clear that, to take full advantage of the investment and to climb the ladder of technological sophistication, any FDI has to be complemented by an increase in indigenous efforts to absorb the imported technology (Blomström and Kokko 2003). This is the part where Malaysia lags behind other countries, especially Japan and Korea, when they were at a comparable stage of development.

Licensing

Firms can obtain technology directly from other firms by licensing. Typically, licensing requires more technical capability than a joint venture, since recipients need to understand the underlying technology in order to use it efficiently (tacit knowledge; Hobday 1995b). As with the studies of reverse engineering, much of the literature concerning licensing comes from detailed case studies of firms or countries—especially Japan and Korea—that successfully used licensing agreements to obtain key technologies in the past. For instance, the initial designs of passenger cars and automotive technology were licensed from Europe, and licenses were obtained from Texas Instruments and Fairchild when attempts to reverse engineer failed in key areas of the semiconductor industry in Japan (Kimura 1997). A number of Korean firms also relied on licensing for key technologies. Goldstar relied heavily on licensing from Hitachi for basic 4- and 16-megabit DRAMs. Hyundai also depended on licensing to acquire semiconductor technology. Taiwan (China) actively purchased foreign licenses, numbering more than 3,000 between 1952 and 1988 (Hobday 1995a).

The Japanese government, especially MITI, was heavily involved in the licensing of foreign technology on behalf of domestic firms.[25] MITI's strat-

25. Overall, Japan's heavy reliance on licensing may have resulted from historically weak protection of intellectual property. Because of this weak protection, FDI leads to a much quicker erosion of the MNC's advantages through "inventing around" compared to licensing. Licensing ensures legal protection, in the form of restrictive usage of the licensed knowledge (McDaniel 2000). However, Japan reformed her patent system in 1988, expanding the scope of patent protection to prevent "inventing around." It will be interesting to see if such reform was successful. And if so, then we would see a decline in licensing and an increase in inward FDI in Japan, controlling for other determinants if McDaniel's claim is correct.

egy was to purchase licenses for domestic firms in order to prevent competing firms from bidding up the license price. This allowed low-cost licensing agreements in the 1950s and early 1960s (Chiang 2000; Wakasugi 1997).[26] In fact, MITI refused to give Sony the necessary foreign exchange to obtain a license for transistors from Western Electric after it had assigned Hitachi, Mitsubishi Electric, and Toshiba as the recipients of a licensing agreement with RCA (Sony Corporation 2003).

Easily quantifiable data are more readily available for analyzing the effects of technology licenses on the economy than for analyzing the effects of other channels of technology transfer. The methodology is similar to using patent and citation data to track the movement of knowledge. Branstetter and Chen (2001) investigate the effect of technology licensing on productivity growth in Taiwan (China). They find that the licensing substantially affected productivity growth in the Taiwanese (Chinese) electronics industry. The magnitude of this effect is close to that of domestic R&D efforts.

However, simply obtaining a license does not guarantee the successful adoption of new technology. Lack of experience in technology, equipment, and operating environments forced Japanese engineers to undertake much experimentation to master the new technology after obtaining licenses in the early years of development (Kimura 1997). Arora (1996) concurs that the purchase of a license itself is not enough, especially for developing countries that tend to lack knowledge of and experience with the underlying technology. These firms require the technology supplier to bundle tacit knowledge (technology service) along with the blueprints for effective assimilation. LG Electronics initially relied on Hitachi to train its engineers and supervise the plant when it signed a licensing agreement to produce black-and-white television sets (Kang 2001). Alternatively, where such bundling of tacit knowledge is not offered, local R&D efforts (learning) are required to understand how the technology works and how it should be used in the local environment in order to reap the maximum benefit (Forbes and Wield 2000).

Furthermore, a limitation arises as firms successfully upgrade their own capabilities. It becomes increasingly more difficult to find suitable licensors once a licensee gains its own capabilities.[27] The situation that Samsung faced is a good illustration (Kim 1997). Samsung contacted many firms in the United States, seeking semiconductor technology. The only licensing agree-

26. The strategy was put into place partly to conserve the scarce supply of foreign exchange (Wakasugi 1997). However, MITI increasingly allowed multiple firms to import foreign technology as it realized that imported technology strengthened the competitiveness of Japanese industry in the global market.

27. Another difficulty is the limited transferability of licenses, which can be a stumbling block during a merger process, as occurred in Korea ("Dead Deal Walking" 2002).

ment that it was able to reach was with Micron, which at the time was facing severe financial problems. Without such luck on Samsung's part, and misfortune on Micron's, Samsung could not have expected to encounter such a cooperative stance from any firm. Thus the ability of firms to license is directly related to the sophistication of licensees. The above studies investigate technology transfer from the point of view of importing firms (demand side). However, looking at the supply side, providers have several methods of sharing (or not sharing) their technology. Unfortunately, the lack of supply-side analysis coupled with the demand side is a shortcoming in this area of research.

Original Equipment Manufacturer

As noted in the previous section, a firm may face a difficult time digesting the information specified in licenses, especially with regard to tacit knowledge. The advantage of producing for an OEM is that the technology supplier provides technical know-how and service to ensure that affiliated firms can produce quality components meeting the supplier's exact specifications (Kim 1991; Kim and Lee 2002). In addition, through the OEM route, firms can expand exports rapidly, even though the profit margins may be thin (Kim and Lee 2002). This greatly enhances the ability of firms to fully assimilate technology and contribute to future innovative activities. For instance, Samsung gained a foothold in the world market for microwave ovens through its links with General Electric (Cyhn 2000). This mode might be especially useful for small- and medium-scale enterprises, since they typically do not have sufficient human and financial resources to acquire technology. Instead, they can achieve this by participating in global production networks led by MNCs, as described in chapters 2 and 4 of this volume (Hayashi 2002).

Subcontracting and OEM were the major sources of technology transfer in Taiwan's (China's) electronics industry, which was populated mainly by small- and medium-scale enterprises (Hobday 1995b).[28] For example, Tatung of Taiwan (China) developed much of its manufacturing know-how through licensing and contracts with OEMs. Whereas most production embodied little original R&D, much of the product and process technologies were on par with those of more technologically advanced countries (Hobday 1995b).

28. One step forward in terms of sophistication is to become an own-design manufacturer (ODM), under which firms conduct some or all of the product design and process tasks needed for manufacturing based on the general design layout given by the foreign partner. As in OEM, the finished product is sold under the brand name of the buyer. ODM allows the manufacturer to internalize production technologies, component design, and product design capabilities. These are the steps necessary to reach the next stage of own-brand manufacturing (OBM) and product innovation (Hobday 1995a).

Similarly, OEM, mainly with large Japanese and American firms, was the dominant means of production for firms in Hong Kong (China). Like their Taiwanese (Chinese) counterparts, these firms had significant design components of their own, even though most firms were too small to conduct R&D, develop both process and product technologies from the ground up, or market their own products.[29] Thus backward linkages were the important avenues that stimulated supporting industries, such as plastic casting, metal parts and plating, tools, printed circuit board assembly, metal working, materials, and components in these economies (Hobday 1995b).

Singaporean firms gained from OEM arrangements to a lesser extent when compared with Korea and Taiwan (China), which relied heavily on OEM for technology transfer. This is due to the small number of local electronics firms in Singapore compared with the other two economies, although Wearnes obtained technical specifications through an OEM agreement with a multinational enterprise. As a result, Wearnes gained not only technical expertise but also production techniques that greatly enhanced productivity (Hobday 1994).

Thus OEM provides another effective avenue for technology transfer. However, there are three main limitations to this approach. The first limitation is that, to be selected by an OEM, firms typically need to show their potential to deliver high-quality goods at a relatively low cost in the first place. This requires firms to possess a certain level of production skill and technological capability. Whereas this may have worked with Korea and Taiwan (China), the hurdle might be too high for other developing countries wishing to pursue this strategy.

Second, due to their extensive OEM arrangements, Korea and other newly industrialized economies are heavily dependent on Japanese and U.S. technologies and machinery. In 1990 they imported approximately $57 billion of primarily Japanese machinery and capital goods and consistently ran trade deficits with Japan because of their reliance on components and material imports. On the one hand, such arrangements create a dependence on foreign capital goods because the foreign partner typically chooses the capital equipment needed for production. On the other hand, OEM and participation in global production networks does facilitate the training of managers, engineers, and technicians and makes available to participating firms advice on production, financing, and management (Hobday 1995a).

Finally, another limitation of this approach is that there is no guarantee that firms can move out of OEM and into ODM or OBM (own-brand manufacturing). In fact, as alluded to in chapter 1, the risk is that firms may for-

29. An exception to this is Vtech, which can design and market its own products.

ever be trapped in OEM, even after receiving significant technical assistance and training.[30] Even though Wearnes created its own development and design capability, it chose to be an OEM manufacturer for U.S. computer makers, including IBM, instead of seeking to become an OBM (Hobday 1994). Many Korean firms entered the personal computer market through OEM arrangements in the second half of the 1980s. Yet none emerged as a world-class manufacturer of personal computers. Instead they all switched to computer peripherals (Kang 2001). When Acer aspired to become an OBM. In order to do so, it had to change the brand image of personal computers coming from Taiwan (China), which were regarded mainly as inexpensive low-end products (Doz, Santos, and Williamson 2001). Acer's move from OEM to OBM was supported by strong in-house R&D efforts and rapid innovation in software, products, and manufacturing processes, supplemented by foreign-trained engineers as a main source of technology building (Hobday 1995b). However, as noted in chapter 3, the effort proved to be far more difficult than anticipated, although Acer became one of the largest computer firms in 2000 (Doz, Santos, and Williamson 2001), especially in the segment of notebook computers. Firms in Taiwan (China) were able to achieve a full transition from OEM to OBM in the bicycle industry, although the technology involved clearly is far less complex than in electronics (Hobday 1995b).

Labor Mobility

Labor mobility plays a key role in technology transfer, especially of tacit knowledge that cannot be traded like explicit knowledge can, although systematic quantitative analysis is lacking in this area. To gain tacit knowledge, firms can acquire workers from other firms, from the subsidiaries of MNCs, and from foreign sources or indigenous workers with work experience abroad (see Yusuf with others 2003).[31] Labor mobility

30. Sony, with its "pocketable" radio, insisted on using its own brand name when first exporting to the United States, when this kind of practice by Japanese electronics firms was rare. At that time, most Japanese electronics manufacturers were own-design (but not own brand) manufacturers outside of Japan, even though they had their own brands established in Japanese markets. (Sony Corporation 2003).

31. Fosfuri, Motta, and Ronde (2001) develop a theoretical model in which technology spillover occurs through worker mobility. They consider a case in which a MNCs must train local workers for its subsidiaries. In this scenario, it is possible for a trained worker to be hired away by a local firm in order to gain his or her "embedded" knowledge. To prevent this erosion of its technology, the MNCs can offer a higher wage to retain the worker. In the end, the effect of worker mobility will depend on competition in the product market. If the competition is fierce, the MNCs will retain the worker. Otherwise, local firms will be able to attract the trained worker away from the MNCs, leading to the transfer of technology embedded in trained workers.

was instrumental in enabling Korean firms to reverse engineer many consumer products (Kim 1997).

Inter-organizational mobility of workers was a key contributor to Korean firms' rapid accumulation of technological capabilities.[32] In Korea 70 percent of consumer electronics producers enter the market with the help of experienced workers poached from other firms (Kim 1995). Yet this external mobility is not the only way for a firm to draw on the pool of experience. For large firms, such as *chaebols* in Korea, intra-firm migration can help to increase their acquisition of technology. When Hyundai entered the automobile industry, it gathered both internal and external personnel to form a task force, creating the base of prior and tacit knowledge required to initiate automobile production (Kim 1998). Similarly, when Samsung entered the telecommunications equipment industry, many engineers were drawn from Samsung Electronics and other parts of the group to compensate for the lack of engineers experienced in this particular field (Hobday 1995b). Every time Samsung wished to move up the product ladder, it would hire experienced engineers from other firms or from abroad to increase its tacit knowledge before embarking on a new project.[33] Clearly, the use of mobile, experienced workers contributed substantially to the development of technological capabilities of Korean firms (Kim 1998). Mobility of workers, especially from large, established firms, also can foster the establishment of small- and medium-scale enterprises supplying parts to former employers, thus adding additional layers to the process of technology transfer (Kim and Lee 2002).

From overseas, many Korean firms hired Koreans and Korean Americans, who are educated in the fields of science and engineering and trained in the United States, to improve their technological capabilities. The students' return to Korea, permanently or temporarily, provides firms with access to technologies developed in the United States (Kim and Yi 1997; Lim 1999). Samsung provides a good illustration of this practice. The company was able to acquire the first semiconductor firm in Korea, founded by Dr. Ki-Dong Kang, a Korean American scientist with prior experience at Motorola.[34] Along with technical specifications, Dr. Kang was able to pass on much tacit knowledge to Samsung engineers. Samsung also recruited Korean American scientists and engineers for the development of semiconductor chips.[35] They

32. For the Japanese experience, see Kusunoki and Numagami (1997).

33. Of course, for any kind of research endeavor, the firm performed prior literature surveys to identify and understand the advances in technology.

34. He obtained his doctorate from Ohio State University.

35. These are Ph.D. holders with working experience in leading semiconductor firms.

were mainly responsible for establishing and furnishing the laboratories necessary to undertake semiconductor research (Lim 1999). Not only were these workers able to bring in the technology, they also brought in American know-how in R&D (as opposed to the Japanese know-how on which Korea had been dependent previously). In many instances in Samsung's history, foreign-educated and experienced scientists and engineers have contributed significantly to the company's ability to advance up the technological ladder (Kim 1997).

A similar international flow of workers is seen in Taiwan (China). The Taiwanese (Chinese) government has invested heavily in education and training, especially in the fields of engineering and science, often in response to the urging of the business community. During the 1980s, thousands of Taiwanese (Chinese) went abroad to study and to work for foreign corporations, mostly in Japan and the United States. From the mid-1980s, returning locals trained in foreign firms became a direct source of technology and new skills (Smith 2000). For instance, the founders of Microelectronics Technology of Taiwan (China) all had work experience in the United States, including at Hewlett-Packard and other leading technology firms. Using their experience, these firms were able to identify and acquire technologies much more easily than firms without such ties to the United States (Hobday 1995a, 1995b).[36] Similarly, Acer regularly taps into a pool of Taiwanese (Chinese) engineers in the United States to gain knowledge of existing and emerging technologies in computer industries (Doz, Santos, and Williamson 2001).

Sending students abroad for training has also provided access to advanced technologies from international sources, and the returnees contribute to the faster creation of knowledge (Kim 1998; Yusuf with others 2003). Kim (1998) finds that one standard deviation increase in the number of students abroad in industrial countries will increase GDP growth per capita by 0.6 of a percentage point. Compared with the 0.75 of a percentage point increase gained by investment in physical capital, this is quite a significant gain. He also finds that effects on economic growth differ depending on the field of study undertaken. As one would expect, scientific fields (natural sciences, engineering, and medical sciences) contribute the most to economic growth. However, only about 1,200 of 15,000 Koreans who earn a doctoral degree in science and engineering ever return. This is a stark contrast to the more than 20 percent of students returning to Taiwan,

36. They also produced under OEMs with main inputs imported from Japan and the United States. Later, these firms moved to ODM and OBM, signifying its learning and technological capabilities.

China (Smith 2000). In very recent years, however, the pace of return migration to both Taiwan (China) and Korea has accelerated.

Sending students to study abroad is a contentious issue for developing countries, which are sensitive to the threat of brain drain. Yet brain drain may not be as detrimental as one might expect. Even if some students do not return to their home country, providing the opportunity to emigrate creates incentives to accumulate more human capital in the home country than otherwise.[37] Beine, Docquier, and Rapoport (2001) present some evidence on "brain gain" in line with this theoretical model. Even if these students do not return, many typically keep close connections to their home country, which can assist technology transfer in the future and provide links with global production networks (see Yusuf with others 2003). Samsung has successfully used this network to increase its technological capabilities in the semiconductor industry.[38]

Labor mobility, both domestically and internationally, seems to be instrumental to the success of technology transfer in Korea and Taiwan (China). Highly educated and experienced workers were catalysts of successful technology transfer. However, since these firms were relying on workers poached from other firms, a question arises concerning the firms' incentive to provide on-the-job training.[39] It is commonly held that U.S. firms do not provide as extensive training as do Japanese firms. This is because, with the possibility of lifetime employment and strong company loyalty in Japanese firms, the probability of trained workers leaving the firm is fairly low. Higher labor turnover in the United States discourages outlay on training (Morita 2001). Yet this begs the question: Is it formal training at firms that matters? Or the tacit knowledge accumulated through experience? It seems that the accumulation of experience matters the most for increased tacit knowledge.

37. Stark, Helmenstein, and Prskawetz (1997, 1998) argue that brain drain can be translated into brain gain in certain cases. Their theoretical approach models the decision to emigrate. The probability of successful emigration depends on the accumulation of human capital. Because of this dependency, workers in the source country invest in the development of human capital in the hope of future emigration to other countries. Since the probability of successful emigration is small, the source country can actually gain by offering the possibility of emigration (as opposed to shutting its door to the outside world). However, the welfare implication is not clear if public education is provided.

38. Labor mobility from foreign to local firms in Taiwan (China) is investigated in Hou and Gee (1993), as cited in Hill and Athukorala (1998).

39. In Korea the sentiment against "poaching" is quite high, although rigorous investigation was not done to verify this claim. The Korean government put forth several measures (including training levies) to encourage firms to provide in-house training, but these measures were largely ineffective (Gill and Ihm 2000).

Even though the mobility of highly educated and experienced workers has been shown to play an important role, empirical studies based on labor mobility are difficult to implement as a result of their high data requirements.

Domestic R&D

Given the limited resources available, should developing countries actively engage in R&D effort? Given that technology can be transferred through imported capital goods, can firms rely solely on imported capital goods without their own R&D efforts? Santarelli and Sterlacchini (1994) find that for small- and medium-scale enterprises in Italy, the acquisition of capital goods is the most important way for firms to accumulate technological capabilities. Since small- and medium-scale enterprises are not well equipped to conduct their own R&D, they need to rely on the R&D efforts conducted by others, especially their suppliers. Panizzolo (1998) obtains similar results using survey data of machine tool industries in Italy. In both studies, customer-buyer-supplier relationships seem to play a crucial role in enabling firms to gether information and in motivating them to adopt new technology.

These two studies suggest that imported machinery can serve as a substitute for in-house R&D efforts, contributing to the debate about whether technology imports substitute or complement domestic R&D efforts.[40] However, in many studies, technology imports and firms' own R&D efforts complement each other and enable the firm to fully adopt the imported technologies. In the context of technology transfer, R&D efforts are geared more toward learning, aimed at imitation and continuous improvements in processes and products, than toward conducting basic research. Therefore, R&D efforts by importing countries should be classified more as "learning" (Aggarwal 2000).

The difference in the degree of technology assimilation can best be illustrated by the following example. China imported turnkey plants to quickly acquire or enlarge productive capacity, as any developing country does at an early stage of development.[41] However, 90 percent of imported plants in China during the 1960s and 1970s failed to yield a reasonable level of production. Many were closed down even before becoming fully oper-

40. For an emphasis on policy issues, see Katrak (1997).

41. During the early years of the Meiji era, Japan imported whole iron and textile plants as "test" plants run by the government to showcase the "new" technology available outside Japan (Hayashi 1990). Korea also relied on importing turnkey plants (Kim 1991), especially when large investment was necessary (typically in the chemical, steel, paper, and cement industries).

ational (Andreosso-O'Callaghan and Qian 1999). This is a stark contrast to Korean firms, which were able to achieve full capacity rather quickly after an intensive learning process and, in some cases, even produced more than the designed capacity (Park 1995).[42] Television production in China resulted in only 25 percent realization of full capacity, indicating a low degree of technical assimilation, although part of the reason for low capacity utilization lies in lower demand for each factory's product because of a general overproduction of television sets within China (Tang 1998).

Moreover, in the recent past China still emphasized the importation of turnkey plants to improve its productive capacity quickly, while neglecting the importance of tacit knowledge. In sheer numbers, initiatives to import technical equipment (including turnkey plants) still dominated China's technology transfer projects during the 1990s (Chen and Sun 2000). This was characterized by the poor assimilation of imported technologies compared with other cases in East Asia. While the importation of light manufacturing industries in the 1980s created a surge in foreign technologies in China that upgraded the technological capabilities of existing enterprises, learning was limited, and many of the inputs (parts, components, and raw materials) were imported to support manufacturing (Ding 1998).[43]

In contrast, Japan and Korea both spend substantial resources assimilating imported technologies (Tang 1998; Xu 1998).[44] Japan is said to devote 30 percent of its R&D to fully understand imported technology (von Tunzelmann 1998).[45] Electronics firms in Taiwan (China) spend 27 percent of R&D on product improvements and another 26 percent on process innovation once licenses are obtained (Chen, Chen, and Chu 2001). Therefore, it seems inevitable that importing firms engage in some R&D efforts of their own. This will lead to more indigenous innovative capacity down the road when they have accumulated enough experience, knowledge, and talents (Bosworth 1996; Dowling and Ray 2000; Lau 1997; Park 1995). When firms wish to catch up to technology leaders, their own R&D efforts are

42. Korea and Taiwan (China) still lag behind the United States in productivity. This is so even when comparing the productivity level of the United States in the 1960s and 1970s to that of Korea and Taiwan (China) in the 1990s, when their capital-labor ratios were comparable to that of the United States (Timmer 2002).

43. Apart from the concern that China was completely relying on foreign technology without improving indigenous technological capabilities, this method also incurs substantial costs in terms of foreign exchange that many developing countries lack.

44. For every $1 that is spent to import technologies, Japan and Korea spend about $3 to understand and assimilate the technology, while China spends only $0.50.

45. During the 1950s, the scientists needed to assimilate foreign technology were mainly former defense engineers from World War II who had substantial experience in research activities (Goto and Odagiri 1997).

the centers of learning in organizing, preparing, and building the capacity to identify and learn from foreign technology (Forbes and Wield 2000). Kim (1997) stresses the importance of the efforts exerted by the recipient of the technology transfer. Samsung engineers spent many months studying the relevant literature on the semiconductor industry prior to contacting experts in the United States and prospective firms that would offer them the technology. If they had not built prior knowledge, Samsung's absorption of technology would have been much more difficult. For that very reason, Korean firms invest substantially in in-house R&D, especially Samsung (Lim 1999).[46] The limited learning achieved by China is most apparent when compared with Korea, using capacity utilization as a measure. This is not to say that R&D effort in China is lacking. Realizing that own effort is necessary to fully internalize the foreign technology, Chinese firms are increasingly engaging in R&D. Using data on Chinese industries, Hu and others (2003) find that technology purchases are effective only when Chinese firms engage in R&D. Malaysia was successful in attracting FDI to start up its electronics and other high-tech industries. However, these activities were mainly limited to assembly, with little value added. Moreover, local design and research skills were weak, because of the lack of a skilled work force and private R&D spending. In 1995 only 0.85 percent of GDP was spent on R&D, of which 80 percent came from public sources (Tidd and Brocklehurst 1999), compared with Japan, where most R&D spending came from the private sector.[47]

Potentially firms can participate in R&D consortia to pool their resources. The success of Japanese R&D consortia has led other countries, both developing and developed, to adopt similar kinds of collaborative R&D schemes to facilitate innovation in an industry.[48] In East Asia, Korea and Taiwan (China) have actively pursued this strategy to assist firms in their R&D

46. Park (1995) sheds light on how the Korean government and firms adopted new foreign technology that enabled them to become world-class producers. The case study focuses on four firms (Daewoo Heavy Industry, Hanyang Chemical, Kolon, and Pohang Iron and Steel [POSCO]) from the mid-1960s to the mid-1980s. The Korean government was heavily involved in selecting the technology to be imported as well as in setting the terms under which these purchases took place. POSCO adopted technology relatively quickly, judging by the speed with which it was able to bring its factory to full capacity and kept adopting new technology as it became available. By looking at the different installment times of technology, Park shows that Korean firms' participation always increased compared with the previous installments. This is indicative of the Korean workers' assimilation of the imported technology.

47. The number of graduates with science and technical backgrounds has also been declining in recent years in both absolute and percentage terms (Tidd and Brocklehurst 1999).

48. The most famous example of such consortia is the VLSI project in 1975. See also Branstetter and Sakakibara (1998, 2002) for an in-depth study of Japanese R&D consortia.

efforts. In the case of Taiwan (China), the collaborative R&D was aimed more at diffusing new technologies than at conducting research per se (Mathews 2002). In the case of Korea, R&D consortia were aimed directly at collaborative research to enhance the competitiveness of domestic firms. However, compared with the Japanese case, Korean R&D consortia were much less effective because of several factors. First, fewer firms were involved in Korea than in Japan. Only 3.4 firms, on average, participated in such collaboration in Korea, whereas 14.8 firms participated in Japan. In addition, Korean consortia mainly targeted applied technologies that could be introduced in the market. This stifled cooperation among Korean firms, whereas Japa-nese consortia typically aimed at developing future generations of technology (Sakakibara and Cho 2002).

One advantage that developing countries have is that the technologies they target have already been filtered and refined for them. They are not exploring uncharted waters. This greatly reduces the uncertainty surrounding the usefulness of imported technology.[49] Nevertheless, domestic R&D efforts are still needed to acquire tacit knowledge through trial and error, especially when the frontier is continuously being extended (Forbes and Wield 2000).

Even though technology embodied in capital goods imports has propelled East Asian countries to higher technological complexity, it is the capabilities of these countries in selecting, adopting, assimilating, and internalizing this imported technology that has helped them to achieve this feat (Lau 1997). For this, own R&D spending geared toward learning is indispensable.

R&D Facilities Abroad

Rather than relying on FDI, firms can actively engage in R&D efforts closer to the so-called "center of excellence" to reap the benefits of the spillover of host countries' expertise in specific technology. Because spillover effects tend to be localized, the R&D facilities tend to be located in close proximity to where new innovations and ideas are generated.

Broadly speaking, there are two reasons for an MNC (or a potential MNC) to establish an R&D facility abroad: "exploitation" of own technology in the host country and "sourcing" of foreign technology

49. However, there is still uncertainty associated with the successful adoption and adaptation of such technology by importing countries.

from the host country.[50] In the first case, the main purpose of "exploitation" R&D facilities is to adapt products and technology to local environments. Therefore, they rely heavily on R&D efforts conducted at home. If, in contrast, the purpose of the facility is sourcing, one would expect this facility to learn from the host country.[51] Since the "exploitation" of own technology in general does not lead to technology transfer, this section focuses on the "sourcing" type of FDI in R&D.

The overseas R&D expenditure of MNCs is highly concentrated in only a few countries, reflecting the availability of advanced technologies, R&D infrastructure, and resources (Kumar 2001). Industrial countries together host more than 90 percent of the overseas R&D of American MNCs, whereas developing countries host only 10 percent. Of these developing economies, the relatively more advanced ones—Brazil, Hong Kong (China), Israel, Mexico, Singapore, and Taiwan (China)—account for close to 90 percent of the share, although R&D facilities in industrializing countries are more associated with adapting technologies to suit local conditions than with engaging in research (see also Kumar 1996; Odagiri and Yasuda 1997).

Among industrial countries, the United States is a popular destination for R&D facilities by MNCs of various nationalities. According to the U.S. Department of Commerce, there are 251 Japanese R&D facilities in the United States, followed by 107 German, 103 British, and 44 French facilities. R&D facilities established by Japanese firms in the United States outnumber U.S. establishments overseas. Korea has increased its R&D presence, with 32 establishments in 1998 compared with approxi-

50. Fors (1997) casts doubt on the effectiveness of subsidiaries' R&D efforts in leading to the sourcing of technology from host countries to home. He finds that the contribution of subsidiaries' R&D efforts has not been used as an input to Swedish plants at home. Most of the technology utilized by home plants and subsidiaries originates in R&D facilities in the home country. Anand and Kogut (1997) also find that FDI in the United States was not motivated by the acquisition of technology. However, in a subsequent study, Fors (1998) finds that Swedish firms locate a higher share of their R&D expenditures in host countries that have technological leadership in the industries that these firms operate. Similarily, Japanese R&D subsidiaries in the United States and Europe seem to be attracted by their technological leadership, whereas R&D subsidiaries in Asia are more motivated by the opportunity to "exploit" (Odagiri and Yasuda 1997). In addition, Frost (2001) uses patent data to identify the origin of ideas. He finds that larger subsidiaries tend to draw their ideas more from the host country's base of knowledge (measured by the citation of patents in the host country) than on the base of knowledge at home.

51. One way to differentiate these two kinds of facilities is to determine if the R&D labs are attached to manufacturing units or if they stand alone. Stand-alone labs tend to be of the sourcing type. One needs to distinguish between the two when analyzing whether an MNC is tapping into the center of excellence in order to test whether overseas R&D facilities are conduits for knowledge transfer.

mately a dozen in 1992. The survey also indicates that the motivation behind foreign firms' investment in U.S. R&D facilities is predominantly to access as well as acquire technologies that complement their technological capabilities (Dalton and Serapio 1999).[52]

An illustration of the use of R&D facilities for sourcing technology is Samsung's development of the semiconductor industry. Samsung established an outpost in Silicon Valley in 1983 aimed at upgrading its semiconductor technology. This R&D facility was staffed with five Korean American scientists, educated in the United States and with relevant experience in major U.S. semiconductor firms. Samsung simultaneously established another R&D facility in Korea. Through training, joint research, and consulting, Samsung was able to transfer much-needed information from the outpost to the home R&D facility (Kim 1997). Another valuable service that an R&D outpost can provide, besides sourcing of new technologies, is information scanning. That is, when a firm wishes to import new technology, it must research what kind of technology is available and who is willing to share it with others. Samsung's recruitment of many Korean American scientists and engineers enabled the company to tap into the frontier of research and was an important component of the R&D outpost (Lim 1999).

However, merely establishing R&D units does not automatically result in the successful observation and assimilation of new developments. These facilities need to be involved in the community and to participate actively in networking. They can achieve this by employing local personnel and developing effective networks with local agents, firms, and research institutions (Zanfei 2000).

Furthermore, the degree of embeddedness of these subsidiaries in local technological development is contingent on the reputation of parent firms in technological arenas (Nobel and Brikinshaw 1998). The more these subsidiaries are embedded in local technological development, the more firms can learn and apply knowledge to their own advantage. To this end, increasing interaction with local firms is crucial. This interaction tends to be more intense when the host country's level of human capital is high (Zanfei 2000). In order to be able to take full advantage of an R&D outpost, a firm needs to have a highly educated work force and an established reputation for technological capabilities. These two requirements seem to be quite daunting for firms in East Asia at the current stage, except for Japan and Korea; unfortunately only a handful of firms in East Asia would be able to engage in this kind of activity.

52. Florida (1997) arrives at similar results.

MAIN FINDINGS AND DISCUSSION

Several common themes emerge from this selective review of the recent literature on technology transfer. First, to understand imported technology, firms must actively engage in R&D, although R&D in this context should be best thought of as learning rather than creating new knowledge, which may come later. That is, the diffusion of technology is not automatic. It is one thing to import new capital equipment that may have several advanced features. Understanding how it actually works and being able to improve on such machinery are another matter. Some level of effort on the firms' part—sometimes great effort—is necessary to fully understand the technology being transferred. New technology is not given; it must be learned. Therefore, importing (or to a lesser degree, just licensing) new technologies is a first step, but a second step is needed to follow it through.

In order to understand new technologies, a firm must possess some capabilities. Essentially this capability is represented by the quality of managers and employees at the firm. If the average level of education and skills is low, the likelihood of mastering technologies, let alone improving on them, is small. Thus, for a firm to assimilate technology successfully, it must employ a high-quality work force at the minimum. Of course, depending on the country's situation, obtaining such workers with advanced knowledge may be hard or close to impossible. A firm can supplement the domestic supply of a highly educated work force (relative to others) by recruiting globally. The cases of Acer, Samsung, and others illustrate the benefits of such an approach. In addition, these engineers and scientists can bring with them tacit knowledge that is much harder to obtain through imported capital goods or licensing.

Because technologies mainly flow from developed to developing countries, firms in developing countries must be outward oriented and ready to exploit the advantages of licensing or of global value chains or imported capital goods with advanced technologies. Global linkage to other firms is increasingly essential. As discussed in earlier chapters, the links may take the form of buyer-supplier relationships or collaboration in R&D efforts. Samsung accumulated valuable experience and technical capabilities through OEM, as did numerous Taiwanese (Chinese) firms. Also, to gain access to tacit knowledge, recruiting engineers and scientists from abroad can be helpful. Thus in some instances, firms must be able to tap the global marketplace for both technologies and potential inputs (capital goods or human capital) in order to successfully transfer foreign technologies.

Role of Human Capital

The appropriate title of Rosenzweig's 1995 article, "Why Are There Returns to Schooling?" puts forth the fundamental question to consider when analyzing the role of education in economic growth. There are two ways in which education can enhance productivity: by improving access to available information and by improving the ability to digest this new information. Considering these, education should play an important role in economic growth, especially when economies are presented with greater opportunities for learning (Rosenzweig 1995). Technology transfer offers such learning opportunities, and, therefore, the level of human capital should influence its success or failure. An example is the green revolution in India. Foster and Rosenzweig (1996) show that the more educated farmers are, the larger the return to using high-yield variety crops, suggesting that educated farmers are better equipped to take full advantage of new technology.

Through the review of literature, it is clear that recipient firms need to possess at least some minimum absorptive capacity—some level of human capital—in order to adopt imported technology (Kim 1997, 1998).[53] Without such capacity, combined with active R&D efforts, any attempt to transfer advanced technology—advanced from the perspective of the host country—will inevitably produce unsatisfactory results (see Lan 1996 on China). At the initial stage of imported technology, workers need to have only a basic educational level (primary and secondary).[54] In the latter part of technology adoption and assimilation, workers need to have a certain level of tertiary education to understand and follow scientific and technological developments available around the world (Hayashi 2002; Kim 1997). Among developing countries, lack of skilled workers and researchers is a typical impediment to successful assimilation and use of foreign technology (Yee and Higuchi 1999). This implies that countries which invest in human capital have a better chance of achieving technological sophistication compared with others.

53. So far, the absorptive capacity is of an absolute kind. Many researchers have concentrated on the recipient firms' capabilities to absorb foreign technologies. However, Lake and Lubatkin (1998) argue that relative absorptive capacity matters more than absolute capacity. In defining their use of "relative," they measure how similar the firms are in terms of prior base of knowledge and organizational structure. On the one hand, Lake and Lubatkin argue that, the more similar they are to each other, the more they can learn (compared to just using the absolute measure of absorptive capacity). On the other hand, growth theory advocates that the larger the gap, the faster the convergence will be.

54. The extent to which secondary education contributes to economic growth in developing countries is hotly debated (see Brist and Caplan 1999).

Korea provides a good example of developing the necessary human capital for technology transfer as part of the country's development strategies. Korea had the highest number of scientists and engineers per 10,000 population among developing countries in 1994. This number was almost equal to that of France and the United Kingdom and helped Korea to move up the technology ladder faster than other countries at comparable levels of per capita income (Kim 1998). Many Korean firms also engage in R&D efforts, aimed mainly at an incremental improvement of existing technology rather than at significant product innovation (Park 1995). In Singapore's case, the government not only emphasized education but also provided subsidies for in-house training to facilitate the learning experience and set up institutes to encourage research in higher-value-added activities (Hobday 1994).[55] In contrast, Malaysia has consistently faced a shortage of skilled workers. This persistent shortage hampers Malaysia's progress in the technology dimension and its ability to attract further investment by MNCs, on which the country depends for industrial development (see chapter 8 of this volume; Hobday 1996).

Moreover, an increasingly skilled labor force is critical to developing countries since most new technologies tend to be more skill intensive in nature.[56] It follows that technologies developed in industrial countries will be used relatively inefficiently in developing countries, with lower associated productivity. Two solutions are available to solve this problem. One is to develop technologies more suitable for developing countries; the other is to increase the level of skills and supply of skilled workers in developing countries (Acemoglu 2001). Given that innovations and development of new technologies are motivated by an anticipation of profits by private agents, relying on or forcing the first solution on developed countries would be impossible to implement. This leads to the conclusion that the most effective way to achieve successful technology transfer is to ensure that the work force in these countries is equal to the challenge posed by new technologies.

This said, empirical evidence linking human capital to economic growth is mixed. In the microeconomic literature, positive returns to schooling are

55. However, the personnel output of Singapore leaned heavily toward technical engineers, not research engineers, lagging behind other industrial countries, including Korea. This will have profound implications for the ability of Singaporean firms to innovate in the future.

56. Not only developing countries but also developed countries face similar problems. In their study of Canadian firms, Baldwin and Lin (2002) find that 29 percent of establishments in the Canadian 1993 Survey of Innovation and Advanced Technology reported shortages of skills, lack of training, or contract issues as impediments to adopting advanced technology. See also Baldwin and Sabourin (2002).

well documented.[57] Individuals are paid according to their productivity, suggesting that education is more than just signaling (Jones 2001). Yet Bils and Klenow (2000) find little evidence that initial schooling affects economic growth significantly, and some studies even find negative coefficients on the level of human capital. Recent studies have tried to reconcile the difference between microeconomic and macroeconomic findings by incorporating institutional differences (Chuang 2000) and quality of education (Dessus 2001; Hanushek and Kimko 2000) and by using better data sets (de la Fuente and Doménech 2002; Krueger and Lindahl 1999). Using any of these corrections results in significant positive coefficients for human capital, broadly confirming the findings in the microeconomic literature. In addition, Barro (1999) finds that secondary and tertiary education greatly affect the growth rate of an economy. He attributes this to the ability of workers to absorb new technology.[58]

Education not only enables workers to learn more efficiently, it also provides them with more frequent opportunities to learn. Tan and Gill (2000) find that in Malaysia, the more educated a worker is, the more likely it is that he or she will receive in-house training, since he or she is likely a better learner than an uneducated worker. This supports Rosenzweig's (1995) argument that education fosters better learning skills. Furthermore, the more a firm engages in R&D activities, the more likely it is that it will offer training to both nonproduction and production workers. Since R&D activities are often used to gauge a firm's sophistication, it is safe to assume that such a firm either employs more advanced technology or at least introduces more advanced technology than firms that do not engage in R&D activities. This correlation between training and advanced technologies signifies that either the educational system (both vocational and general) is inadequate in providing necessary skills or that the technology employed is firm specific so that additional training is necessary, even if the educational system is adequate in general. It may follow that, given firms' propensity to train workers, general education should be more important than vocational education, especially when higher and higher skill levels are required.[59] This is especially true in Malaysia, where more low- and mid-skilled workers mainly receive vocational training (Tan and Gill 2000).

57. See Krueger and Lindahl (1999) for a survey of the microeconomic literature.

58. However, he finds no systematic relationship between educational attainment (years of schooling) and investment, contrary to the hypothesis that a higher ratio of human to physical capital should facilitate the accumulation of capital.

59. Even if governments try to increase the number of graduates from vocational schools, if the social demand for such education is low, any governmental efforts are doomed to fail, as in the case of Korea (Gill and Ihm 2000).

Given the well-established microeconomic findings and more encouraging results from the growth literature—even though analysts still may not fully understand the exact process in which human capital influences growth—human capital seems to play a pivotal role in technology transfer.

POLICY DIRECTIONS

An important issue surrounding technology transfer pertains to the role of government. To resolve this, one needs to understand the kinds of impediments that local firms face. Typically, these are a lack of sufficient information concerning the availability of technology; limited financial resources; lack of technical, legal, and commercial expertise to evaluate the technology;[60] constraints imposed by government regulations (such as permits or regulations); and scarcity of skilled labor (Lau 1997; Lau and Higuchi 1999).[61]

A government can actively disseminate information concerning the availability and development of new technologies and licensors around the world as the Japanese, Korean, and Singaporean governments have done. Advances in information and communications technology greatly reduce the cost of gathering and disseminating these kinds of information. Often, both suppliers and recipients of technology cite inter-governmental assistance and incentives as playing a crucial role in technology transfer (Lau 1997). Lack of skills can be overcome by increasing the supply of skilled workers through education and immigration. Beyond this, as Hill also describes in chapter 8, governments can facilitate the successful adoption of foreign technologies by setting up research labs and institutes,[62] creating and nurturing incubators, assisting in licensing, and forming R&D consortia. In the early stages of development, the Korean government actively assisted firms in acquiring foreign technologies, often placing many re-

60. Wakasugi (1997) finds that the ability to search out and absorb foreign technology has a positive effect on supply. This ability is generated by the indigenous R&D efforts.

61. Other factors include religious and traditional customs, high royalty payments, and restricted conditions imposed on the acquisition of technology.

62. For instance, the Industrial Technology Research Institute in Taiwan (China) has facilitated the diffusion of technology (Mathews 2002; Mathews and Cho 2000). However, the effectiveness of such institutes is unclear. See Kimura (1997) for Japan; Kim and Yi (1997) for Korea; Hsu and Chiang (2001) for Taiwan (China).

strictions on technology suppliers.[63] Korean firms used these requirements to their advantage, and, as a consequence, some firms emerged as successful exporters.[64]

Arguably the scale and efficacy of technology transfer critically depend on the level of human capital.[65] Once a country achieves a level of technological sophistication and is willing and able to be an *innovator* rather than an *imitator*, high-quality research capabilities are needed. Therefore, strengthening tertiary education is vital.[66] To this end, FDI can bring yet another benefit, aside from being a conduit for technology transfer. MNCs often demand skilled workers—generally more skilled than those demanded by domestic industries—and this translates into higher demand for a more educated work force (Blomstrom and Kokko 2003).

Related to the supply of human capital is the mobility of skilled workers. Technologies are embedded not only in imported machinery but often in skilled workers in the form of tacit knowledge. The more the skilled workers circulate among different firms, the more they spread such tacit knowledge. For this to function effectively, a fluid labor market in terms of both domestic labor and international flow of skilled workers is an advantage.

CONCLUSION

This chapter has reviewed various channels of technology transfer available to firms wishing to upgrade their technological capabilities. The literature does not offer any clear-cut answers on which methods a firm (or a country as a policy direction) should take. On the one hand, this is a shortcoming of the literature, and further research is needed on the relative

63. The Korean government's requirements included (1) acquisition of patents, designs, and know-how, (2) training of engineers and managers abroad or on-site, (3) speedy replacement of expatriates with Korean workers, and (4) access to improvements in products and processes in the future.

64. The effects of this policy can be seen in the iron and steel industries, where there were four different installments of imported foreign technology. As Korean workers became acquainted with the technology, their participation in every stage increased. They also took advantage of access to improvements and installed them as more capacity was needed (Park 1995).

65. The educational level of host countries matters to close the gap between the technological capabilities of the parent and subsidiaries (Urata and Kawai 2000).

66. Graduate education in science and engineering in Korea has been neglected in the past. This has forced Korean firms to recruit from abroad where such education is provided (Kim 1995).

effectiveness of different pathways of technology transfer. On the other hand, no firm will choose only one avenue of accessing new technologies developed elsewhere. More likely successful firms will use multiple channels of technology transfer, depending on the nature of the technology in question and the constraints they face. For instance, manufacturing firms tend to rely on the direct purchase of capital goods as well as licensing, subcontracting, and equity acquisition, whereas insurance and financial firms prefer licensing and human resource cultivation (Lau and Higuchi 1999).

However, the literature does suggest two key lessons: openness is essential, and own effort matters. Needless to say, in order for technology transfer to succeed, technologies need to be able to flow from one place to the other. In most cases, this means the flow of ideas from developed to developing countries. As explained in detail in this chapter, the technology can flow through trade, especially on capital goods, spillovers from FDI, and circulation of skilled workers. These all point to the importance of openness to trade, investment, and the flow of people. When any one of these channels is restricted, the potential benefit of technological development outside the country is dimished.

Openness also provides an incentive to firms to upgrade their technological capabilities. Without such incentive, firms are unlikely to embark on the risky and costly exercise of technology upgrading. The contrast between East Asia and Latin America seems to offer the clearest difference in competitive pressures felt by firms in these two regions. Both regions have invested in human capital development, but firms in Latin America have tended to be shielded from competitive pressures and to be inward looking, with other domestic firms as their main competitors as the result of import substitution. In contrast, firms in East Asia have tended to be more outward oriented and to face more competitive pressures in the global market rubbing shoulders with leading firms from developed countries (de Ferranti and others 2003). Because of this pressure and outward orientation, firms in East Asia have felt compelled to upgrade their technological capabilities. Their participation in global production networks has facilitated this goal greatly.

Even though openness facilitates the flow of ideas through multiple channels and provides incentives to upgrade, technology transfer is not automatic. Even if new ideas flow, if firms are ill prepared to take advantage of them, the payoff from such flow is much smaller. In order to fully comprehend and use the new technologies, firms must actively engage in R&D, mostly focusing on learning rather than on developing new technologies. To do so firms must have some requisite human capital, depending on the complexity of the technology. Hence an abundant supply of educated workers is necessary to the technological development of firms.

Regardless of the mode of technology transfer, the efforts of East Asian economies to master new technologies will remain indispensable to profitability in the current and expected global environment. Continued innovation is fundamental to achieving this goal. In most instances, the ability to innovate must be learned from experience, and it is a slow and tedious process.[67] This experience is accumulated through technology transfers of various kinds and domestic R&D efforts that allow firms to fully comprehend foreign technology. By redoubling their efforts to absorb technology from elsewhere—coupled with ensuring a high level of education, domestic R&D efforts, and a fluid labor market—and developing their own technologies, firms can ensure their profitability and growth under conditions of an increasingly competitive global marketplace.

REFERENCES

The word *processed* describes informally reproduced works that may not be commonly available through libraries.

Acemoglu, Daron. 2001. "Directed Technical Change." NBER Working Paper 8287. National Bureau of Economic Research, Cambridge, Mass. Processed.

Aggarwal, Aradhna. 2000. "Deregulation, Technology Imports, and In-House R&D Efforts: An Analysis of the Indian Experience." *Research Policy* 29(9):1081–93.

Ahuja, Gautam, and Riitta Katila. 2001. "Technological Acquisitions and the Innovation Performance of Acquiring Firms: A Longitudinal Study." *Strategic Management Journal* 22(3):197–220.

Aitken, Brian, Gordon H. Hanson, and Ann E. Harrison. 1997. "Spillovers, Foreign Investment, and Export Behavior." *Journal of International Economics* 43(1–2):103–32.

Aitken, Brian J., and Ann E. Harrison. 1999. "Do Domestic Firms Benefit from Direct Foreign Investment? Evidence from Venezuela." *American Economic Review* 89(3):605–18.

Anand, Jaideep, and Bruce Kogut. 1997. "Technological Capabilities of Countries, Firm Rivalry, and Foreign Direct Investment." *Journal of International Business Studies* 28(3): 445–65.

Andreosso-O'Callaghan, Bernadette, and Wei Qian. 1999. "Technology Transfer: A Mode of Collaboration between the European Union and China." *Europe-Asia Studies* 51(1):123–42.

Arora, Ashish. 1996. "Contracting for Tacit Knowledge: The Provision of Technical Services in Technology Licensing Contracts." *Journal of Development Economics* 50(2):233–56.

Baldwin, John R., and Zhengxi Lin. 2002. "Impediments to Advanced Technology Adoption for Canadian Manufacturers." *Research Policy* 31(1):1–18.

67. Looking at the electronics industry in Singapore, Hobday (1994) concludes that leapfrogging may not be possible. From Korean experience in memory chips, Lee and Lim (2001) conclude that some minor leapfrogging (skipping some intermediate steps) is possible, but major leapfrogging is not.

Baldwin, John R., and David Sabourin. 2002. "Advanced Technology Use and Firm Performance in Canadian Manufacturing in the 1990s." *Industrial and Corporate Change* 11(4):761–89.

Baptista, Rui. 2000. "Do Innovations Diffuse Faster within Geographical Clusters?" *International Journal of Industrial Organization* 18(3):515–35.

Barro, Robert J. 1999. "Human Capital and Growth in Cross-Country Regressions." *Swedish Economic Policy Review* 6(2):237–77.

Bayoumi, Tamim, David T. Coe, and Elhanan Helpman. 1999. "R&D Spillovers and Global Growth." *Journal of International Economics* 47(2):399–428.

Beine, Michel, Frederic Docquier, and Hillel Rapoport. 2001. "Brain Drain and Economic Growth: Theory and Evidence." *Journal of Development Economics* 64(1):275–89.

Belderbos, Rene. 2001. "Overseas Innovations by Japanese Firms: An Analysis of Patent and Subsidiary Data." *Research Policy* 30(2):313–32.

Belderbos, Rene, Giovanni Capannelli, and Kyoji Fukao. 2000. "The Local Content of Japanese Electronics Manufacturing Operations in Asia." In Taketoshi Ito and Anne O. Krueger, eds., *The Role of Foreign Direct Investment in East Asian Economic Development.* Chicago, Ill.: University of Chicago Press.

———. 2001. "Backward Vertical Linkages of Foreign Manufacturing Affiliates: Evidence from Japanese Multinationals." *World Development* 29(1):189–208.

Bils, Mark, and Peter J. Klenow. 2000. "Does Schooling Cause Growth?" *American Economic Review* 90(5):1160–83.

Blalock, Garrick, and Paul Gertler. 2002. "Technology Acquisition in Indonesian Manufacturing: The Effect of Foreign Direct Investment." Background paper prepared for East Asian Prospects Project, World Bank, Washington, D.C. Processed.

Blomström, Magnus, and Ari Kokko. 2003. *Human Capital and Inward FDI.* CEPR Discussion Paper 3762. London: Centre for Economic Policy Research.

Blomström, Magnus, and Fredrik Sjoholm. 1999. "Technology Transfer and Spillovers: Does Local Participation with Multinationals Matter?" *European Economic Review* 43(4–6):915–923.

Bosworth, Derek. 1996. "Determinants of the Use of Advanced Technologies." *International Journal of the Economics of Business* 3(3):269–93.

Braconier, Henrik, Karolina Ekholm, and Karen-Helene Midelfart-Knarvik. 2000. *Does FDI Work as a Channel for R&D Spillovers? Evidence Based on Swedish Data.* CEPR Discussion Paper 2469. London: Centre for Economic Policy Research.

Branstetter, Lee G., and Jong-Rong Chen. 2001. "The Impact of Technology Transfer and R&D on Productivity Growth in the Taiwanese Electronics Industry: Microeconometric Analysis Using Plant-Level Data." Unpublished ms. University of California, Davis. Processed.

Branstetter, Lee G., and Mariko Sakakibara. 1998. "Japanese Research Consortia: A Microeconometric Analysis of Industrial Policy." *Journal of Industrial Economics* 46(2): 207–33.

———. 2002. "When Do Research Consortia Work Well and Why? Evidence from Japanese Panel Data." *American Economic Review* 92(1):143–59.

Brist, Lonnie E., and Arthur J. Caplan. 1999. "More Evidence on the Role of Secondary Education in the Development of Lower-Income Countries: Wishful Thinking or Useful Knowledge?" *Economic Development and Cultural Change* 48(1):155–75.

"Can Proton Deliver? After Roaring Back from the Brink, the Company Faces Stiffer Competition." 2003. *Business Week*. February 3.

Caselli, Francesco, and Wilbur John Coleman II. 2001. "Cross-Country Technology Diffusion: The Case of Computers." *American Economic Review* 91(2):328–35.

Chen, Tain-Jy, Been-Lon Chen, and Yun-Peng Chu. 2001. "The Development of Taiwan's Electronics Industry." In Poh-Kam Wong and Chee-Yuen Ng, eds., *Industrial Policy, Innovation, and Economic Growth*, pp. 245–82. Singapore: Singapore University Press.

Chen, X., and C. Sun. 2000. "Technology Transfer to China: Alliances of Chinese Enterprises with Western Technology Exporters." *Technovation* 20(7):353–62.

Chiang, Jong-Tsong. 2000. "Institutional Frameworks and Technological Paradigms in Japan: Trageting Computers, Semiconductors, and Software." *Technology in Society* 22(2):151–74.

Chuang, Yih-chyi. 2000. "Human Capital, Exports, and Economic Growth: A Causality Analysis for Taiwan, 1952–1995." *Review of International Economics* 8(4):712–20.

Coe, David T., and Elhanan Helpman. 1995. "International R&D Spillovers." *European Economic Review* 39(5):859–87.

Coe, David T., Elhanan Helpman, and Alexander W. Hoffmaister. 1997. "North-South R&D Spillovers." *Economic Journal* 107(440):134–49.

Cyhn, Jin W. 2000. "Technology Development of Korea's Electronics Industry: Learning from Multinational Enterprises through OEM." *European Journal of Development Research* 12(1):159–87.

Dalton, Donald H., and Manuel G. Serapio Jr. 1999. "Globalizing Industrial Research and Development." Washington, D.C.: U.S. Department of Commerce.

"Dead Deal Walking." 2002. *Economist*, February 7.

de Ferranti, David, Guillermo E. Perry, Indermit S. Gill, J. Luis Guasch, William F. Maloney, Carolina Sánchez-Páramo, and Norbert Schady. 2003. *Closing the Gap in Education and Technology*. Washington, D.C.: World Bank.

de la Fuente, Angel, and Rafael Doménech. 2002. *Human Capital in Growth Regressions: How Much Difference Does Data Quality Make? An Update and Further Results.* CEPR Discussion Paper 3587. London: Centre for Economic Policy Research.

Dessus, Sebastian. 2001. "Human Capital and Growth: The Recovered Role of Educational System." Policy Research Working Paper 2632. World Bank, Washington, D.C. Processed.

Ding, Jingping. 1998. "Using Imported Technology to Transform Existing Enterprises in China." In Charles Feinstein and Christopher Howe, eds., *Chinese Technology Transfer in the 1990s.* Lyme Regis, U.K.: Edward Elgar.

Dowling, Malcolm, and David Ray. 2000. "The Structure and Composition of International Trade in Asia: Historical Trends and Future Prospects." *Journal of Asian Economics* 11(3):301–18.

Doz, Yves, Jose Santos, and Peter Williamson. 2001. *From Global to Metanational: How Companies Win in the Knowledge Economy.* Boston, Mass.: Harvard Business School Press.

Eaton, Jonathan, and Samuel Kortum. 1999. "International Technology Diffusion: Theory and Measurement." *International Economic Review* 40(3):537–70.

———. 2001. "Trade in Capital Goods." *European Economic Review* 45(7):1195–235.

Florida, Richard. 1997. "The Globalization of R&D: Results of a Survey of Foreign-Affiliated R&D Laboratories in the USA." *Research Policy* 26(1):85–103.

Forbes, Naushad, and David Wield. 2000. "Managing R&D in Technology-Followers." *Research Policy* 29(9):1095–109.

Fors, Gunnar. 1997. "Utilization of R&D Results in the Home and Foreign Plants of Multinationals." *Journal of Industrial Economics* 45(3):341–58.

———. 1998. "Locating R&D Abroad: The Role of Adaptation and Knowledge-Seeking." In Pontus Braunerhjelm and Karolina Ekholm, eds., *The Geography of Multinational Firms*, pp. 117–344. Boston, Mass.: Kluwer Academic Publishers.

Fosfuri, Andrea, Massimo Motta, and Thomas Ronde. 2001. "Foreign Direct Investment and Spillovers through Workers' Mobility." *Journal of International Economics* 53(1): 205–22.

Foster, Andrew D., and Mark R. Rosenzweig. 1996. "Technical Change and Human-Capital Returns and Investments: Evidence from the Green Revolution." *American Economic Review* 86(4):931–53.

Frantzen, Dirk. 2000. "Innovation, International Technological Diffusion, and the Changing Influence of R&D on Productivity." *Cambridge Journal of Economics* 24(2):193–210.

Frost, Tony S. 2001. "The Geographic Sources of Foreign Subsidiaries' Innovations." *Strategic Management Journal* 22(2):101–23.

Gill, Indermit S., and Chon-Sun Ihm. 2000. "Republic of Korea." In Indermit S. Gill, Fred Fluitman, and Amit Dar, eds., *Vocational Education and Training Reform*. New York: Oxford University Press.

Goto, Akira, and Hiroyuki Odagiri. 1997. *Innovation in Japan*. Oxford: Clarendon Press.

Haddad, Mona, and Ann Harrison. 1993. "Are There Positive Spillovers from Direct Foreign Investment? Evidence from Panel Data for Morocco." *Journal of Development Economics* 42(1):51–74.

Hanushek, Eric A., and Dennis D. Kimko. 2000. "Schooling, Labor-Force Quality, and the Growth of Nations." *American Economic Review* 90(5):1184–208.

Hayashi, Mitsuhiro. 2002. "The Role of Subcontracting in SME Development in Indonesia: Micro-Level Evidence from the Metalworking and Machinery Industry." *Journal of Asian Economics* 13(1):1–26.

Hayashi, Takeshi. 1990. *The Japanese Experience in Technology*. Maastricht, the Netherlands: United Nations University.

Hill, Hal, and Prema-Chandra Athukorala. 1998. "Foreign Investment in East Asia: A Survey." *Asian-Pacific Economic Literature* 12(2):23–50.

Hobday, Michael. 1994. "Technological Learning in Singapore: A Test Case of Leapfrogging." *Journal of Development Studies* 30(4):831–58.

———. 1995a. "East Asian Latecomer Firms: Learning the Technology of Electronics." *World Development* 23(7):1171–93.

———. 1995b. *Innovation in East Asia: The Challenge to Japan*. Aldershot, U.K., and Brookfield, Vt.: Edward Elgar.

———. 1996. "Innovation in South-East Asia: Lessons for Europe?" *Management Decision* 34(9):71–81.

Hou, Chi-Ming, and San Gee. 1993. "National Systems Supporting Technical Advance in Industry: The Case of Taiwan." In Richard R. Nelson, eds., *National Innovation Systems: A Comparative Analysis*, pp. 384–413. New York: Oxford University Press.

Hsu, Chiung-Wen, and Hsueh-Chiao Chiang. 2001. "The Government Strategy for the Upgrading of Industrial Technology in Taiwan." *Technovation* 21(2):123–32.

Hu, Albert G. Z., Gary H. Jefferson, Guan Xiaojing, and Qian Jinchang. 2003. "R&D and Technology Transfer: Firm-Level Evidence from Chinese Industry." Working Paper 582. William Davidson Institute, Ann Arbor, Mich. Processed.

Jones, Patricia. 2001. "Are Educated Workers More Productive?" *Journal of Development Economics* 64(1):57–79.

Kang, Hojin. 2001. "The Development Experience of South Korea: Government Policies and Development of Industries—The Case of Electronics Industries." In Poh-Kam Wong and Chee-Yuen Ng, eds., *Industrial Policy, Innovation, and Economic Growth*, pp. 397–430. Singapore: Singapore University Press.

Katrak, Homi. 1997. "Developing Countries' Imports of Technology, In-House Technological Capabilities, and Efforts: An Analysis of the Indian Experience." *Journal of Development Economics* 53(1):67–83.

Keller, Wolfgang. 1998. "Are International R&D Spillovers Trade-Related? Analyzing Spillovers among Randomly Matched Trade Partners." *European Economic Review* 42(8):1469–81.

———. 2002. "Geographic Localization of International Technology Diffusion." *American Economic Review* 92(1):120–42.

Kim, Linsu. 1991. "Pros and Cons of International Technology Transfer: A Developing Country's View." In Tamir Agmon and Mary Ann von Glinow, eds., *Technology Transfer in International Business*, pp. 223–39. New York: Oxford University Press.

———. 1995. "Absorptive Capacity and Industrial Growth: A Conceptual Framework and Korea's Experience." In Bon-Ho Koo and Dwight H. Perkins, eds., *Social Capability and Long-Term Economic Growth*, pp. 266–87. New York: St. Martin's.

———. 1997. "The Dynamics of Samsung's Technological Learning in Semiconductors." *California Management Review* 9(4):506–21.

———. 1998. "Crisis Construction and Organizational Learning: Capability Building in Catching Up at Hyundai Motor." *Organization Science* 9(4):506–21.

Kim, Linsu, and Gihong Yi. 1997. "The Dynamics of R&D in Industrial Development: Lessons from the Korean Experience." *Industry and Innovation* 4(2):167–82.

Kim, Youngbae, and Byungheon Lee. 2002. "Patterns of Technological Learning among the Strategic Groups in the Korean Electronic Parts Industry." *Research Policy* 31(4):543–67.

Kimura, Yui. 1997. "Technological Innovation and Competition in the Japanese Semiconductor Industry." In Akira Goto and Hiroyuki Odagiri, eds., *Innovation in Japan*, pp. 121–58. Oxford: Clarendon Press.

Krueger, Alan B., and Mikael Lindahl. 1999. "Education for Growth in Sweden and the World." *Swedish Economic Policy Review* 6(2):289–339.

Kumar, Nagesh. 1996. "Intellectual Property Protection, Market Orientation, and Location of Overseas R&D Activities by Multinational Enterprises." *World Development* 24(4):673–88.

———. 2001. "Determinants of Location of Overseas R&D Activity of Multinational Enterprises: The Case of U.S. and Japanese Corporations." *Research Policy* 30(1):159–74.

Kusunoki, Ken, and Tsuyoshi Numagami. 1997. "Intrafirm Transfers of Engineers in Japan." In Akira Goto and Hiroyuki Odagiri, eds., *Innovation in Japan*, pp. 173–203. Oxford: Clarendon Press.

Lake, Peter J., and Michael Lubatkin. 1998. "Relative Absorptive Capacity and Interorganizational Learning." *Strategic Management Journal* 19(5):461–77.

Lan, Ping. 1996. "Role of IJVs in Transferring Technology to China." In Roger Baran, Yigang Pan, and Erdener Kaynak, eds., *International Joint Ventures in East Asia.* New York: International Business Press.

Lau, Sim Yee. 1997. "Technology Transfer in East Asia and Its Implications for Regional Cooperation." *Global Economic Review* 26(4):65–88.

Lau, Sim Yee, and Yoichiro J. Higuchi. 1999. "Technology Transfer in the East Asian Six: A Multivariate Analytical Inquiry." *Asia-Pacific Development Journal* 6(1):19–34.

Lawrence, Robert Z., and David E. Weinstein. 2001. "Trade and Growth: Import-Led or Export-Led? Evidence from Japan and Korea." In Joseph E. Stiglitz and Shahid Yusuf, eds., *Rethinking the East Asian Miracle*, pp. 379–408. New York: Oxford University Press.

Lee, Keun, and Chaisung Lim. 2001. "Technological Regimes, Catching up, and Leapfrogging: Findings from the Korean Industries." *Research Policy* 30(3):459–83.

Lim, Youngil. 1999. *Technology and Productivity: The Korean Way of Learning and Catching Up.* Cambridge, Mass.: MIT Press.

Madden, Gary, and Scott J. Savage. 2000. "R&D Spillovers, Information Technology, Telecommunications, and Productivity in Asia and the OECD." *Information Economics and Policy* 12(4):367–92.

Markusen, James R., and Anthony J. Venables. 1999. "Foreign Direct Investment as a Catalyst for Industrial Development." *European Economic Review* 43(2):335–56.

Mathews, John A. 2002. "The Origins and Dynamics of Taiwan's R&D Consortia." *Research Policy* 31(4):633–51.

Mathews, John A., and Dong Sung Cho. 2000. *Tiger Technology: The Creation of a Semiconductor Industry in East Asia.* Cambridge Asia-Pacific Studies. Cambridge, U.K.: Cambridge University Press.

McDaniel, Christine A. 2000. "Inventing around and Impacts on Modes of Entry in Japan: A Cross-Country Analysis of U.S. Affiliate Sales and Licensing." Working Paper. U.S. International Trade Commission, Washington, D.C. Processed.

Morita, Hodaka. 2001. "Choice of Technology and Labour Market Consequences: An Explanation of U.S.-Japanese Differences." *Economic Journal* 111(468):29–50.

Narayanan, Suresh. 1999. "Factors Favouring Technology Transfer to Supporting Firms in Electronics: Empirical Data from Malaysia." *Asia-Pacific Development Journal* 6(1):55–72.

Narayanan, Suresh, and Lai Yew Wah. 2000. "Technological Maturity and Development without Research: The Challenge for Malaysian Manufacturing." *Development and Change* 31(2):435–57.

NASDA (National Space Development Agency of Japan). 1997. *Japanese Rocket Development History.* Report 56. Tokyo.

Navaretti, Giorgio Barba, and Isidro Soloaga. 2001. "Weightless Machines and Costless Knowledge: An Empirical Analysis of Trade and Technology Diffusion." Policy Research Working Paper 2598. World Bank, Washington, D.C. Processed.

Nobel, Robert, and Julian Brikinshaw. 1998. "Innovation in Multinational Corporations: Control and Communication Patterns in International R&D Operations." *Strategic Management Journal* 19(5):479–96.

Odagiri, Hiroyuki, and Hideto Yasuda. 1997. "Overseas R&D Activities of Japanese Firms." In Akira Goto and Hiroyuki Odagiri, eds., *Innovation in Japan*, pp. 204–28. Oxford: Clarendon Press.

Panizzolo, Roberto. 1998. "Managing Innovation in SMEs: A Multiple Case Analysis of the Adoption Implementation of Product and Process Design Technologies." *Small Business Economics* 11(1):25–42.

Park, Woo Hee. 1995. "Technology Transfer and Absorption in Korea: Methodology and Measurement." *Seoul Journal of Economics* 8(2):195–226.

Rosenzweig, Mark R. 1995. "Why Are There Returns to Schooling?" *American Economic Review* 85(2):153–58.

Saggi, Kamal. 2002. "Trade, Foreign Direct Investment, and International Technology Transfer: A Survey." *World Bank Research Observer* 17(2):191–235.

Sakakibara, Mariko, and Dong Sung Cho. 2002. "Cooperative R&D in Japan and Korea: A Comparison of Industrial Policy." *Research Policy* 31(5):673–92.

Santarelli, Enrico, and Alessandro Sterlacchini. 1994. "Embodied Technological Change in Supplier-Dominated Firms: The Case of Italian Traditional Industries." *Empirica* 21(3):313–27.

Shan, Weijian, and Jaeyong Song. 1997. "Foreign Direct Investment and the Sourcing of Technological Advantage: Evidence from the Biotechnology Industry." *Journal of International Business Studies* 28(2):267–84.

Shi, Yizheng. 2001. "Technological Capabilities and International Production Strategy of Firms: The Case of Foreign Direct Investment in China." *Journal of World Business* 36(2):184–204.

Smith, Heather. 2000. *Industry Policy in Taiwan and Korea in the 1980s: Winning with the Market.* Cheltenham, U.K., and Northampton, Mass.: Edward Elgar; distributed by American International Distribution Corporation, Williston, Vt.

Sony Corporation. 2003. *Sony History.* Available at http://www.sony.co.jp/en/Fun/SH/.

Stark, Oded, Christian Helmenstein, and Alexia Prskawetz. 1997. "A Brain Gain with a Brain Drain." *Economics Letters* 55(2):227–34.

———. 1998. "Human Capital Depletion, Human Capital Formation, and Migration: A Blessing or a 'Curse'?" *Economics Letters* 60(3):363–67.

Tan, Hong W., and Indermit S. Gill. 2000. "Malaysia." In Indermit S. Gill, Fred Fluitman, and Amit Dar, eds., *Vocational Education and Training Reform*, pp. 218–60. New York: Oxford University Press.

Tang, Shiguo. 1998. "Sino-Japanese Technology Transfer and Its Effects." In Charles Feinstein and Christopher Howe, eds., *Chinese Technology Transfer in the 1990s.* Lyme Regis, U.K.: Edward Elgar.

Tidd, Joseph, and Michael Brocklehurst. 1999. "Routes to Technological Learning and Development: An Assessment of Malaysia's Innovation Policy and Performance." *Technological Forecasting and Social Change* 62(3):239–57.

Timmer, Marcel P. 2002. "Climbing the Technology Ladder Too Fast? New Evidence on Comparative Productivity Performance in Asian Manufacturing." *Journal of the Japanese and International Economies* 16(1):50–72.

Urata, Shujiro, and Hiroki Kawai. 2000. "Intrafirm Technology Transfer by Japanese Manufacturing Firms in Asia." In Taketoshi Ito and Anne O. Krueger, eds., *The Role of Foreign Direct Investment in East Asian Economic Development.* Chicago: University of Chicago Press.

van Pottelsberghe de la Potterie, Bruno, and Frank Lichtenberg. 2001. "Does Foreign Direct Investment Transfer Technology across Borders?" *Review of Economics and Statistics* 83(3):490–97.

von Tunzelmann, Nick. 1998. "The Transfer of Process Technologies in Comparative Perspective." In Charles Feinstein and Christopher Howe, eds., *Chinese Technology Transfer in 1990s.* Lyme Regis, U.K.: Edward Elgar.

Wah, Lai Yew, and Suresh Narayanan. 1997. "The Quest for Technological Competence Via MNCs: A Malaysian Case Study." *Asian Economic Journal* 11(4):407–22.

Wakasugi, Ryuhei. 1997. "Technological Importation in Japan." In Akira Goto and Hiroyuki Odagiri, eds., *Innovation in Japan*, pp. 20–38. Oxford: Clarendon Press.

Xu, Bin. 2000. "Multinational Enterprises, Technology Diffusion, and Host-Country Productivity Growth." *Journal of Development Economics* 62(2):477–93.

Xu, Jiangping. 1998. "China's International Technology Transfer: The Current Situation, Problems, and Future Prospects." In Charles Feinstein and Christopher Howe, eds., *Chinese Technology Transfer in the 1990s.* Lyme: Edward Elgar Publishing.

Yee, Lau Sim, and Yoichiro J. Higuchi. 1999. "Technology Transfer in the East Asian Six: A Multivariate Analytical Inquiry." *Asia-Pacific Development Journal* 6(1):19–34.

Ying, Fan. 1996. "Research on Joint Ventures in China: Progress and Prognosis." In Roger Baran, Yigang Pan, and Erdener Kaynak, eds., *International Joint Ventures in East Asia.* New York: International Business Press.

Yusuf, Shahid, with M. Anjum Altaf, Barry Eichengreen, Sudarshan Gooptu, Kaoru Nabeshima, Charles Kenny, Dwight H. Perkins, and Marc Shotten. 2003. *Innovative East Asia: The Future of Growth.* New York: Oxford University Press.

Zanfei, Antonello. 2000. "Transnational Firms and the Changing Organisation of Innovative Activities." *Cambridge Journal of Economics* 24(5):515–42.

CHAPTER 10

AN INVESTIGATION OF FIRM-LEVEL R&D CAPABILITIES IN EAST ASIA

Gary H. Jefferson and Zhong Kaifeng

For developing nations, a necessary precondition for catch-up with the world's most advanced economies is the capacity to innovate new technologies and sustain technological progress. This chapter focuses on the role of the firm as the key actor on which successful innovation depends. The observation of Nelson and Rosenberg (1993, p. 4) that firms are the entities that "master and get into practice product designs and manufacturing processes" underscores the central role of the firm in the innovation process.

Although firms may represent the most fundamental unit of the national innovation system, they are but a piece of the whole.[1] Institutional conditions, including laws and regulations, government policy and programs, factor and product markets, research institutes, universities, and research and development (R&D) networks together shape the environment within which firms struggle to achieve competitive advantage. This chapter is based on an extraordinary survey of these conditions as reported by a large sample of firms in 10 Asian metropolitan economies. With these survey data, the World Bank has created a rich database that documents firm-level R&D resources and performance as well as the broad institutional endowment of the Asian metropolitan economies within which these firms conduct their R&D operations.

The authors very much appreciate the comments of Simon Evenett, Dale Jorgenson, Kaoru Nabeshima, Shahid Yusuf, and three anonymous referees on earlier versions of this chapter as well as the comments and suggestions of participants at the conference on East Asia's future economy, Harvard University, October 1–2, 2001. They also appreciate support from the National Science Foundation, grant 9905259.

1. See Nelson and Rosenberg (1993) and Jefferson and Hu's (2000) study of China's industrial innovation system.

This chapter investigates two major questions regarding the factors that affect the innovation capabilities of firms. The first question concerns the general impact of R&D on firm-level performance. Within the sample of Asian firms, what is the evidence that R&D affects firm performance, and is it possible to identify the channels through which R&D operates, such as product or process innovation?

The second question concerns the range of institutional factors that influence the effectiveness of R&D. These factors fall into four categories—international exposure, human capital, the R&D network, and the policy setting—for which survey data have been collected at the firm level. A number of these factors are explored in Yusuf with others (2003, ch. 4). Can we identify a subset of these institutional factors that enhance the effectiveness of firm-level R&D?

We now begin to aggregate our firm-level findings to investigate differences in overall R&D capabilities across the metropolitan economies in our sample. We use our firm-level data to investigate the following question: Is it possible to identify and measure differences in the institutional endowment of metropolitan economies that can potentially account for differences in their R&D performance?

We investigate the association between the returns to R&D at the metropolitan level and differences in the measured institutional endowments of these economies. That is, is there an identifiable set of institutional attributes that systematically enhances the effectiveness of R&D activity across metropolitan economies?

This chapter is organized as follows. The following section outlines the research agenda and method of the paper. Because the firm is the basic unit of observation on which we construct metropolitan economy–wide measures of R&D performance, we characterize the optimizing problem of the firm that causes it to allocate R&D resources for deliberate innovation. This is done in the second section. The third section identifies the institutional attributes that potentially affect firm and metropolitan area–wide R&D capabilities; that section also identifies literature that has highlighted the importance of these factors. The fourth section reports on the first stage of the applied analysis—the impact of R&D on firm performance and the channels through which R&D operates. The fifth section identifies the institutional attributes that interact with R&D to enhance firm performance. The sixth section identifies the incidence of these key institutional attributes across the 10 metropolitan economies included in this phase of our analysis. For the six metropolitan economies for which the necessary data are available, the seventh section investigates the empirical relationship between the R&D institutional endowment of these metropolitan economies and our estimates of returns to R&D. A final section discusses our findings and draws conclusions from the overall analytical exercise.

RESEARCH AGENDA AND METHOD

Our research agenda is shaped by the opportunities and limitations presented by the World Bank survey introduced above, which includes 1,826 firms distributed across 11 cities: Bangkok, Jakarta, Kuala Lumpur, Manila, Seoul, Singapore, and five Chinese cities—Beijing, Chengdu, Guangzhou, Shanghai, and Tianjin. For each of these 11 metropolitan economies, the sample of firms spans 10 industries, consisting of five manufacturing and five service industries. The distribution of the 1,826 firms across the 11 metropolitan areas and 10 industries is shown in table 10.1.

This survey collected data on 23 institutional characteristics that potentially enhance the productivity and profitability of firm-level R&D operations. These characteristics are listed in table 10.2. We aggregate these data up to the metropolitan level to develop economy-wide measures of international exposure, human capital, R&D networks, and the policy setting for each metropolitan area. We then measure whether these institutional differences can explain differences in overall returns across the economies surveyed.

The basic strategy is to use these firm-level data to construct economy-wide measures of R&D capabilities and performance. We achieve this using the following four-step procedure: assessing the impact of R&D on firm productivity; identifying factors that enhance the effectiveness of R&D; differentiating the R&D capabilities of metropolitan areas; and associating returns to R&D with R&D capabilities.

Assessing the Impact of R&D on Firm Productivity

To examine the impact of R&D on firm performance, we employ a three-equation approach that separates R&D into two distinct processes. The first equation is a knowledge production function that explores the relationship between R&D effort and specific innovation outputs, including counts of product and process innovations. The second equation characterizes the impact of specific innovation outputs—that is, the product and process innovations that are produced through the knowledge production function—on firm performance, including productivity and profitability. Finally, in order

Table 10.1 Distribution of Firms, by Metropolitan Area

Location	Manufacturing industries					Service industries					
	Apparel and leather goods	Electronic equipment	Electronic components	Consumer products	Vehicles and vehicle parts	Information technology services	Communication services	Accounting and related services	Advertising and related services	Business logistics services	Total
China											
Beijing	49	41	43	21	44	25	11	23	20	23	300
Tianjin	42	36	41	35	46	29	11	23	19	18	300
Shanghai	40	40	40	40	40	20	20	20	20	20	300
Guangzhou	46	40	39	33	42	30	12	18	11	29	300
Chengdu	45	35	40	36	44	24	17	20	19	20	300
Total	222	192	203	165	216	128	71	104	89	110	1,500
Other East Asia											
Jarkata	21	0	4	0	5	11	6	10	4	15	76
Seoul	9	15	18	8	5	8	7	2	5	8	85
Kuala Lumpur	4	7	9	4	4	7	2	4	4	11	56
Manila	18	9	5	0	5	7	4	4	4	9	65
Singapore	0	1	3	0	2	2	0	3	2	5	18
Bangkok	8	1	4	1	2	2	0	1	1	6	26
Total	60	33	43	13	23	37	19	24	20	54	326

Source: World Bank survey data.

Table 10.2 Institutional Attributes that Potentially Enhance the Effectiveness of R&D

Classification	Attribute
Openness and competition	Share of foreign ownership
	Number of competitors
	Imported equipment
	Export sales ratio
	Engagement in activities with a foreign firm located abroad (other than providing design or R&D services)
	Firm's market share
	Import market share
Human capital	Percentage of workers using the Internet
	Percentage of work force with foreign experience
	Management's level of education
R&D network	Information technology assets as a percentage of total fixed assets
	Purchase of outside technology
	Purchase of a foreign license
	Receipt of external R&D assistance
	Provision of design or R&D services
	Purchase of externally performed R&D services
Institutions	Share of foreign ownership
	Share of public ownership
	Industrial park or export processing zone
	Government assistance in identifying a foreign relationship
	Membership of a business association
	Useful functions of a business association
	Constraints on growth in the domestic market

Source: World Bank survey data.

to explore the direct impact of R&D on firm performance, we substitute the knowledge production function for the innovation impact function. The resulting reduced form provides estimates of the elasticity of output and profit with respect to R&D and thereby establishes the basis for estimating returns to R&D. This three-equation exercise is intended to establish the efficacy of R&D among all or a portion of the firms within our sample and to identify the channels through which the innovation process operates.

For the purpose of estimating the three equations in this first phase of the research agenda, we use the full sample. That is, we pool the firm-level data across the 11 participating metropolitan economies.

Identifying Factors that Enhance the Effectiveness of R&D

Having examined the impact of R&D inputs on productivity and profitability, our next task is to identify from among the 23 institutional attributes those that affect firm-level performance. These attributes and the classification that we have imposed on them are shown in table 10.2.

Our interest is in testing whether these attributes operate through the R&D channel to enhance firm productivity or, if not, whether they have a direct impact on firm performance. We anticipate that some of these attributes, such as R&D networks and investment in information technology, improve firm performance by enhancing the effectiveness of the firm's R&D operation. Other factors, such as imported equipment and the share of imports in the firm's relevant market, may affect the overall productivity of the firm, but not the quality of its R&D operation. By interacting R&D effort with this broad set of institutional attributes, we identify those characteristics of the institutional framework that serve to enhance R&D performance at the firm level.

Differentiating the R&D Capabilities of Metropolitan Areas

Once we have identified the set of attributes that enhance innovation capabilities at the firm level, our next challenge is to aggregate the measures of these attributes into a metropolitan economy–wide measure. Among the 23 attributes, we identify 12 that enhance innovation capabilities and five that raise productivity outside the R&D channel. For these 12 R&D-enhancing attributes, we create a composite measure, which we use to compare the overall R&D institutional environment among the metropolitan economies studied.

Although our data are typically metropolitan-level data, not national data, data from the major metropolitan areas in each country are likely to be highly correlated with national data that rank the relative technological capabilities of the nations represented in the Bank survey. These national rankings, prepared by the United Nations Development Programme (UNDP), are shown in table 10.3. The rankings show the Republic of Korea and Singapore at the top, followed by Malaysia. Thereafter, Thailand, the Philippines, and China are closely grouped, with Indonesia lagging somewhat behind.

Because comparisons of the R&D capabilities of the metropolitan areas in our sample require that we separate and analyze the data by metropolitan area, we implicitly assume that each metropolitan area individually includes a substantial number of representative firms. However, the Singapore portion of the survey includes only 18 firm-level observations. Because this sample is particularly small and the results from these observations exhibit certain counterintuitive results, for the purpose of creating comparable measures of metropolitan area R&D capabilities, we drop Singapore from the exercise.

Table 10.3 National-Level Innovation Intensities and Capabilities, by Country

Country	UNDP technology achievement index	Scientists and engineers in R&D (per 100,000 population)	High-tech exports (as a percentage of total in 1999)	Ratio of R&D to GDP (percent)	R&D expenditure in business (as a percentage of total)
Korea, Rep. of	0.666	2,193	33	2.8	84.0
Singapore	0.585	2,318	58	1.1	62.5
Malaysia	0.396	93	52	0.2	8.3
Thailand	0.337	103	30	0.1	12.2
Philippines	0.300	157	26	0.2	1.9
China	0.299	454	21	0.7	—
Indonesia	0.211	182	7	0.1	76.4

—Not available.

Source: UNDP (2001, pp. 48–55).

Associating Returns to R&D with R&D Capabilities

The final step in our research agenda is to estimate the composite returns to R&D for each metropolitan area and to test the relationship between our estimates of composite R&D returns and the composite measures of the metropolitan R&D institutional environment. In order to compare returns to R&D across metropolitan economies, we have to adjust the exchange rate to the nominal estimates of returns to R&D across the national economies. For only six of our metropolitan economies—the five Chinese metropolitan economies and Seoul—are the exchange rate comparisons plausible. For these six metropolitan economies, our composite measures of the R&D institutional environment and returns to R&D exhibit a surprisingly robust association.

Before examining each of these issues, we create a conceptual framework for the analysis.

A CONCEPTUAL FRAMEWORK: THE FIRM'S PROBLEM

We situate this research agenda within a theoretical perspective. Like Romer (1990), we augment the basic Solow neoclassical growth model with human capital and the possibility for deliberate technical change. However, rather than viewing these dynamics from an economy-wide perspective, we allow the firm's manager to solve an optimizing problem, which determines the amount of human capital to dedicate to the firm's R&D operation. The solution to this problem involves the decision to divide retained earnings between physical investment and investment in human capital.[2]

The firm starts with an intensive production function of the form

$$y(t) = E(t)^{\phi} k(t)^{\alpha} \tag{10.1}$$

in which y is output (value added) per worker, E is a measure of innovation outputs, such as counts of new product or process innovations, and k is capital per worker.

Innovation outputs are driven by the knowledge production function shown in equation 10.2:

$$E(t) = aR(t)^{\gamma} \tag{10.2}$$

where γ is the elasticity of counts of new innovation with respect to R, the stock of R&D personnel.

Finally, for the purpose of measuring the direct impact of R&D effort on labor productivity, we substitute the knowledge production function, equation 10.2, into equation 10.1. The result is shown in equation 10.3:

$$y(t) = aR(t)^{\delta} k(t)^{\alpha} \tag{10.3}$$

where equation 10.3 is the "reduced form," which describes the direct impact of R&D on labor productivity.

Equation 10.3 shows that the accumulation of physical or human capital results in higher output per worker. Where rising productivity translates into the growth of retained earnings, new earnings spur investment in both physical and human capital. Hence the introduction of human capital and R&D creates the possibility for continuous growth of productivity and for the endogenous growth of output per worker.

An essential purpose of this chapter is to evaluate the ability of various institutional conditions to enhance the performance of R&D. To represent the impact of any one of the 23 attributes that potentially enhance the efficiency of the firm's R&D operation, we add the variable Z_i, $i = 1 \ldots 23$. By interacting each of the Z_i institutional attributes with R, our measure of R&D input, we are able to estimate the complementary effect of the institutional attributes as it enhances the effectiveness of R&D. That is, to test the complementary effect of an institutional attribute with R&D personnel, we estimate the equation

2. The model described here is partially based on Aghion and Howitt (1999).

$$\ln y(t) = a + \gamma \ln R(t) + \theta \ln R(t) * \ln Z_i(t) + \ln k(t) \quad (10.4)$$

where θ captures the elasticity of labor productivity with respect to the interactive term, which includes the *i*th institutional attribute.

CATEGORIES OF ANALYSIS AND LITERATURE REVIEW

We classify the list of 23 attributes that are measured in the World Bank survey into four broad categories. The categories, which approximate those used by Hill (chapter 8 of this volume), are institutional exposure, human capital, the R&D network, and the policy setting. Other studies employ similar categories in their attempt to summarize firm capabilities. Kumar and Chadee (2002), for example, formulate measures of three internal factors—technology and information and communications technology, human resources, and organizational structure—and two measures of external factors—the role of government policy and the role of finance and capital. They do not, however, interact these attributes with R&D. We review highlights of the literature that relate to each of the four categories.

International Exposure

The recent growth literature generally agrees that there is no systematic evidence to support the notion of a convergence of living standards among the world's rich and poor economies (for example, DeLong 1988).[3] The growing acceptance of the stylized fact of nonconvergence has motivated researchers to identify conditions that distinguish metropolitan areas that exhibit convergence from those that do not. One such study of conditional convergence demonstrates the tendency toward convergence among the world's most open developing economies. Creating an index of openness, Sachs and Warner (1995) demonstrate that the 30 countries among those most open to trade between 1965 and 1997 show convergence in income per capita. Hill, in chapter 8 of this volume, articulates one explanation of the importance of openness: "More than 90 percent of the world's R&D is undertaken in the OECD (Organisation for Economic Co-operation and Development) economies, and thus openness to global R&D capacity is

3. Sala-i-Martin (2002) demonstrates that this condition of nonconvergence depends critically on the unit of analysis. While the finding of nonconvergence holds for the universe of all nations, it does not hold when the unit of analysis is the individual.

critical for borrowers and latecomers." His observation emphasizes openness as a precondition to successful international technology transfer.

The World Bank survey collected data on a wide range of measures that relate to economic openness. Among these measures are the firm's share of the domestic market, the share of the domestic market supplied by imports, exports as a share of sales, the share of foreign direct investment in total equity, and the number of competitors identified by each firm. In the fifth section, we identify whether these attributes affect firm performance and whether their influence operates through the R&D channel.

Human Capital

In their influential paper, Mankiw, Romer, and Weil (1992) demonstrate that the "fit" of the Solow model with the convergence hypothesis could be improved by extending the model to include human capital. The basic measure that we use to capture R&D intensity—R&D personnel—simply counts the number of persons assigned to the firm's R&D function; it may not capture differences in the quality of human capital engaged in the R&D process across different firms.

Among the measures included in the World Bank survey that serve to augment basic R&D personnel and differential quality are the foreign work experience of the work force, the proportion of workers using the Internet, and the educational level of management. The econometric analysis presented later in this chapter examines the extent to which each of these enhances the effectiveness of the firm's R&D operations.

The R&D Network

During the decades following the publication of Solow's neoclassical growth model, which underscores the reliance of long-run growth of living standards on technological progress, economists implicitly or explicitly assumed that technology was a public good. As a public good, technology could, at little or no cost, be shared or transferred among agents.

In recent years, researchers have become more interested in the barriers to technology transfer. These include formal, intentional barriers, such as intellectual property rights law, and practical impediments. The study by the British Commission on Intellectual Property Rights notes that, in certain critical areas, the strict enforcement of intellectual property rights in the international system has become a burden on the development and diffusion of technology (CIPR 2002). Practical impediments, such as the lack of access to computers with which to access the Internet, effectively limit the public good component of technology (UNDP 2001, box 2.3, p. 35).

In their examination of the geographic concentration, agglomeration, and co-location of university research and industrial R&D, Agrawal and Cockburn (2002) find strong evidence of the co-location of upstream and downstream constituents of "local innovation systems." This idea of neighborhood innovation systems is further reinforced by the notion of interfirm networks of know-how, such as those examined by Carter (1987) and Von Hippel (1987). They find that, while these know-how networks serve as channels for the diffusion of technology, they also limit their memberships to users with the capacity to reciprocate.

Finally, market structure may also affect the motivation for R&D networking. Goyal and Moraga (2000) examine the effects of collaboration on the R&D effort of individual firms. They find that, in individual markets, R&D effort increases with the level of collaborative activity, whereas, if firms are Cournot competitors, individual effort declines with the level of collaborative activity.

Our characterization of R&D networking consists of the six measures shown in table 10.2. These include commercial transactions, such as the purchase of externally performed R&D services and foreign technology licenses. A second broad measure of networking relates to institutional linkages, such as collaborations with R&D institutions, including universities and research institutes, and the receipt of either external R&D assistance or final R&D products. We examine the extent to which these network measures augment the effectiveness of basic R&D personnel later in the chapter.

The Policy Setting

The World Bank survey collected data that focus on the role of public ownership in firm governance. These data include asset shares in public and foreign ownership. They also measure the importance of government-sponsored industrial and technology research parks as well as the intensity of use of government services intended to assist with establishing a range of relationships with foreign firms. Finally, we include measures of the incidence of participation in industry trade associations as a measure of the extensiveness of private industry associations outside the scope of government. As with the other categories of attributes, these measures of institutional quality are interacted with R&D personnel to identify the extent to which they augment R&D effectiveness.

We use econometric methods to identify the extent to which firms located within each of the 10 cities identify these institutional attributes as important or not important to their R&D operations. These measures are then aggregated to the metropolitan level for each of the four categories of institutional attributes; the four categories are subsequently aggregated to create composite measures of institutional quality for each of the participating metropolitan economies.

R&D, INNOVATION OUTPUT, AND THEIR IMPACTS

In this section, we estimate the knowledge production function (equation 10.2), the innovation impact function (equation 10.1), and the reduced form (equation 10.3). The results are shown, respectively, in tables 10.4, 10.5, and 10.6.

Knowledge Production

In order to estimate the impact of R&D personnel on innovation outputs, we use equation 10.2 as the core of our estimation equation. To distinguish between the impact of R&D resources and the size of the firm, we include sales as a control variable; to capture fixed area and industry effects, we also include dummy variables for the metropolitan area and industry composition.

The knowledge production function estimates, which control for firm size (lnSALES), industry (IND), and metropolitan area (MA), are shown in table 10.4. These results show that, for most of the five measures of innovation output, knowledge production is highly associated with R&D intensity. For both the full sample that includes reported levels of zero activity for the relevant innovation measure and for subsamples that drop the zero observations, the role of R&D personnel is highly statistically significant for each of the five measures of innovation.

Innovation Impact

In order to estimate the impact of the innovation outputs on firm performance, we use equation 10.1. Estimates of the impact of innovation on firm performance—both productivity and profitability—are shown in table 10.5. The introduction of new products, new management techniques, and quality controls all exhibit statistically significant relationships with productivity and profitability. New process innovation, while not highly statistically

Table 10.4 Knowledge Production

$\ln E_i = \alpha_0 + \gamma \ln(R + 0.0001) + \alpha_2 \ln SALES + \Sigma\beta_j \ln MA + \Sigma\beta_k \ln IND + \varepsilon$

$(i = 1, 2, \ldots 5)$

	New products		New business lines		Process innovations		New management techniques		New quality controls	
Variable	All observations	Observations > 0 only	All observations	Observations > 0 only	All observations	Observations > 0 only	All observations	Observations > 0 only	All observations	Observations > 0 only
Constant	−1.527	−0.230	−2.177	−1.217	−0.677	−0.631	−0.769	−0.645	−0.735	−0.336
	(−3.54)	(−0.33)	(−4.30)	(−1.58)	(−1.54)	(−0.91)	(−1.91)	(−0.90)	(−1.80)	(−0.47)
ln*R**	0.159	0.299	0.136	0.359	0.183	0.161	0.113	0.184	0.123	0.189
	(7.47)	(3.46)	(5.71)	(3.93)	(8.16)	(1.91)	(5.48)	(2.15)	(5.88)	(2.20)
lnSALES	0.156	0.136	0.096	0.106	0.110	0.137	0.127	0.185	0.148	0.192
	(4.90)	(2.60)	(2.72)	(0.055)	(3.38)	(2.64)	(4.31)	(3.45)	(4.90)	(3.55)
Jakarta	0.378	0.704	0.136	1.060	−0.306	−.319	−1.141	−1.457	−.388	−0.708
	(0.91)	(0.84)	(0.28)	(1.26)	(−0.58)	(−0.35)	(−2.51)	(−1.65)	(−0.95)	(−0.86)
Seoul	−1.573	−1.499	−1.065	−1.446	−0.766	−1.090	−2.670	−3.102	−2.141	−2.698
	(−4.36)	(−3.17)	(−2.73)	(−2.84)	(−2.11)	(−2.31)	(−6.90)	(−5.96)	(−5.98)	(−5.42)
Kuala Lumpur	−0.292	−0.823	−0.134	−1.679	1.051	−0.023	−0.608	−1.851	−1.083	−2.671
	(−0.76)	(2.58)	(−0.33)	(−1.55)	(2.77)	(−0.03)	(−1.73)	(−2.44)	(−2.76)	(−3.01)
Manila	−1.087	−0.686	−1.139	−0.485	−0.091	−0.123	−1.574	−1.709	−1.355	−1.175
	(−2.55)	(−0.85)	(−2.10)	(−0.51)	(−0.22)	(−0.15)	(−3.87)	(−2.04)	(−3.40)	(−1.48)
Bangkok	0.221	−0.291	0.478	—	−0.551	—	−2.637	—	—	—
	(0.37)	(−0.22)	(0.75)		(−0.70)		(−2.50)			
Tianjin	−0.324	0.089	−0.395	−0.132	−0.302	−0.102	−0.215	−0.161	−0.278	−0.460
	(−1.45)	(0.24)	(−1.53)	(−0.34)	(−1.27)	(−0.28)	(−1.09)	(−0.44)	(−1.37)	(−1.25)

(continued)

Table 10.4 Knowledge Production (*continued*)

$\ln E_i = \alpha_0 + \gamma \ln(R + 0.0001) + \alpha_2 \ln SALES + \Sigma\beta_j \ln MA + \Sigma\beta_k \ln IND + \varepsilon$

$(i = 1, 2, \ldots 5)$

Variable	New products		New business lines		Process innovations		New management techniques		New quality controls	
	All observations	Observations > 0 only	All observations	Observations > 0 only	All observations	Observations > 0 only	All observations	Observations > 0 only	All observations	Observations > 0 only
Shanghai	0.564	0.817	–0.235	–0.274	0.199	0.030	–0.141	–0.279	–0.064	–0.444
	(2.71)	(2.58)	(–1.01)	(–0.88)	(0.92)	(0.10)	(–0.73)	(–0.94)	(–0.33)	(–1.46)
Guangzhou	–0.352	–0.066	–0.284	0.018	0.123	0.161	0.325	0.466	0.264	0.151
	(–1.67)	(–0.22)	(–1.21)	(0.06)	(0.58)	(0.53)	(1.71)	(1.48)	(1.37)	(0.48)
Chengdu	0.331	0.548	0.124	0.204	0.504	0.591	0.249	0.303	0.235	0.129
	(1.65)	(1.91)	(0.57)	(0.71)	(2.44)	(2.06)	(1.35)	(1.05)	(1.25)	(0.44)
Industry	Yes	Yes	Yes	Yes	Yes	Yes	Yes	Yes	Yes	Yes
Adjusted R^2	0.168	0.108	0.102	0.067	0.170	0.089	0.105	0.109	0.118	0.101
Number of observations	1,458	572	1,458	569	1,458	569	1,459	569	1,442	569

—Not available.

Note: R = average R&D personnel (1998–2000); 0 observations for R&D personnel have been converted to 1. Estimation equation includes a dummy variable for 0 observations on E that have been converted to 0.0001. No dummy estimates are statistically significant at the 5 percent level.

Source: Authors' calculations based on World Bank survey data.

Table 10.5 Impact of Innovation on Firm Performance

$\ln \text{PERF}_i = \alpha_0 + \alpha_1 \ln K + \alpha_2 \ln L + \phi \ln E_i + \Sigma\beta_j \ln \text{MA} + \Sigma\beta_k \ln \text{IND} + \varepsilon$ $(i = 1, 2, \ldots 5)$**

Variable	Productivity ln (VA)					Profitability ln (Profit)				
	New product	New business line	Process innovations	New management techniques	New quality controls	New product	New business line	Process innovations	New management techniques	New quality controls
Constant	1.920	1.887	1.894	1.866	1.846	0.753	0.728	0.734	0.697	0.678
	(7.19)	(6.99)	(7.02)	(6.94)	(6.86)	(2.14)	(2.05)	(2.08)	(1.98)	(1.92)
ln*K*	0.438	0.450	0.447	0.445	0.442	0.547	0.557	0.552	0.548	0.549
	(13.66)	(13.90)	(13.80)	(13.76)	(13.66)	(13.20)	(13.40)	(13.27)	(13.23)	(13.20)
ln*L*	0.464	0.467	0.465	0.463	0.468	0.326	0.328	0.325	0.320	0.328
	(9.05)	(8.98)	(8.93)	(8.94)	(9.06)	(4.95)	(4.94)	(4.90)	(4.84)	(4.96)
New product	0.434	n.a.	n.a.	n.a.	n.a.	0.404	n.a.	n.a.	n.a.	n.a.
	(4.20)					(3.02)				
New business line	n.a.	0.116	n.a.	n.a.	n.a.	n.a.	0.152	n.a.	n.a.	n.a.
		(1.02)					(1.05)			
New process inncvation	n.a.	n.a.	0.157	n.a.	n.a.	n.a.	n.a.	0.291	n.a.	n.a.
			(1.50)					(2.12)		
New management technique	n.a.	n.a.	n.a.	0.235	n.a.	n.a.	n.a.	n.a.	0.392	n.a.
				(2.40)					(3.11)	
New quality controls	n.a.	n.a.	n.a.	n.a.	0.255	n.a.	n.a.	n.a.	n.a.	0.301
					(2.61)					(2.40)
Metropolitan area	Yes	Yes	Yes	Yes	Yes	Yes	Yes	Yes	Yes	Yes
Industry	Yes	Yes	Yes	Yes	Yes	Yes	Yes	Yes	Yes	Yes
Adjusted R^2	0.68	0.67	0.67	0.67	0.67	0.60	0.60	0.60	0.60	0.60
Number of observations	797	797	797	797	797	730	730	730	730	730

n.a. Not applicable.

Note: The estimation equation includes a dummy variable for "0" observations on *E* that have been converted to 0.0001; no dummy estimates were statistically significant at the 5 percent level.

Source: Authors' calculations based on World Bank survey data.

significant in its impact on productivity, is significantly associated with profitability. That the introduction of new business lines does not exhibit such a connection with performance may result from the contemporaneous disruption and subsequent break-in time required for radical product innovation to translate into improved firm performance.

Arguably, the five innovation outputs shown in table 10.5 are correlated, causing the estimation equations to suffer from misspecification of the omitted variables and a degree of upward bias in the reported estimates. If the five innovation output variables were included in a single estimation equation, the same correlation would lead to multicollinearity and a tendency to underestimate the statistical significance of the impact of R&D on productivity and profitability. Indeed, the fact that the five elasticities on the innovation count in the productivity equation and the profitability equation sum to 1.2 and more than 1.5, respectively, suggests the presence of upward bias in the reported estimates of the coefficients on innovation counts.

The Reduced Form

This structural model may capture only a portion of the impact of R&D on firm performance. A firm's R&D activity may affect firm performance through channels other than those formally measured. R&D personnel, for example, may expend effort on improvements in the quality of existing products, on the installation and efficient use of new machinery, or on other incremental tasks that substantially improve firm performance but are not captured in counts of innovation outputs.

In their study of R&D in Chinese industry, Jefferson and others (forthcoming) demonstrate that conventional measures of innovation activity account for but a fraction of overall measured returns to R&D. Using data on counts of patent applications and the share of sales accounted for by new products, Jefferson and his colleagues find that, among China's large and medium firms, these measures account for only 16 percent of the returns to R&D personnel.[4] While patents account for just 5 percent of total returns, new product sales represent 11 percent of overall returns.

4. The share of measured returns to R&D expenditure accounted for by patents and new products is substantially higher: 73 percent. The authors argue that the reason for this disparity is the tendency for R&D to be substantially more capital intensive in large firms than in smaller firms. Within the typical large firm, the model is one of an R&D lab in which R&D expenditures are focused on the production of patentable innovations and measurable new products. By comparison, the typical medium-size enterprise maintains a far more labor-intensive R&D operation in which R&D teams appear to focus on incremental innovation, such as product quality and small process improvements.

Table 10.6 reports the elasticities of productivity and profitability with respect to R&D personnel. Both elasticities are statistically significant. R&D personnel in Seoul, Shanghai, and Guangzhou—in descending order—exhibit the largest impacts on productivity and profitability.

As a three-year average over the period 1998–2000, our R&D variable is a kind of stock measure representing both the cumulative and lagged impact of R&D on firm performance. While at least a portion of the stock measure may be viewed as predetermined, there is still cause to be concerned about the endogeneity of R&D personnel in the performance equations. For example, firms that consistently exhibit above-average levels of profitability may be well placed to support comparatively large R&D operations. This

Table 10.6 Effect of R&D on Firm Performance (Reduced Form)
$\ln PERF_i = \alpha_0 + \alpha_1 \ln K + \alpha_2 \ln L + \delta \ln R + \Sigma\beta_j \ln MA + \Sigma\beta_k \ln IND + \varepsilon$

Variable	Productivity ln (VA)	Profit ln (Profit)
Constant	3.094	1.897
	(7.67)	(3.51)
ln*K*	0.373	0.484
	(8.01)	(7.91)
ln*L*	0.270	0.139
	(3.71)	(1.48)
ln*R**	0.325	0.276
	(6.05)	(3.85)
Jakarta	Dropped	Dropped
Seoul	5.753	4.742
	(9.43)	(5.09)
Kuala Lumpur	–0.614	–2.112
	(–0.68)	(–1.88)
Manila	1.372	1.064
	(1.16)	(0.73)
Bangkok	Dropped	Dropped
Tianjin	–00178	0.031
	(–0.07)	(0.09)
Shanghai	0.831	0.804
	(4.81)	(3.45)
Guangzhou	0.525	0.456
	(2.90)	(1.85)
Chengdu	–0.058	0.068
	(–0.34)	(0.29)
Adjusted R^2	0.688	0.580
Number of observations	408	359

Note: R = average R&D personnel (1998–2000); 0 observations for R&D personnel have been converted to 1.

Source: Authors' calculations based on World Bank survey data.

is likely because, as Jefferson and others (forthcoming) show, retained earnings are an important source of R&D finance. Nonetheless, they also demonstrate that efforts to correct for possible endogeneity do not alter substantially the robust impact of R&D effort on firm performance. Here we do not attempt to correct for possible endogeneity associated with fixed effects.

ATTRIBUTES THAT DETERMINE FIRM-LEVEL INNOVATION CAPABILITIES

In order to investigate the impact of a range of institutional attributes on the efficiency of R&D personnel, we estimate equation 10.4 (not in intensive form) for each of the 23 attributes. The results are shown in table 10.7, which reports results for the 12 attributes that do exhibit statistically significant interactions with R&D personnel. Table 10.8 reports results for five of the remaining 11 attributes that fail to exhibit significant interactions with R&D personnel but do exhibit a statistically significant independent effect on firm performance, that is, productivity or profitability.

The first two factors in table 10.7 demonstrate the impact of foreign versus state ownership on firm-level R&D capabilities. While foreign ownership significantly enhances the effectiveness of R&D, public ownership limits the returns to a firm's R&D operation. This result may reflect the fact that FDI creates a channel for internal technology transfer that may provide a fertile environment for innovation within the firm. Public ownership may, by comparison, create a relatively insulated environment in which firms are relatively protected from import competition and therefore less motivated to innovate.

The results in table 10.7 also show that R&D personnel who are Internet users tend to operate relatively successful R&D operations. As a source of technical information and a channel for the exchange of know-how, the Internet provides an invaluable R&D network for personnel. Similarly, firms that invest heavily in information technology are creating the physical infrastructure needed for the firm to participate effectively in the Internet-based R&D network.

The finding that both of these characteristics—the proportion of workers using the Internet and the investment in information technology assets—are important drivers of R&D efficiency probably indicates that, to a substantial degree, these are complementary inputs to an Internet-based R&D network. In principle, we should include each of these attributes as well as other characteristics that we believe enhance the effectiveness of the

Table 10.7 Effect of R&D Interacted with Firm Attributes
$\ln X = \alpha_0 + \alpha_1 \ln K + \alpha_2 \ln L + \alpha_3 \ln R + \alpha_4(\ln R^* \ln Z) + \Sigma\alpha_1 \ln LOC + \Sigma\alpha_1 \ln IND + \varepsilon$

Attribute and variable	X_1 = Value added	X_2 = Profit
Share of foreign ownership		
ln *R*	0.189	0.221
	(5.19)	(4.79)
ln *R**	0.009	0.013
	(2.49)	(2.86)
ln (foreignsh)[1]	0.189	0.221
	(5.19)	(4.79)
R^2	0.662	0.591
Number of observations	876	801
Share of public ownership		
ln *R*	0.073	0.085
	(2.12)	(1.92)
ln *R** ln (publicsh)[1]	–0.014	–0.013
	(4.11)	(2.97)
Adjusted R^2	0.667	0.593
Number of observations	875	800
Percentage of workers using the Internet		
ln *R*	0.107	0.108
	(3.44)	(2.66)
ln *R** ln (netsh)[1]	0.023	0.031
	(4.63)	(4.77)
Adjusted R^2	0.678	0.596
Number of observations	866	798
Information technology assets as a percentage of total fixed assets		
ln *R*	0.227	0.318
	(4.69)	(4.38)
ln *R** ln (ITsh)[1]	0.047	0.055
	(4.04)	(3.61)
Adjusted R^2	0.718	0.636
Number of observations	502	482
Number of competitors		
ln *R*	0.226	0.276
	(4.58)	(4.58)
ln *R** ln (compete)	–0.032	–0.045
	(2.43)	(2.70)
Adjusted R^2	0.668	0.580
Number of observations	837	723
Industrial park or export processing zone		
ln *R*		
ln *R**zone	0.099	0.184
	(2.33)	(3.39)
Adjusted R^2	0.674	0.596
Number of observations	881	808

(continued)

Table 10.7 Effect of R&D Interacted with Firm Attributes (*continued*)

$\ln X = \alpha_0 + \alpha_1 \ln K + \alpha_2 \ln L + \alpha_3 \ln R + \alpha_4(\ln R^* \ln Z) + \Sigma\alpha_1 \ln LOC + \Sigma\alpha_1 \ln IND + \varepsilon$

Attribute and variable	X_1 = Value added	X_2 = Profit
Management's level of education		
ln *R*	0.496	0.474
	(6.15)	(5.03)
ln *R**educ	0.391	0.360
	(4.67)	(3.69)
Adjusted R^2	0.678	0.607
Number of observations	742	653
Purchase of a foreign license		
ln *R*	0.131	0.147
	(4.05)	(3.54)
ln *R** ln (forlicence)	0.179	0.174
	(2.88)	(2.19)
Adjusted R^2	0.672	0.583
Number of observations	857	790
Purchase of outside technology		
ln *R*	0.092	0.113
	(2.60)	(2.52)
ln *R** ln (purch_tech)	0.135	0.111
	(3.17)	(2.05)
Adjusted R^2	0.673	0.593
Number of observations	862	793
Provision of design or R&D services		
ln *R*	0.131	0.123
	(3.66)	(2.81)
ln *R** *D*(provide_RD)	0.016	0.032
	(1.38)	(2.27)
Adjusted R^2	0.659	0.588
Number of observations	642	540
Receipt of external R&D assistance		
ln *R*	0.095	0.076
	(2.45)	(1.54)
ln *R***D*(extRDasst)	0.114	0.177
	(2.70)	(3.31)
Adjusted R^2	0.660	0.581
Number of observations	779	705
Percentage of work force with foreign experience		
ln *R*	0.420	0.647
	(2.42)	(2.93)
ln *R**(% for_exper)	0.057	0.111
	(1.92)	(2.95)
Adjusted R^2	0.889	0.811
Number of observations	65	57

Source: Authors' calculations based on World Bank survey data.

R&D process, into a single regression equation. Otherwise, owing to their high correlation, estimating the effect of each of these attributes alone is likely to create an upward bias in estimates of their contribution to the efficiency of R&D. However, the inclusion of multiple measures of R&D quality is likely to create multicollinearity, thereby frustrating any attempt to distinguish between the separate contributions of highly complementary inputs. In this chapter, we examine the impact of each attribute without controlling for complementary factors, other than the usual controls for conventional capital and labor.

We next examine the impact of the number of competitors on returns to R&D. The negative sign most likely reflects the tendency of competition to limit the markup on innovations. Lower markups—and prices—depress measures of both profit and productivity. This result seems to support the view of Schumpeter (1950), who emphasizes the advantages of market power in innovation, rather than that of Arrow (1962), who emphasizes the benefit of competitive markets in motivating innovation.

Each of the next four attributes shown in table 10.7 relates to explicit external R&D-related transactions of the firm. The purchase of outside technology, including the purchase of foreign licenses, enhances the effectiveness of R&D. Likewise, the receipt of R&D assistance from an external source raises R&D efficiency. Finally, firms that provide design or R&D services to a foreign firm exhibit relatively efficient R&D operations. The results suggest a consistent pattern in which R&D networking—external commercial transactions and reciprocal exchanges—enhances the effectiveness of R&D.

Table 10.8 identifies a number of attributes that are associated with higher firm productivity and profitability, but their effect does not, at least in a statistically significant sense, operate through the firm's formal R&D channel. The attributes each create positive intercept shifts in the measure of firm performance.

One variable that we might expect to operate through the R&D operation, but seemingly does not, is the external R&D variable. For this variable, firms were asked, "Did you have a contractual or long-standing relationship with any of the following to perform R&D for your plant?" The possible sources of R&D services are a local university, government research institute, private research institute, or private company. That this attribute significantly affects the firm's performance without enhancing its R&D program indicates that contracting for external services may represent a substitute for internal R&D capabilities. The services improve firm performance, but, unlike other forms of R&D transactions and networking, the avenue of impact does not operate through the firm's R&D operation.

Table 10.8 Direct Impact of Firm Attributes on Value Added and Profit[a]

$\ln X = \alpha_0 + \alpha_1 \ln K + \alpha_2 \ln L + \alpha_3 \ln R + \alpha_4 \ln Z + \Sigma\alpha_1 \ln LOC + \Sigma\alpha_1 \ln IND + \varepsilon$

Factor and variable	X_1 = Value added	X_2 = Profit
Imported equipment		
Constant	2.448	1.388
	(9.32)	(4.07)
ln *R*	0.142	0.155
	(4.60)	(3.93)
D(import_equip)[1]	0.217	0.269
	(2.01)	(1.89)
Adjusted R^2	0.672	0.590
Number of observations	884	813
Firm's market share		
Constant	2.193	1.392
	(6.88)	(3.52)
ln *R*	0.123	0.136
	(3.48)	(3.12)
D(firm_mktsh)	0.156	0.120
	(4.01)	(2.45)
Adjusted R^2	0.678	0.602
Number of observations	631	570
Receipt of external R&D assistance		
Constant	2.330	1.270
	(8.94)	(3.77)
ln *R*	0.128	0.110
	(3.93)	(2.65)
D(RDnet)	0.163	0.483
	(1.34)	(3.13)
Adjusted R^2	0.061	0.604
Number of observations	877	806
Import market share		
Constant	2.559	1.358
	(8.89)	(3.63)
ln *R*	0.135	0.150
	(4.11)	(3.49)
ln (import_mktshare)	0.026	0.134
	(2.92)	(1.21)
Adjusted R^2	0.675	0.582
Number of observations	738	692
Government assistance in identifying a foreign relationship		
Constant	2.183	1.132
	(7.81)	(3.06)
ln *R*	0.127	0.137
	(3.86)	(3.21)
D(gov't_assistance)	0.263	0.367
	(2.11)	(2.21)
Adjusted R^2	0.676	0.580
Number of observations	726	662

a. That is, no impact through the R&D channel.

Source: Authors' calculations based on World Bank survey data.

The other factors—imported equipment, market share, the market share of imports, and government assistance in locating a foreign client, supplier, or investor relationship—all enhance firm performance. Like externally performed R&D services, each of these attributes affects firm performance through channels other than the firm's R&D operation.

THE DISTRIBUTION OF ATTRIBUTES BY METROPOLITAN ECONOMY

Having identified firm-level attributes that enhance firm performance, both directly and indirectly by enhancing the firm's R&D function, we now examine how these attributes are distributed over the 10 metropolitan economies covered in the World Bank survey.

To identify the incidence of these attributes in each of the 10 metropolitan economies, we regress each attribute on a set of metropolitan area dummies as well as a set of industry dummies. When the survey data provide us with a continuous measure of the attribute, we use a log-linear estimation equation. When the attribute is an on-off measure, such as imported equipment, we use a logit model.

As shown in tables 10.9 through 10.12, we have created two composite measures. The first is the number of estimates that are statistically significant at the 95 percent level ($t \approx 1.95$). The incidence of each attribute, relative to Beijing, the reference, may be positive or negative. The second composite measure is the sum of the t-statistics. While summing the t-statistics is a somewhat arbitrary approach, as a measure of the robustness of the relevant estimates, the ranking should, as a general principle, be useful for distinguishing gradations of international exposure, human capital, the R&D network, and the policy setting.[5]

We report two summations of the t-statistics. The first, shown in the next-to-last grouping in tables 10.9 through 10.12, represents the summation of the t-statistics for all of the institutional attributes that have an impact on firm performance, whether they operate indirectly through the R&D channel or directly on performance only. The second summation, shown in the bottom grouping, includes only those institutional attributes that operate on performance through the R&D channel.

5. Djankov and Murrell (2002) use this approach to develop aggregate weights of findings that are reported in a wide range of research literature using similar estimation equations.

International Exposure

Our measure of economy-wide international exposure includes five of the 23 measures included in tables 10.7 and 10.8. These are shown in table 10.9. We find that the greater the number of competitors, the lower the firm's productivity and profitability. Also the larger the firm's own market share, the greater the firm's productivity and profitability. In creating the composite measure of international exposure as a source of enhanced firm performance, we reverse the sign on the number of competitors. We anticipate that a reduced markup associated with heightened competition erodes measures of productivity and profitability. Where competition is relatively low, measured productivity and profitability should be relatively high.

Viewing table 10.9, we find that Manila, Jakarta, and Kuala Lumpur exhibit the highest degree of international exposure, followed by Shanghai. The second tier consists of Seoul and Bangkok, followed by Guangzhou. The measures of international exposure for Tianjin, Chengdu, and Beijing lag considerably behind those of the other metropolitan areas.

Our composite measure for the last indicator in table 10.9, relating to attributes that affect the effectiveness of R&D, suggests four tiers, with Manila and Shanghai occupying the top tier; Jakarta, Kuala Lumpur, and Bangkok occupying the second tier; Seoul, Beijing, and Guangzhou occupying the third tier; and Chengdu lagging well behind.

Human Capital

Table 10.10 shows the distribution of human capital over the 10 economies. Since all of the eligible attributes operate through the R&D channel, we compute a single composite measure. Seoul stands out as the metropolitan economy with the highest measured intensity of human capital. Only with respect to foreign work experience does Seoul lag behind any of its peers. This finding is consistent with the relatively closed character of the Seoul economy. While all four of the ASEAN (Association of South East Asian Nations) economies lag behind Seoul, their composite human capital measures are notably higher than those of the next tier—the Chinese cities not including Tianjin. At least with respect to our measure of the quality of human capital, Tianjin lags considerably behind all of the other nine metropolitan areas.

The R&D Network

In table 10.11, we examine the distribution of five measures of R&D institutions and networks across our 10 economies. All but one of these meas-

Table 10.9 International Exposure

Indicator	Jakarta	Kuala Lumpur	Manila	Bangkok	Seoul	Tianjin	Shanghai	Guangzhou	Chengdu	Adjusted R^2	Number of observations
Import market share	1.112	2.888	0.260	0.508	2.822	0.320	1.192	1.099	−1.038	0.122	1,316
	(1.29)	(3.14)	(0.32)	(0.38)	(3.75)	(0.61)	(2.22)	(2.06)	(2.08)		
Share of foreign ownership	0.979	1.804	5.382	2.398	−2.356	−0.061	2.656	0.578	−2.001	0.131	1,399
Percent of foreign direct investment in total capital	(0.76)	(1.57)	(5.19)	(1.33)	(2.76)	(0.12)	(5.34)	(1.16)	(4.15)		
Imported equipment	0.985	1.201	1.724	0.411	0.562	−0.316	0.632	0.703	−0.445	0.179	1,530
	(2.81)	(3.28)	(4.73)	(0.75)	(1.94)	(1.49)	(3.09)	(3.49)	(2.14)		
Number of competitors[a]	−0.876	−0.632	−0.803	−0.570	−0.827	0.187	−0.184	−0.116	0.096	0.114	1,345
	(3.11)	(2.14)	(2.78)	(1.20)	(3.52)	(1.13)	(1.10)	(0.67)	(0.60)		
Firm's market share	2.038	0.697	1.198	1.430	0.764	−0.362	0.323	−0.260	−0.097	0.161	1,269
	(9.31)	(3.08)	(4.50)	(3.97)	(4.20)	(2.81)	(2.48)	(1.83)	(−0.71)		
All institutional attributes											
Number for which t > 2	3	4	4	1	4	0	4	2	0	n.a.	n.a.
Number for which t < −2	0	0	0	0	1	1	0	0	3	n.a.	n.a.
Summation of t-statistics	16.28	13.21	17.52	7.63	10.65	−2.68	14.23	5.55	−9.68	n.a.	n.a.
R&D-only attributes[b]											
Number for which t > 2	1	1	2	0	1	0	1	0	0	n.a.	n.a.
Number for which t < −2	0	0	0	0	1	0	0	0	1	n.a.	n.a.
Summation of t-statistics	3.87	3.71	7.97	2.53	0.76	−1.25	6.44	1.83	−4.75	n.a.	n.a.

n.a. Not applicable.

a. Because a large number of competitors is found to depress the measure of productivity and profitability (probably through reduced markups), in creating the composite measures in the last row we reverse the sign of the reported *t*-statistics.

b. R&D only represents the composite measure for those attributes that affect firm performance through the R&D channel.

Source: Authors' calculations based on World Bank survey data.

Table 10.10 Human Capital

Indicator	Jakarta	Kuala Lumpur	Manila	Bangkok	Seoul	Tianjin	Shanghai	Guangzhou	Chengdu	Adjusted R^2	Number of observations
R&D personnel as a percentage of total workers	–0.203	–1.141	–1.623	–0.510	1.897	–1.220	0.095	0.046	0.605	0.258	1,530
	(0.51)	(2.67)	(3.99)	(0.80)	(5.38)	(5.02)	(0.39)	(0.19)	(2.55)		
Percentage of workers using the Internet	2.111	3.142	1.485	3.395	4.422	–0.995	.464	0.673	–1.165	0.185	1,492
	(3.04)	(4.20)	(2.09)	(3.04)	(7.18)	(2.31)	(1.06)	(1.56)	(2.79)		
Percentage of workers with foreign work experience	0.148	0.227	0.862	0.842	0.495	0.842	0.005	–0.254	–0.139	0.245	369
	(0.47)	(0.73)	(2.67)	(1.75)	(1.83)	(2.39)	(0.02)	(0.94)	(0.52)		
Management's level of education	0.073	0.243	0.307	0.294	0.243	–0.034	–0.002	–0.018	0.028	0.259	1,155
	(1.78)	(6.91)	(6.00)	(3.37)	(6.91)	(1.36)	(0.09)	(0.75)	(1.20)		
All institutional attributes											
Number for which t > 2	1	2	3	2	3	1	0	0	1	n.a.	n.a.
Number for which t < –2	0	1	1	0	0	2	0	0	1	n.a.	n.a.
Summation of t-statistics	5.34	9.17	6.77	7.36	21.30	–6.30	1.38	1.06	0.44	n.a.	n.a.
R&D only attributes[a]											
Number for which t > 2	1	2	3	2	3	1	0	0	1	n.a.	n.a.
Number for which t < –2	0	1	1	0	0	2	0	0	1	n.a.	n.a.
Summation of t-statistics	5.34	9.17	6.77	7.36	21.30	–6.30	1.38	1.06	0.44	n.a.	n.a.

n.a. Not applicable.

a. R&D only represents the ccmposite measure for those attributes that affect firm performance through the R&D channel.

Source: Authors' calculations based on World Bank survey data.

Table 10.11 The R&D Network

Indicator	Jakarta	Kuala Lumpur	Manila	Bangkok	Seoul	Tianjin	Shanghai	Guangzhou	Chengdu	Adjusted R^2	Number of observations
Purchase of externally performed R&D services	1.070	0.026	−0.446	n.a.	0.423	−0.788	0.019	−0.124	0.657	0.070	1,502
	(3.07)	(0.06)	(0.94)		(1.32)	(2.71)	(0.08)	(0.49)	(2.96)		
Purchase of external R&D assistance	−0.245	0.670	0.656	0.850	0.630	−0.380	0.625	0.206	0.327	0.092	1,314
	(0.55)	(1.75)	(1.46)	(1.43)	(1.97)	(1.60)	(2.88)	(0.90)	(1.54)		
Purchase of a foreign license	1.204	1.308	0.940	n.a.	−0.869	−1.366	0.558	−0.032	−0.152	0.109	1,462
	(2.22)	(2.70)	(1.80)		(1.13)	(2.39)	(1.59)	(0.08)	(0.40)		
Purchase of outside technology	0.242	0.095	0.189	1.010	0.207	−0.811	0.680	0.287	1.189	0.096	1,473
	(0.42)	(0.18)	(0.33)	(1.45)	(0.51)	(2.16)	(2.45)	(0.97)	(4.64)		
Provision of R&D services for a foreign firm	1.008	2.058	1.430	0.345	0.647	−0.469	0.580	0.756	−0.321	0.080	1,170
	(1.44)	(3.97)	(2.68)	(0.32)	(1.20)	(0.94)	(1.46)	(1.95)	(0.67)		
All institutional attributes											
Number for which t > 2	2	2	1	0	1	0	2	1	2	n.a.	n.a.
Number for which t < −2	0	0	0	0	0	3	0	0	0	n.a.	n.a.
Summation of t-statistics	6.60	8.66	5.33	3.20	3.87	−7.64	8.46	3.25	8.07	n.a.	n.a.
R&D-only attributes[a]											
Number for which t > 2	1	2	1	0	1	0	2	1	1	n.a.	n.a.
Number for which t < −2	0	0	0	0	0	2	0	0	0	n.a.	n.a.
Summation of t-statistics	3.53	8.60	6.27	3.20	2.55	−7.09	8.38	3.74	5.11	n.a.	n.a.

n.a. Not applicable.

a. R&D only represents the composite measure for those attributes that affect firm performance through the R&D channel.

Source: Authors' calculations based on World Bank survey data.

ures affect the quality of R&D operations. Among the 10 economies, Kuala Lumpur, Shanghai, and Chengdu exhibit the highest total composite measures of R&D networking. Again, Tianjin lags far behind. Closer examination shows that Chengdu's surprisingly high ranking results in part from two measures: the purchase of externally performed R&D services and the purchase of outside technology.

When we examine the four attributes that affect R&D alone, Chengdu loses its leadership status, while Kuala Lumpur and Shanghai continue to exhibit strength. Tianjin continues to appear at the bottom.

The Policy Setting

Table 10.12 covers five measures of the policy setting of the firms included in the Bank survey. One of these—share of foreign-owned assets—is double counted with openness. For this measure of policy setting, the five non-Chinese metropolitan economies typically outperform the Chinese metropolises. These differences reflect the relative scarcity of public ownership in the cities outside China. The four ASEAN cities dominate Seoul largely due to the low incidence of FDI in Korean companies and the relatively low level of government assistance in establishing links with foreign firms. The relatively low incidence of FDI in Seoul is consistent with the characterizations of Seoul's industrial development strategy as emphasizing the role of human capital in reverse engineering imported goods and equipment rather than the role of FDI (Chapter 9, Westphal 1990). Seoul's emphasis on human capital is confirmed by the results reported in table 10.10.

For the only composite measure of R&D, Manila and Kuala Lumpur exhibit the highest measures, while Shanghai, Jakarta, and Seoul follow. Bangkok, Guangzhou, and Tianjin exhibit comparable levels of institutional quality, followed by Beijing and, by an even wider margin, by Chengdu.

Summary of Attributes

In table 10.13 we construct a composite measure of the four individual country-city attributes. Extending the approach used to construct the composite measures for each of the four broad categories, we construct single composite measures of competitiveness and R&D capabilities.

The rankings in table 10.13 suggest several basic findings. We focus on the composite measure of R&D capability. We broadly identify two clusters that stand out at the tails of the distribution. These are Kuala Lumpur, Manila, and Seoul at the high end of the distribution, with Tianjin and

Table 10.12 The Policy Setting

Indicator	Jakarta	Kuala Lumpur	Manila	Bangkok	Seoul	Tianjin	Shanghai	Guangzhou	Chengdu	Adjusted R^2	Number of observations
Share of public ownership[a]	−4.483	−3.916	−4.179	−4.431	−3.930	−0.598	−0.465	−1.336	1.472	0.085	1,397
	(3.51)	(3.35)	(3.91)	(2.30)	(4.53)	(1.19)	(−0.92)	(2.64)	(3.00)		
Share of foreign ownership	0.979	1.804	5.382	2.398	−2.356	−0.061	2.656	0.578	−2.001	0.131	1,399
	(0.76)	(1.57)	(5.19)	(1.33)	(−2.76)	(0.12)	(5.34)	(1.16)	(4.15)		
Industrial park or export processing zone	0.759	2.170	1.339	−0.457	1.509	0.219	0.702	0.141	0.196	0.135	1,456
	(2.02)	(5.84)	(3.89)	(0.58)	(4.98)	(0.95)	(3.11)	(0.61)	(0.86)		
Government assistance in identifying a foreign relationship	0.245	0.721	0.589	2.476	−0.910	−0.384	0.101	−0.062	0.686	0.065	1,253
	(0.59)	(1.80)	(1.38)	(4.11)	(−1.65)	(1.29)	(0.37)	(−0.22)	(2.93)		
Information technology assets as a percentage of total fixed assets	3.711	−0.047	−0.117	0.810	0.300	0.536	−0.088	0.073	0.118	0.430	523
	(2.32)	(−0.08)	(−0.27)	(1.01)	(0.52)	(1.86)	(−0.42)	(0.33)	(0.57)		
All institutional attributes											
Number for which t > 2	3	2	3	2	2	0	2	1	1	n.a.	n.a.
Number for which t < −2	0	0	0	0	1	0	0	0	2	n.a.	n.a.
Summation of t-statistics	9.20	12.48	14.10	11.37	5.62	2.59	9.32	4.52	−2.79	n.a.	n.a.
R&D-only attributes[b]											
Number for which t > 2	2	2	3	1	2	0	2	1	0	n.a.	n.a.
Number for which t < −2	0	0	0	0	1	0	0	0	2	n.a.	n.a.
Summation of t-statistics	8.61	10.68	12.72	5.22	7.27	3.88	8.95	4.74	−5.72	n.a.	n.a.

n.a. Not applicable.

a. The sign of public ownership share is reversed in the tallies shown in the last row.

b. R&D only represents the composite measure for those attributes that affect firm performance through the R&D channel.

Source: Authors' calculations based on World Bank survey data.

Table 10.13 Composite Measure of Country and City Attributes

Metropolitan area	Openness	Human capital	R&D network	Institutional quality	Composite measure of competitiveness	Composite measure of R&D capability
Jakarta						
Number of all institutional attributes for which t > 2	1	1	2	3	7	5
Number of all institutional attributes for which t < –2	0	0	0	0	0	0
Summation of t-statistics	4.86	5.34	6.60	9.20	26.00	20.72
Kuala Lumpur						
Number of all institutional attributes for which t > 2	2	2	2	2	8	7
Number of all institutional attributes for which t < –2	0	1	0	0	1	1
Summation of t-statistics	7.99	9.17	8.66	12.48	36.30	32.16
Manila						
Number of all institutional attributes for which t > 2	2	3	1	3	9	9
Number of all institutional attributes for which t < –2	0	1	0	0	1	1
Summation of t-statistics	10.24	6.77	5.33	14.10	36.44	33.04
Bangkok						
Number of all institutional attributes for which t > 2	0	2	0	2	4	3
Number of all institutional attributes for which t < –2	0	0	0	0	0	0
Summation of t-statistics	2.46	7.36	3.20	1.37	24.39	18.31
Seoul						
Number of all institutional attributes for which t > 2	2	3	1	2	8	7
Number of all institutional attributes for which t < –2	1	0	0	1	2	2
Summation of t-statistics	2.93	21.30	3.87	5.62	31.33	31.88

Beijing						
Number of all institutional attributes for which t > 2	0	0	0	0	0	0
Number of all institutional attributes for which t < –2	0	0	0	0	0	0
Summation of t-statistics	0.00	0.00	0.00	0.00	0.00	0.00
Chengdu						
Number of all institutional attributes for which t > 2	0	1	2	1	4	2
Number of all institutional attributes for which t < –2	3	1	0	2	6	4
Summation of t-statistics	–8.37	0.44	8.07	–2.79	–2.65	–7.92
Guangzhou						
Number of all institutional attributes for which t > 2	2	0	1	1	4	2
Number of all institutional attributes for which t < –2	0	0	0	0	0	0
Summation of t-statistics	6.71	1.06	3.25	4.52	15.54	6.63
Shanghai						
Number of all institutional attributes for which t > 2	3	0	2	2	7	5
Number of all institutional attributes for which t < –2	0	0	0	0	0	0
Summation of t-statistics	10.65	1.38	8.46	9.32	27.97	25.15
Tianjin						
Number of all institutional attributes for which t > 2	0	1	0	0	0	1
Number of all institutional attributes for which t < –2	0	2	3	0	3	5
Summation of t-statistics	–0.93	–6.30	–7.64	2.59	–12.28	–10.26

Source: Authors' calculations based on World Bank survey data.

Chengdu at the low end of the distribution. A second notable result is the considerable variation among the Chinese cities. The difference between Shanghai and Tianjin is about twice the difference between the highest- and lowest-ranking non-Chinese Asian cities. Finally, we find considerable variation across the four attributes. Measured by openness, Seoul ranks among the lowest of the 10 metropolitan areas, whereas no other economy comes close to Seoul in its composite measure of human capital. Our chosen index, for which we add together the composite t-statistics for each of the four categories, suggests a high degree of substitutability among the four components that shape R&D capability. For example, rather than capture spillovers for foreign investment, Seoul employs a highly trained corps of scientists and engineers to reverse engineer imported equipment, which underscores the point made by Nabeshima in chapter 9 of this volume.

RETURNS TO R&D AND THE INSTITUTIONAL CONTEXT FROM AN ECONOMY-WIDE PERSPECTIVE

In this section, we develop estimates of firm-level and composite metropolitan-level returns to R&D and explore the association of composite returns with the composite measures of R&D capability, shown in table 10.14. Our estimates of the metropolitan-level marginal products of R&D

Table 10.14 Association between Composite Measures of R&D Capability and Performance in Seoul and the Five Chinese Metropolitan Areas

	Composite measures of R&D capability[a]			Composite measures of performance		
Metropolitan area	Number of t-statistics for which $t > 2$ (1)	Number of t-statistics for which $t < -2$ (2)	Summation of t-statistics (3)	Marginal productivity of R&D personnel (U.S. dollars) (4)	Wage of R&D personnel (U.S. dollars) (5)	Ratio of column 4 to column 5 (6)
Seoul	7	2	31.18	37,639	20,847	1.81
Shanghai	5	0	25.15	24,086	5,655	4.26
Guangzhou	2	0	6.63	14,984	3,249	4.62
Beijing	0	0	0.00	13,479	3,494	3.86
Chengdu	2	4	−7.92	9,676	3,102	3.12
Tianjin	1	5	−10.76	8,818	1,569	5.62

a. The estimates are from table 10.13; the Korean won is converted using the average exchange rate as of May 2001 (that is, 1,130); the Chinese yuan is converted using 8.28, the average for that period.

Source: Authors' calculations based on World Bank survey data.

personnel are performed using the estimate of the output elasticity of R&D personnel shown in table 10.6: that is, 0.325. We then construct an average product for R&D personnel as a ratio of the sum of total value added and total R&D personnel for all of the firms included in each city sample. Finally, we convert these estimates of marginal productivity to common U.S. dollar measures using average exchange rates for May 2001, the period that matches or closely approximates the date of the survey work.

This exercise of estimating the marginal products of R&D personnel for each of the 10 metropolitan economies substantially narrows the field of included metropolitan areas. Apart from Seoul and the five Chinese cities, we find that a substantial proportion of firms tend to use different units of account. Although this irregularity is generally not a problem for estimation work using logarithmic transformations of reported values, it does become problematic when we reconvert the data to the metropolitan level. The problems of small samples, missing observations, and inconsistent use of units are severe enough to eliminate the samples other than Seoul and the Chinese cities from the metropolitan-level comparisons. Our estimates of the marginal productivity of R&D labor for the six metropolitan economies are reported in column 3 of table 10.14.

Using these figures, we estimate the relationship between the composite marginal productivity of R&D labor and the composite measure of R&D capabilities. Regressing the observations in column 3 on column 2, we report the results in table 10.15. Our composite measure of R&D capability exhibits surprising explanatory power. The adjusted R^2 for the regression is 0.910. Figure 10.1 shows a scatter plot using these data, with the observations for Seoul and Shanghai anchoring the upper end of the regression line. Our conclusion is that the composite measure of the institutional attributes of metropolitan economies provides a robust indicator of the returns to

Table 10.15 Impact of Firm, Market, and Institutional Factors (Summarized by the Composite R&D t-statistic) on the Composite Marginal Productivity of R&D Personnel

Variable	Estimate
Constant	13,543
	(8.10)
β	602.96
	(6.35)
R^2	0.910
Number of observations	6

Source: Authors' calculations based on World Bank survey data

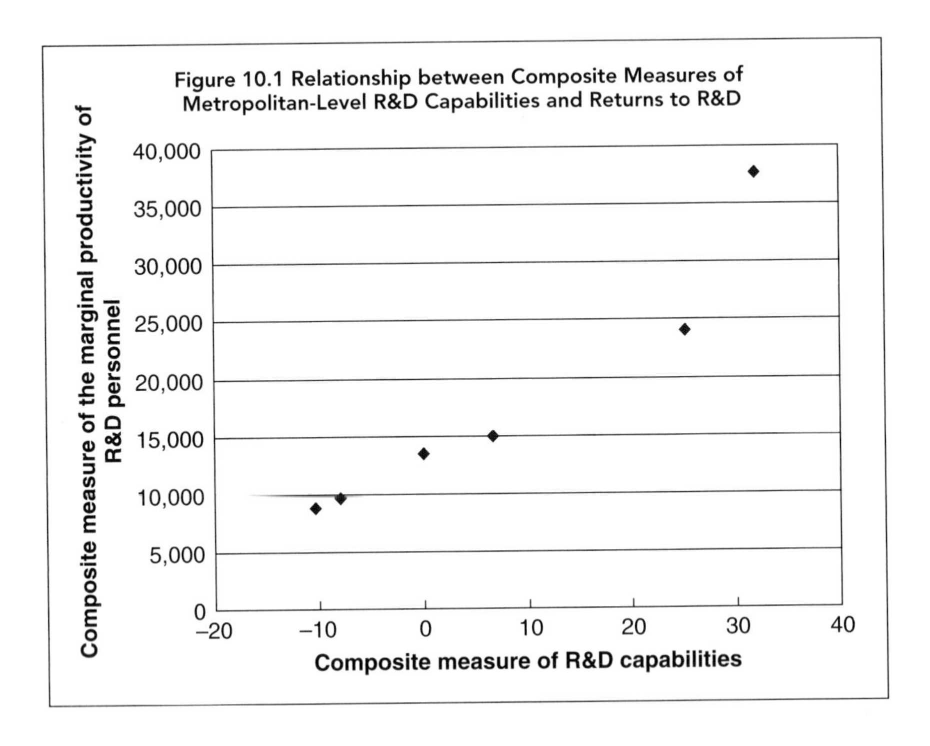

Figure 10.1 Relationship between Composite Measures of Metropolitan-Level R&D Capabilities and Returns to R&D

R&D across these economies. This result provides substantial support for the notion that the innovation function has to be viewed within the broad context of the public goods that have been created in large metropolitan economies, including an open and competitive economy, access to human capital, R&D networks, and public policies that support innovation.

Before we conclude this section, we use the analytical framework developed above to try to explain a recent phenomenon: that is, evidence that many large Japanese firms are relocating significant portions of their R&D operations to China, particularly to the Shanghai region. At least two sources have commented on the relative R&D cost advantage that has emerged in China. Writing in the *New York Times*, James Brooke (2002) suggests that Japanese investment in China's R&D sector is contributing to China's transformation from "the factory of the world" to the "design laboratory of the world." Brooke catalogues the R&D investments of a number of Japanese multinationals in China, virtually all of which are centered on Shanghai. Spurring the investments are, according to Brooke, the low wages of Chinese engineers, the growing Chinese market for computer chips, and the expectation that China's entry into the World Trade Organization will bring protection for patents. Pursuing the same theme in the *Far Eastern Economic Review*, David Kruger (2002) reports that one large Japanese firm, having already moved its technology operations, is shifting its headquarters to Shanghai, where, according to Kruger, costs are 10 times lower than in Japan.

We use the data shown in table 10.14 to explain this phenomenon of R&D sourcing. The data in table 10.14 show the average reported wage for R&D personnel for each of the six metropolitan economies. In the last column, we compute the ratio of the marginal productivity to the wage of R&D personnel. Before comparing the ratios, we note that all of the ratios are considerably greater than unity, the result that we might expect if firms hired R&D personnel up to their marginal productivity. That the ratios significantly exceed unity may result from one or both of two conditions. First, fringe benefits, not reported in the survey, may account for a significant fraction of the base wage, so that, in practice, the ratio of marginal product to total compensation may be somewhat less than shown. Second, it is also possible that the estimate of the output elasticity of R&D personnel, shown in table 10.6, is biased upward. We have discussed this problem earlier. If this estimate were biased, the use of the "correct" estimate would serve to reduce our estimates of marginal products and the ratios shown in table 10.14, but the changes would be proportionate, so the relative size of the ratios would remain unchanged.

The striking feature of the ratios shown in the last column of table 10.14 is the large disparity between Seoul and the Chinese metropolitan areas. In the Chinese metropolitan economies, the ratio of marginal product over the wage typically is larger than that of Seoul by a factor of two or three. While the articles that describe the relocation of Japanese R&D operations to China make no mention of Korea, our finding that R&D in the five Chinese metropolitan areas offers substantially larger producer surplus than does R&D in Seoul provides an explanation for the direct relocation of these R&D operations to China.

Our finding is that Chinese metropolitan economies, particularly Shanghai, have begun to create the institutional attributes that boost the productivity of R&D. While the compensation for R&D personnel has begun to rise to capture these productivity gains, wages still lag far behind productivity and behind the salaries of R&D personnel in more prosperous Asian economies, notably Korea (or Seoul and Tokyo) and Japan. This gap between productivity and cost appears to be making China, particularly Shanghai, an attractive focus for foreign investment and technology. Our findings in table 10.10 suggest that these factors, which are flowing to Shanghai from abroad, are precisely the factors that are raising the productivity of R&D in Shanghai and elsewhere. It may be that the wages of R&D personnel rise to reduce this productivity-wage gap. Alternatively, the gap may create the potential for a "virtuous circle" in which a large productivity-wage spread attracts investment and technology from abroad, which in turn enlarges or sustains the productivity-wage spread that initially attracted overseas investment.

DISCUSSION AND CONCLUSIONS

Using a set of survey data spanning firms from 10 East Asian metropolitan economies, we have investigated the factors that determine the productivity and profitability of R&D operations at the firm level. Our analysis builds up to a test of the robustness of the relationship between the R&D institutional endowment and the return to R&D across metropolitan economies within our sample.

The following conclusions stand out from our analysis. We identify a substantial number of institutional variables that affect R&D capabilities and firm performance. These can be grouped into four categories: international exposure, human capital, the R&D network, and the policy setting. Among the 23 institutional attributes that we investigate, 12 operate on firm performance through the R&D channel. Five others operate outside the R&D channel to enhance firm performance. Not surprisingly, all of the three human capital variables work to enhance the effectiveness of R&D; five of the six R&D network variables also serve to improve R&D effectiveness.

Our results lend support to the emerging view of the importance of clusters for high-performance R&D. Among the cluster attributes that enhance R&D effectiveness are concentrations of industrial park users, concentrations of information technology investments, the intensity of Internet users, extensive R&D network relationships, and concentrations of competitors. We follow up in annex A with a comparison of measures of FDI and R&D intensity among the industrial and technology parks of Beijing, Shanghai, and Taiwan (China). The evidence collected in the annex suggests that the industrial parks in Shanghai are emerging as the dominant clusters in the region.

The 10 participating metropolitan economies exhibit considerable differences in their level of international exposure, human capital, R&D network, and policy setting. Several comparisons stand out. One is the strikingly low measure of Seoul's international exposure and, by comparison, its strikingly high measure of human capital. We surmise that the relatively low participation of FDI in Seoul is compensated for by a rich supply of high-quality domestic scientists and engineers. A second striking finding is, within China, the large variation in composite measures of metropolitan institutional R&D capabilities. In particular, Shanghai stands out for its achievements, which make its measures of overall institutional endowment more comparable to those of Seoul and the ASEAN metropolitan economies than to Beijing, Tianjin, and Chengdu.

Recently, China's *Economic Daily* ("Ranking China's Cities" 2001) reported rankings of aggregate competitiveness for 10 Chinese cities prepared by the Management School of China South-East University (CSEU). While the CSEU study omits Chengdu, it does include our other four metropolitan areas in China. They match our composite ranking. Below, we summarize for each Chinese city the findings of our study and compare these with the survey results reported in the CSEU survey.

Shanghai

According to our World Bank survey data, Shanghai stands out among the five participating Chinese metropolitan areas; it dominates the other Chinese cities along all dimensions. Shanghai exhibits the highest incidence of FDI by a wide margin. A high incidence of Shanghai firms also report being located in industrial parks—a substantially higher incidence than in any of the other four cities, but less than in several of the other metropolitan economies.[6]

According to the CSEU study, Shanghai ranks first in aggregate competitiveness. Shanghai enjoys the most advantageous location and best infrastructure. It is the financial center of China. Shanghai's government is well organized, efficient, flexible, and creative. Shanghai ranks first in the competitiveness of capital, technology, location, social order, and management; it ranks second in the competitiveness of human capital and culture.

Guangzhou

Among the five Chinese metropolitan economies in our survey, Guangzhou trails only Shanghai in international exposure. Like Shanghai, it registers high shares of imports and a high incidence of imported equipment. By a substantial margin, Guangzhou registers the lowest incidence of public ownership.

Guangzhou is the first city opened to the Western world in China's modern history. The CSEU survey emphasizes Guangzhou's mercantile tradi-

6. We do not know the proportion of firms in each city that reside in industrial or technology parks, but the robust performance of Shanghai (and possibly other cities) may result from oversampling of firms situated in such parks.

tion and its active entrepreneurship. Guangzhou is the key city in China's southern economy; it enjoys close proximity to Hong Kong (China). The CSEU survey ranks Guangzhou's economic structure as the most competitive among Chinese cities, citing its strong manufacturing industry and strong service sector. Guangzhou ranks second in the competitiveness of its enterprise management. According to the CSEU survey, Guangzhou's municipal government is efficient, but the city needs to improve its physical environment and social order.

Beijing

The World Bank survey does not rank Beijing highly in its human capital and R&D network measures. However, because human capital is the one area in which Beijing is not at a disadvantage relative to Shanghai and Guangzhou, human capital may be viewed as one of Beijing's *comparative* advantages. Our survey shows that Beijing's R&D network is not as robust as that of Shanghai, the leader, or those of Guangzhou and Chengdu. Our finding of Beijing's relatively low ranking with respect to the purchase of foreign technology and linkages for external R&D assistance may account for CSEU's urging for institutional reform in Beijing in support of technology transfer. Nonetheless, the CSEU survey seems more enthusiastic about Beijing than our own results would suggest, ranking Beijing first in the competitiveness of human capital and internal R&D research.

Chengdu

According to our survey results, among all 10 metropolitan economies—both those in China and those in the other Asian countries—Chengdu ranks the lowest in international exposure; it is the only economy in which all three measures of openness are significantly negative. This relatively low ranking in international exposure is not surprising; unlike the other 10 economies, Chengdu has no coastal area. With respect to the other four Chinese cities, Chengdu not only rates lowest with respect to FDI, it also rates highest on public ownership. Notwithstanding its clear competitive disadvantage with respect to openness and institutional quality, Chengdu ranks comparatively well in certain aspects of human capital and R&D networking. The Chengdu firms in our sample rank highest in R&D intensity; among the 10 economies, it is second only to Seoul. However, measured in terms of Internet use, foreign experience, and management's level of education, the quality of its human capital is not high. Much of Chengdu's R&D

expenditure, it seems, is used to purchase technology or R&D services, generally from domestic sources, rather than to perform R&D internally.

Chengdu and Tianjin represent the two cities for which the findings of the CSEU survey and the World Bank survey are at variance. Chengdu is omitted from the CSEU's top 10 most competitive cities. Our data yield a higher ranking for Chengdu than for Tianjin, which ranks seventh in the CSEU survey. Both areas rank significantly below Beijing.

Tianjin

Results based on the World Bank survey do not rank Tianjin as highly as results based on the CSEU survey. Beijing and Tianjin rank similarly in both openness and institutional quality. The dimension along which Tianjin suffers relative to the other Chinese cities and the other Asian economies is in its measures of human capital and R&D network. Next to Manila, Tianjin ranks lowest in R&D intensity. It has the lowest percentage of workers using the Internet. Tianjin's R&D networking is weak: Tianjin firms tend not to purchase foreign or outside technology or to receive external R&D services.

Tianjin enjoys a strong locational advantage in northern China. According to the CSEU survey, Tianjin has a strong manufacturing industry and good infrastructure. Its investment costs are relatively low, and profitability is relatively high. The CSEU survey finds that many MNCs are switching their production to Tianjin.

Because the World Bank survey spans only five Chinese areas, the list of all 10 cities included in the CSEU survey may be of interest. The reported ranking in 2001 is Shanghai, Shenyang, Guangzhou, Beijing, Xiamen, Wuchang, Tianjin, Dalien, Hangzhou, and Nanjing.

So that we can examine the association between R&D performance and the institutional endowment of the six metropolitan areas for which data are complete (the five Chinese metropolitan economies and Seoul), we construct measures of metropolitan performance, including overall productivity, the value of marginal productivity of R&D, and salaries for R&D personnel. For Seoul and the five Chinese metropolitan economies, where comparisons can be made, these measures of performance are highly associated with our composite metropolitan-wide measures of R&D institutional endowments. We interpret these empirical results as supporting the broad hypothesis developed in Yusuf with others (2003) regarding the critical role of the institutional environment—notably openness, FDI, human capital, and R&D networks—for creating innovative capabilities.

The ratio of R&D productivity to wages is particularly high in the five Chinese metropolitan economies. This productivity-wage spread—a rough measure of producer surplus—appears to be at least partially responsible for the recent acceleration of large and visible flows of overseas investment in R&D operations in China. Whether this spread is narrowed by rapid increases in salaries for R&D personnel or is sustained by continued flows of investment and technology to China will be critical to the evolution of R&D operations throughout Asia.

REFERENCES

The word *processed* describes informally reproduced works that may not be commonly available through libraries.

Aghion, Philippe, and Peter Howitt. 1999. *Endogenous Growth Theory*. Cambridge, Mass.: MIT Press.

Agrawal, Ajay, and Iain Cockburn. 2002. "University Research, Industrial R&D, and the Anchor Tenant Hypothesis." NBER Working Paper 9212. National Bureau of Economic Research, Cambridge, Mass. Processed.

Arrow, Kenneth J. 1962. "Economic Welfare and the Allocation of Resources for Invention." In National Bureau of Economic Research, *The Rate and Direction of Inventive Activity*. Princeton, N.J.: Princeton University Press.

Brooke, James. 2002. "Japan Embraces for a Designed in China World." *New York Times*, April 21, sec. 3, p. 1.

Carter, Anne P. 1987. "Know-how Trading as Economic Exchange." *Research Policy* 18(3):155–63.

CIPR (Commission on Intellectual Property Rights). 2002. "Integrating Intellectual Property Rights and Development Policy." Available at www.iprcommission.org. Processed.

DeLong, J. Bradford. 1988. "Productivity Growth, Convergence, and Welfare." *American Economic Review* 78(5, December):1138–54.

Djankov, Simeon, and Peter Murrell. 2002. "Enterprise Restructuring in Transition: A Quantitative Survey." *Journal of Economic Literature* 40(3):739–92.

Goyal, Sanjeev, and José Luis Moraga. 2000. "R&D Networks." Discussion Paper 00-075/1. Rotterdam: Tinbergen Institute.

Jefferson, Gary H., and Albert Hu. 2000. "China's Industrial Innovation System." Economics Working Paper. Brandeis University. Processed.

Jefferson, Gary, Albert G. Z. Hu, Xiaojing Guan, and Xiaoyun Yu. 2003. "Ownership, Performance, and Innovation in China's Large and Medium-Size Industrial Enterprise Sector." *China Economic Review* 14(1):89–113.

Jefferson, Gary, Huamao Bai, Xiaojing Guan, and Xiaojing Yu. Forthcoming. "R and D Performance in Chinese Industry." *Economic Innovation and New Technology* 13(1–2). Special issue on empirical studies of innovation in the knowledge-driven economy.

Kruger, David. 2002. "Taking a Shine to Shanghai." *Far Eastern Economic Review*, April 25.

Kumar, Rajiv, and Doren Chadee. 2002. "International Competitiveness of Asian Firms: An Analytical Framework." ERD Working Paper 4. Asian Development Bank, Manila. Processed.

Mankiw, Gregory, David Romer, and David Weil. 1992. "A Contribution to the Empirics of Economic Growth." *Quarterly Journal of Economics* 107(2):407–38.

Nelson, Richard R., and Nathan Rosenberg. 1993. "Technical Innovation and National Systems." In Richard Nelson, ed., *National Innovation Systems: A Comparative Analysis.* New York: Oxford University Press.

"Ranking China's Cities." 2001. *Economic Daily.*

Romer, Paul. 1990. "Endogenous Technological Change." *Journal of Political Economy* 98(5, pt. 2):71–102.

Sachs, Jeffrey, and Andrew Warner. 1995. "Economic Reform and the Process of Global Integration." *Brookings Papers on Economic Activity* 1:1–118.

Sala-i-Martin, Xavier. 2002. "The Disturbing 'Rise' of Global Income Inequality." NBER Working Paper 8904. Cambridge, Mass.: National Bureau of Economic Research.

Schumpeter, Joseph. 1950. *Capitalism, Socialism, and Democracy,* 3d ed. New York: Harper and Row.

UNDP (United Nations Development Programme). 2001. *Human Development Report 2001.* New York: Oxford University Press.

Von Hippel, Eric. 1987. *The Sources of Innovation.* New York: Oxford University Press.

Westphal, Larry. 1990. "Industrial Policy in an Export-Propelled Economy: Lessons from South Korea's Experience," *The Journal of Economic Perspectives* 4(3, summer):41–59.

Yusuf, Shahid, with M. Anjum Altaf, Barry Eichengreen, Sudarshan Gooptu, Kaoru Nabeshima, Charles Kenny, Dwight H. Perkins, and Marc Shotten. 2003. *Innovative East Asia: The Future of Growth.* New York: Oxford University Press.

ANNEX

COMPARISON OF TECHNOLOGY PARKS

From table A.1 we see that the quality and productivity of a firm's R&D operation tends to be enhanced by locating in an industrial park or export processing zone. In order to better understand the ways in which a park setting might influence the R&D capabilities of cities, we examine three zones more closely. These are the Haidian Science Park in Beijing, the Pudong New Area and its four constituent parks, and the Hsinchu Science and Technology Park in Taiwan (China). Although Taiwan (China) is not included in our study, we compare the parks in Beijing and Shanghai with Hsinchu because the success of Hsinchu has created a standard for science and technology parks in Asia.

The comparisons in table A.2 show that, in terms of current sales, the Pudong New Area, the youngest of the three parks, already exceeds the scale of its counterparts in Taiwan (China) and Beijing. Total sales and R&D spending by firms in the Pudong New Area overshadow those of Hsinchu and Haidian.

With only 272 companies to account for its voluminous sales, Hsinchu clearly comprises large-size companies. The average annual sales of the Hsinchu firms are approximately $50 million, about 15 times the size of the average firm in the Pudong New Area and more than 40 times the scale of its typical counterpart in the Haidian Science Park. The substantially larger average size of firms in Taiwan (China) most likely reflects two conditions. One is the relative age of the parks. Hsinchu was established in 1980, acquiring a head start over the respective 1988 and 1992 start-up dates for Beijing and Shanghai (see also chapter 7). A second distinctive feature of Hsinchu is that a substantial portion of the firms, approximately 40 percent, were founded by returning expatriates, who probably had far better access to capital than their counterparts in Beijing and Shanghai. Still, both the Beijing and Shanghai parks include substantial FDI—a flow of $123 million

Table A.1 Effect of Firm, Market, and Institutional Factors on Firm Performance

Factor	Productivity	Profit
Operates through the R&D channel		
Share of foreign ownership	+	+
Share of public ownership	–	–
Number of competitors		–
Industrial park or export processing zone	+	+
Percentage of workers using the Internet	+	+
Information technology assets as a percentage of total fixed assets	+	+
Percentage of work force with foreign experience	+	+
Management's level of education	+	+
Purchase of outside technology	+	+
Purchase of a foreign license	+	+
Receipt of external R&D assistance	+	+
Provision of design or R&D services	+	+
Influences directly, not through R&D	+	+
Firm's market share	+	+
Import market share	+	+
Imported equipment	+	+
Purchase of externally performed R&D services	+	+
Government assistance in identifying a foreign relationship	+	+
Exhibits no impact		
Export sales ratio	0	0
Engagement in activities with a foreign firm located abroad (other than providing design or R&D services)	0	0
Member of a business association	0	0
Useful functions of a business association	0	0
Constraints on growth in the domestic market	0	0

in 1998 for the Beijing park and a cumulative investment by 2000 of $34 billion for the Shanghai park.

An overall comparison of the three parks indicates that the Shanghai park is challenging and overtaking the achievements of its counterpart in Taiwan (China) and substantially outperforming the park in Beijing. R&D intensity of its member companies is approaching that of Hsinchu. The fact that the Pudong area has become China's leading center of financial reform and FDI further underscores the vast potential of the Shanghai zone.

Table A.2 Comparison of Technology Parks in China

Indicator	Taiwan (China)	Beijing	Shanghai
Name of park and year founded	Hsinchu Science and Technology Park (1980)	Haidian Science Park (1988)	Pudong New Area (1990)
Total number of technology parks	One	Haidian park is the largest of three parks in the Zhongguancun Science Park	(1) Zhangjiang High-Tech Park, (2) Waigaoqiao free trade zone, (3) Lujiazui finance and trade zone, (4) Jinqiao export processing zone
Purpose	Hsinchu park was created to promote the development of high-tech industries in Taiwan (China)	Also known as China's "Silicon Valley," Haidian park is the country's leading incubator of high-tech businesses and a major cradle of the knowledge-based economy in China	Part of the nation's strategy for economic development, Pudong New Area is intended to build Shanghai into an international center of trade and finance to regenerate the economy of the entire Yangtze River Valley
Scale	In 1998, 272 companies with combined annual sales of $13.7 billion and total employment of 72,623, including more than 3,000 returned expatriates	In 1998, 4,546 companies and annual sales of RMB 45.2 billion; Haidian park had 147,286 employees, including 748 returned expatriates	In 1998, 3,967 companies and annual sales of RMB 135 billion ($16.3 billion); employment in 1998 was 606,100
University linkages	National Tsinghua University and National Chiaotong University	Peking University, Tsinghua University, and China Science and Technology Institute	37 regular institutions of higher learning and four advanced vocational and technical colleges in Shanghai

(continued)

Table A.2 Comparison of Technology Parks in China (*continued*)

Indicator	Taiwan (China)	Beijing	Shanghai
Foreign participation	109 firms, that is, about 40 percent, were founded by returning expatriates	Among the 4,506 firms located in Haidian park in 1998, 19.7 percent were wholly foreign-owned enterprises and joint venture entities, and foreign direct investment totaled $123 million	By 2000 Pudong had attracted 6,635 foreign-invested companies; cumulative investment was $34.4 billion; contracted investment was $14.5 billion
R&D intensity	In 1997 park companies spent 6.2 percent of their sales revenue on R&D ($850 million) compared with only about 1 percent for manufacturing industry in all of Taiwan (China); R&D personnel constituted 11 percent of the work force	Total spending on R&D was RMB 1.07 billion in 1998; the implied R&D was 2.4 percent of sales	Total R&D spending in 2000 was RMB 7.553 billion (that is, $0.91 billion); the implied R&D was 5.6 percent of sales
Patents	More than half of the patents granted in Taiwan (China) are from Hsinchu Science and Technology Park	218 new patents in 1998	—

—Not available.

INDEX

Note: *f* indicates figures, *n* indicates notes, and *t* indicates tables.

ABOUT THE EDITORS

Shahid Yusuf is a research manager in the Development Economics Research Group at the World Bank. He holds a PhD in economics from Harvard University. Dr. Yusuf is the team leader for the World Bank–Japan project on East Asia's Future Economy. He was the director of the *World Development Report 1999/2000: Entering the 21st Century*. Prior to that, he has served the World Bank in several other capacities.

Dr. Yusuf has written extensively on development issues, with a special focus on East Asia. His publications include *China's Rural Development*, with Dwight Perkins (Johns Hopkins University Press 1984); *The Dynamics of Urban Growth in Three Chinese Cities*, with Weiping Wu (Oxford University Press 1997); *Rethinking the East Asian Miracle*, edited with Joseph Stiglitz (Oxford University Press 2001); *Can East Asia Compete? Innovation for Global Markets*, with Simon Evenett (Oxford University Press 2002); and *Innovative East Asia: The Future of Growth*, for which he was the lead author (Oxford University Press 2003). He has also published widely in various academic journals.

M. Anjum Altaf is a senior economist in the Urban Development Sector Unit of the East Asia and Pacific Region at the World Bank. He holds a PhD in engineering–economic systems and an MA in economics, both from Stanford University, and an MS in electrical and computer engineering from Oregon State University. Prior to joining the Bank, Dr. Altaf was a visiting associate professor and Fulbright Scholar at the University of North Carolina at Chapel Hill (1991–93) and an associate professor at the Applied Economics Research Centre, University of Karachi (1985–91). He has worked in the private sector and is presently a Visiting Fellow at the Sustainable Development Policy Institute in Islamabad, Pakistan.

Dr. Altaf's research on poverty, migration, and infrastructure policy and on behavioral, environmental, and urban economics has been published in various academic journals. He was a member of the team that prepared the *World Development Report 1999/2000: Entering the 21st Century* and is one of the authors of *Innovative East Asia: The Future of Growth* (Oxford University Press 2003).

Kaoru Nabeshima is an economist in the Development Economics Research Group at the World Bank. He holds a PhD in economics from the University of California, Davis. He is a team member for the World Bank–Japan project on East Asia's Future Economy and is one of the authors of *Innovative East Asia: The Future of Growth* (Oxford University Press 2003). His research interest lies in the economic development of East Asia, especially in the innovation capabilities of firms.